Saas-Fee Advanced Course 24
Lecture Notes 1994

J. G. Kirk   D. B. Melrose   E. R. Priest

Plasma Astrophysics

Saas-Fee Advanced Course 24
Lecture Notes 1994
Swiss Society for Astrophysics and Astronomy

Edited by A.O. Benz and T.J.-L. Courvoisier

With 139 Figures

Springer-Verlag
Berlin Heidelberg New York
London Paris Tokyo
Hong Kong Barcelona
Budapest

Dr. J. G. Kirk
Max-Planck-Institut für Kernphysik, Postfach 103980, D-69117 Heidelberg, Germany

Professor D. B. Melrose
Department of Theoretical Physics, The University of Sydney, Sydney, NSW 2006, Australia

Professor E. R. Priest
Department of Applied Mathematics, University of St. Andrews, St. Andrews, KY16 9SS, Fife, Scotland, UK

Volume Editors:

Professor A. O. Benz
Institut für Astronomie, ETH-Zentrum, CH-8092 Zürich, Switzerland

Professor T. J.-L. Courvoisier
Observatoire de Genève, ch. des Maillettes 51, CH-1290 Sauverny, Switzerland

This series is edited on behalf of the Swiss Society for Astrophysics and Astronomy:

Société Suisse d'Astrophysique et d'Astronomie
Observatoire de Genève, ch. des Maillettes 51, CH-1290 Sauverny, Switzerland

Cover photograph: Supernovae Remnants Puppis and Vela observed by the ROSAT satellite.
Courtesy of the Max-Planck-Institut für Extraterrestrische Physik, Garching, Germany.

ISBN 3-540-58327-0 Springer-Verlag Berlin Heidelberg New York
ISBN 0-387-58327-0 Springer-Verlag New York Berlin Heidelberg

Library of Congress Cataloging-in-Publication Data

Kirk, J. G. (John G.) Plasma astrophysics / J. G. Kirk, D. B. Melrose, E. R. Priest; edited by A. O. Benz and T. J.-L. Courvoisier. p. cm. – (Saas-Fee advanced course 24 lecture notes; 1994) Includes bibliographical references and index. ISBN (invalid) 3-540-58327-0 (Berlin: alk. paper). – ISBN (invalid) 0-387-58327-0 (New York: alk. paper) 1. Plasma astrophysics. 2. Double stars. 3. Cataclysmic stars. I. Melrose, D. B., 1940- . II. Priest, E. R. (Eric Ronald), 1943- . III. Benz, A. O. IV. Courvoisier, T., 1953- . V. Title. VI. Series: Saas-Fee advanced course ... lecture notes; 1994, 24.QB462.7.K57 1994 94-33108 523.01–dc20 CIP

Typesetting: Camera ready copy from the author/editor
SPIN 10476986 55/3140 - 5 4 3 2 1 0 - Printed on acid-free paper

Foreword

The 1994 Advanced Course of the Swiss Society for Astrophysics and Astronomy dealt with Plasma Astrophysics. This subject is an important tool needed to understand diverse cosmic phenomena such as the corona of the Sun, supernovae remnants, the acceleration of particles to very high energies and therefore the origin of cosmic rays and of the emissions observed in pulsars and active galaxies.

As always in this series of lectures, three leading scientists each present one aspect of the subject. E.R. Priest described the physical foundation of magnetohydrodynamics, D.B. Melrose discussed kinetic plasma theory, and J.G. Kirk concentrated on particle acceleration processes. The three lecturers each gave nine lectures. The speakers not only presented the theoretical aspects of the subject but also discussed many applications to the understanding of current observations. The set of lectures thus gives a well-balanced approach to many facets of Plasma Astrophysics.

These lecture notes are a written version of the three sets of lectures. We are grateful to the three lecturers for preparing their manuscripts with care and in a very short time. We expect that the lecture notes will be useful not only to the participants of the school but also to many researchers and students in various fields of modern astrophysics.

The course took place in Les Diablerets, a small village in the Swiss Alps, from March 21 to 26. The more than 70 participants took an active part in the lectures, asking many questions and using their free time not only for skiing, but also for many lively discussions.

We would like to thank all our colleagues who helped in the organisation of the course and in particular our students P. Buehler and S. Krucker. The financial support of the Swiss Academy of Sciences was extremely helpful and is gratefully acknowledged.

Zurich, Geneva
May 1994

Arnold O. Benz
Thierry J.-L. Courvoisier

Contents

Magnetohydrodynamics

By *Eric R. Priest* (With 87 figures)

Kinetic Plasma Physics

By *Don B. Melrose* (With 26 figures)

List of Previous Saas-Fee Advanced Courses

1994 PLASMA ASTROPHYSICS
J.G. Kirk, D.B. Melrose, E.R. Priest
1993 THE DEEP UNIVERSE (in preparation)
R. Kron, M. S. Longair, A. Sandage
1992 INTERACTING BINARIES
S.N. Shore, M. Livio, E.P. J. van den Heuvel
1991 THE GALACTIC INTERSTELLAR MEDIUM
W.B. Burton, B.G. Elmegreen, R. Genzel
1990 ACTIVE GALACTIC NUCLEI
R. Blandford, H. Netzer, L. Woltjer
1989 THE MILKY WAY AS A GALAXY
G. Gilmore, I. King, P. van der Kruit
1988 RADIATION IN MOVING GASEOUS MEDIA
H. Frisch, R.P. Kudritzki, H.W. Yorke
1987 LARGE SCALE STRUCTURES IN THE UNIVERSE
A.C. Fabian, M. Geller, A. Szalay
1986 NUCLEOSYNTHESIS AND CHEMICAL EVOLUTION
J. Audouze, C. Chiosi, S.E. Woosley
1985 HIGH RESOLUTION IN ASTRONOMY
R.S. Booth, J.W. Brault, A. Labeyrie
1984 PLANETS, THEIR ORIGIN, INTERIOR AND ATMOSPHERE
D. Gautier, W.B. Hubbard, H. Reeves
1983 ASTROPHYSICAL PROCESSES IN UPPER MAIN SEQUENCE STARS
A.N. Cox, S. Vauclair, J.P. Zahn
*1982 MORPHOLOGY AND DYNAMICS OF GALAXIES
J. Binney, J. Kormendy, S.D.M. White
1981 ACTIVITY AND OUTER ATMOSPHERES OF THE SUN AND STARS
F. Praderie, D.S. Spicer, G.L. Withbroe
*1980 STAR FORMATION
J. Appenzeller, J. Lequeux, J. Silk

*1979 Extragalactic High Energy Astrophysics
F. Pacini, C. Ryter, P.A. Strittmatter
*1978 Observational Cosmology
J.E. Gunn, M.S. Longair, M.J. Rees
*1977 Advanced Stages in Stellar Evolution
I. Iben Jr., A. Renzini, D.N. Schramm
*1976 Galaxies
K. Freeman, R.C. Larson, B. Tinsley
*1975 Atomic and Molecular Processes in Astrophysics
A. Dalgarno, F. Masnou-Seeuws, R.V.P. McWhirter
*1974 Magnetohydrodynamics
L. Mestel, N.O. Weiss
*1973 Dynamical Structure and Evolution of Stellar Systems
G. Contopoulos, M. Hénon, D. Lynden-Bell
*1972 Interstellar Matter
N.C. Wrickramasinghe, F.D. Kahn, P.G. Mezger
*1971 Theory of the Stellar Atmospheres
D. Mihalas, P. Pagel, P. Souffrin

Books up to 1989 may be ordered from: SAAS-FEE COURSES
GENEVA OBSERVATORY
ch. des Maillettes 51
CH-1290 Sauverny, Switzerland

Books from 1990 on may be ordered from Springer-Verlag.

* Out of print

Magnetohydrodynamics

Eric R. Priest

Mathematical and Computational Sciences Dept.
The University
St Andrews
KY16 9SS

Prelude

Many fundamental processes in the universe are determined by the magnetic field. The object in which we can study such processes in greatest detail is the Sun, and so it will be natural in this course of lectures to focus on the Sun. In the first lecture I aim to introduce you to it and then in the second to set up the basic theory for magnetic field behaviour. In later lectures this theory is applied to many mysterious processes on the Sun that are only just beginning to be understood. Chapters 3, 4 and 6 develop the theory for the fundamental processes of magnetohydrodynamic force balance, waves and reconnection. Chapter 5 describes the solar wind, bearing in mind that winds occur in many other stars and galaxies. Chapter 7 discusses mechanisms for heating the solar corona, which are likely to be relevant to the coronae of other stars and of accretion discs. Chapter 8 describes prominences, which are evidence of the basic astrophysical process of radiative instability and which occur on other stars. Finally, the flare-like behaviour modelled in Chapter 9 is not restricted to the Sun alone. Thus many of the key processes occur elsewhere in the cosmos, so it is very important to develop a good understanding of what the Sun is telling us about them.

1 The Sun

1.1 Brief History

The Sun has been observed for thousands of years. The Egyptians and Aztecs used to worship it and, even now, many north Europeans trek south and pay homage to it each summer! As long ago as 2000 BC eclipses were recorded and predicted by the Chinese. Sunspots were observed in the fourth century BC by the Greeks and then their existence forgotten in the Middle Ages. Around 1610 AD Galileo and others rediscovered sunspots.

The nineteenth century saw many advances. In 1814 Fraunhofer discovered lines in the solar spectrum, which show up when sunlight is passed through a prism. In 1842 prominences were rediscovered at an eclipse, having previously been mentioned in medieval Russian chronicles. The following year Schwabe suggested the existence of the 11-year sunspot cycle. Then 1851 saw the first photograph of the corona, the outer atmosphere of the Sun, during another eclipse. In 1859 the first solar flare was observed, a brilliant patch of light near a sunspot. Then, in 1868, the emission line of a new element was detected by Secchi at an eclipse and was given the name "helium" after Helios (the Greek sun god). The following year another new line was ascribed to another new element called "coronium", and it was not until 70 years later that Edlen showed that it and other coronal emission lines come instead from normal elements (such as iron) which are many times ionised by the extremely hot corona.

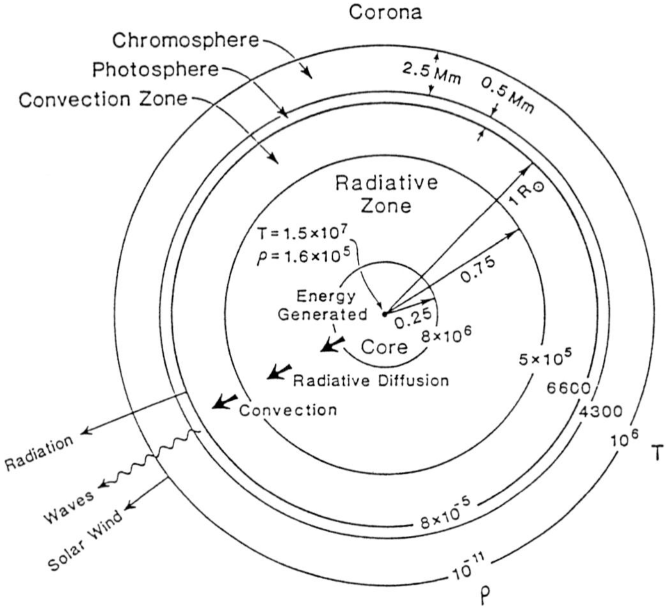

Fig. 1.1. Overall structure of the solar interior and core

Near the beginning of this century (in 1908) it was realised that sunspots possess a strong magnetic field, and in 1930 the coronograph was invented for viewing the corona outside eclipses. In 1938 the carbon - nitrogen and proton - proton chains were proposed by Bethe as a source of solar energy. Then in 1956 Cowling's classic book layed the foundation for the theory of magnetic field behaviour. 1957 gave us the first satellite observations of the interplanetary medium. More recently, during the 1980's and 1990's, there has been a great development of solar theory, which has been complemented by high-resolution observations from ground-based telescopes and space satellites such as Skylab (1973-1974), Solar Maximum Mission (1980-1989) and Yohkoh (1991 - ...). Hopefully, this will be continued as a new suite of world-class telescopes is developed on the Canary Islands and the SOHO satellite is launched in 1995.

1.2 Overall Properties

For astronomers the Sun is a fairly ordinary star (of spectral type G2V) and with an absolute magnitude 4.8. However, its proximity makes it unique and of central importance for understanding other stars and the behaviour of cosmical plasma in general. Her vital statistics are as follows:

Age = 4.5×10^9 years
Mass = 1.99×10^{30} kg (330,000 times the Earth's mass)
Radius = 696,000 km (=696 Mm) (109 times the Earth's radius)
Mean density = 1.4×10^3 kg m^{-3} ($\frac{1}{4}$ that of the Earth)
Mean distance from the Earth = 150×10^6km = $215 R_\odot$
Surface gravity = $274\,\mathrm{m\,s^{-2}}$ (27 times that of Earth)
Escape velocity at surface = 618 km s^{-1}
Radiation emitted (luminosity) = 3.86×10^{26}W
Equatorial rotation period = 26 days
Mass loss rate = $10^9\,\mathrm{kg\,s^{-1}}$
Composition: 90% H, 10% He, 0.1% other elements (C,N,O ...)

The Sun is a ball of hot gas held together by gravity. We can only see its surface, so what is happening inside? We know roughly, but the details depend on models and are highly uncertain. The central temperature and density are so high that nuclear reactions are taking place, supplying energy that continually leaks outwards and is eventually radiated into space (Figure 1.1). In the core, energy is generated by the 5 million tons of hydrogen that are being fused to form helium per second.

The electromagnetic radiation leaks out slowly, but the interior is very opaque, like a thick fog, so that it undergoes many reflections: if it came straight out it would take 2 seconds, but there are so many reflections that it takes 10 million years! Also between $0.71 R_\odot$ and $R_\odot$, the temperature of the gas is decreasing so rapidly that it becomes unstable and produces turbulent convective motions.

1.3 The Atmosphere

The atmosphere consists of three regions (Figure 1.2). The *photosphere* emits most of the Sun's light and consists of a thin skin only $\frac{1}{2}$ Mm thick with a density of 10^{23} particles per cubic metre (one hundredth of that in the Earth's atmosphere). The *chromosphere* is rarer and more transparent with a density of $10^{17}\mathrm{m}^{-3}$. The *corona* is even less dense with a density of $10^{15}\mathrm{m}^{-3}$: it extends to the Earth's orbit (where the density is $10^7\mathrm{m}^{-3}$) and beyond.

At first sight one may expect the temperature to decrease as we go away from the solar surface, and at first it does do so, to a minimum value of 4200 K. However, beyond that it rises slowly through the chromosphere and then rapidly in a narrow transition region to a few million degrees in the corona. As

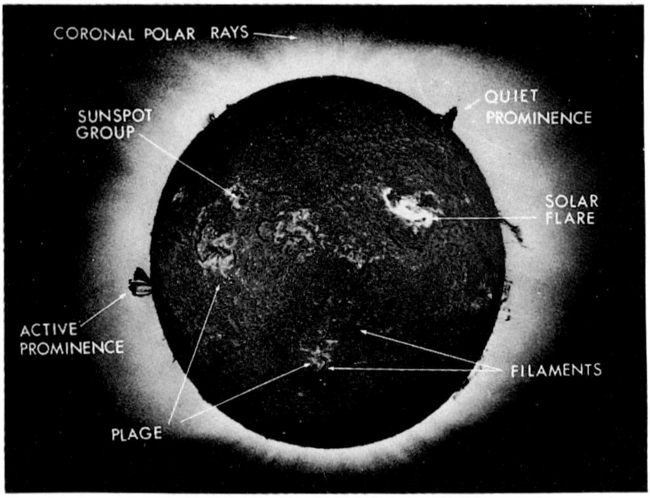

Fig. 1.2. A composite of the chromosphere and corona (S. Martin)

mentioned above, it was only in 1940 that it was realised the corona is so hot. If you put your hand in a bottle of coronal gas, you would find its temperature rise by - only a fraction of a degree, since the coronal density is so small that it contains a minute amount of heat.

The photosphere is the top of the convection zone and is a seething mass of continuously changing material. A granular structure (*granulation*) covers the Sun at this level, with each cell having a diameter of about 0.5 Mm and a lifetime of typically 5 minutes. Also a larger pattern (*supergranulation*) is present with a scale of 30 Mm: material rises in the centre of a cell, moves outward at about 500 $\mathrm{m\,s^{-1}}$ and moves down at the edges. The megametre ($1\mathrm{Mm} = 10^6\mathrm{m}$) proves to be a convenient measure of length for many small-scale solar phenomena. The whole Sun rotates with a period of 25 days near the equator, but 31 days near the poles. Different wavelengths are emitted from different heights in the atmosphere, so observing the Sun through filters of different colours gives pictures of the Sun at different levels. For instance, an Hα filter shows us the *chromosphere*.

The corona is observed at eclipses as a faint halo of very low density and high temperature (Figure 1.3). Low down the magnetic field tends to be closed; further out it is radial, pulled out by the solar wind. On average, there is only one eclipse per year, lasting two minutes, so artificial eclipses have been created by a coronograph, a telescope with a disc that blots out the glare of the photosphere: this is difficult because the corona is only one-millionth of the normal brightness of the Sun, about as bright as the full moon, so normally the corona cannot be seen through the dazzling light of the photosphere.

Fig. 1.3. The solar corona seen at an eclipse (High Altitude Observatory)

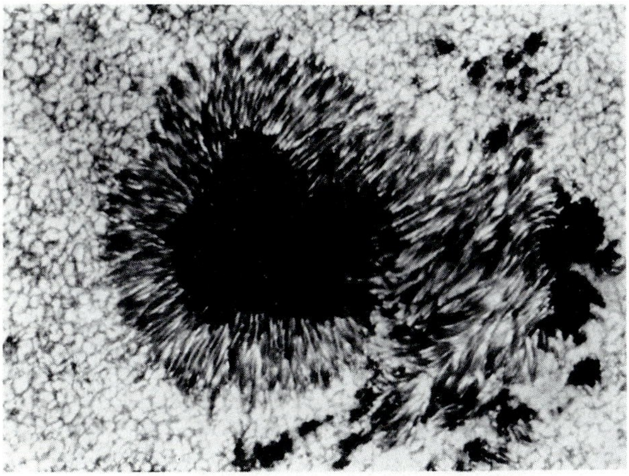

Fig. 1.4. A close-up of a sunspot (R. Muller)

1.4 Solar Activity

Several types of transient activity are observed. *Sunspots* are dark, cool areas in the photosphere with extremely strong magnetic fields, up to 3000 G (Figure 1.4). They often occur in pairs where a large flux tube pokes through the solar surface. They occur in two zones either side of the equator and the number of spots varies with an eleven-year period, the sunspot cycle. A sunspot group is surrounded by a region of moderate field strength (about 100 G), an active region, which is hotter and brighter than its surroundings.

Fig. 1.5. A close-up of a prominence at the limb (Big Bear Observatory)

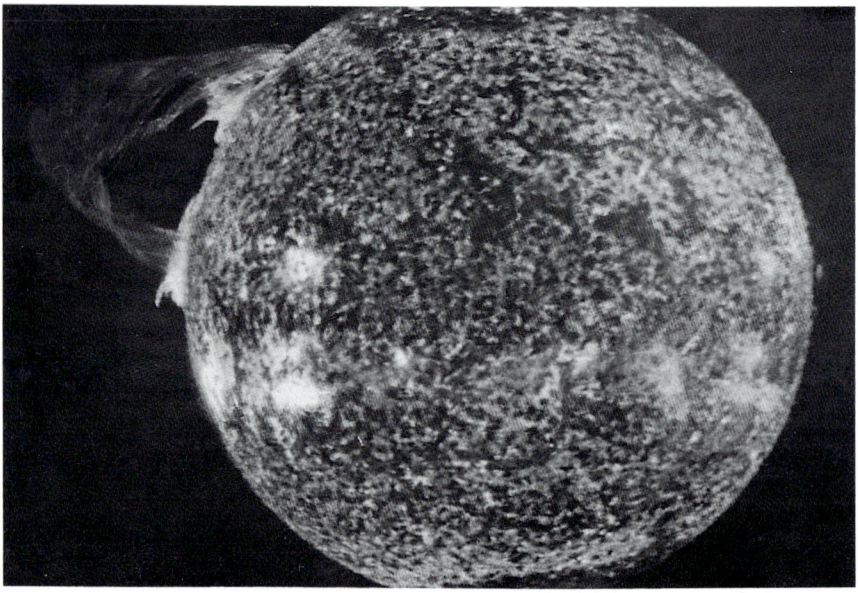

Fig. 1.6. An erupting prominence (Naval Research Lab)

Prominences are vertical sheets of very dense cool material in the corona, observed in projection on the disc as thin, dark filaments (Figure 1.5). Whereas spots usually fade after a few weeks, prominences keep growing for months up to 1000 Mm in length. They are the most stable of all surface features and can endure for nine months. Sometimes a prominence becomes unstable and erupts outwards (Figure 1.6). *Solar flares* are rapid brightenings in the chromosphere and corona near sunspots (see ch 9).

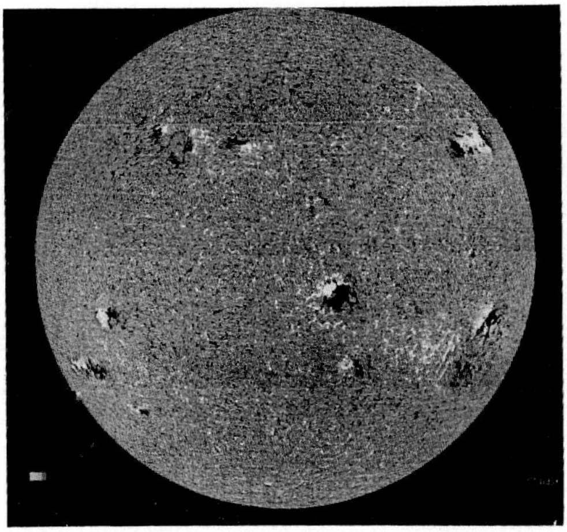

Fig. 1.7. Photospheric magnetic field, with light and dark showing opposite polarity (Kitt Peak Observatory)

1.5 The Solar Revolution

The traditional view of the Sun is of a well understood object with a spherically symmetric atmosphere and a magnetic field that is negligible ($\approx$ 1 G) except in sunspots; the atmosphere is heated by sound waves and an excess pressure drives a spherically symmetric expansion, the solar wind.

Many features of this old view have been completely transformed because of high-resolution observations from the ground of the photosphere and corona and X-ray observations from satellites of the corona. We now realise that the plasma atmosphere of the Sun is highly structured and dynamic and that most of what we see is caused by the magnetic field.

There is a similar change of thinking in astrophysics, where now the magnetic field is realised to be crucial in e.g., star formation, stellar activity (cycles, spots, coronae, flares), magnetospheres of compact objects (white dwarfs, neutron stars, and black holes), jets and accretion discs. However, the Sun continues to be a Rosetta stone for astronomy because here we can study many of the basic physical questions in depth.

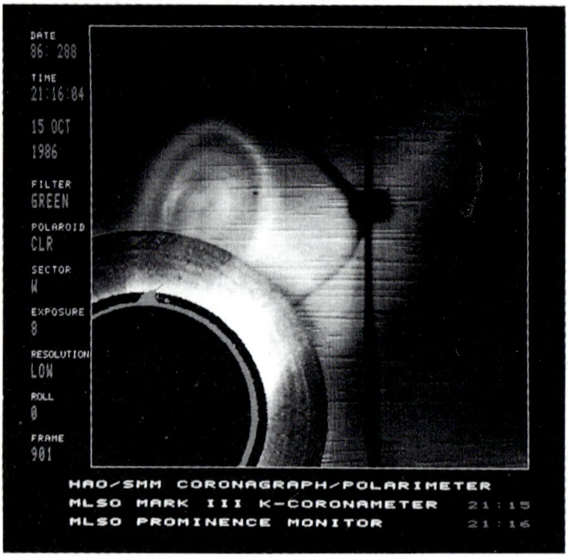

Fig. 1.8. A coronal mass ejection

In what ways has the traditional picture of the Sun changed? Many key topics are not well understood at all, such as the detailed internal structure, coronal heating, the origin of the solar wind, and the causes of eruptions and of solar flares. Also, there have recently been many dramatic discoveries:

(i) Detections of electron neutrinos at Earth give a flux of only 2 SNU (rather than the predicted 6 SNU), but is this because of a smaller than expected central temperature due to mixing or because the electron neutrino has a significant mass which makes two-thirds of them change into μ and τ neutrinos on their way from the Sun to the Earth?

(ii) The Sun is oscillating globally. Over 1000 different normal modes of vibration have been detected and are being used to probe the interior and deduce its structure, just as seismology is used to infer the Earth's internal structure. Preliminary results have given the location of the base of the convection zone as 0.71 $R_\odot$ and have shown surprisingly that the core is probably rotating rigidly while the convection zone has constant angular velocity on radial lines. Also the frequencies change by about 0.1% with the solar cycle.

(iii) The magnetic field outside active regions is concentrated at the edges of supergranule cells to form intense flux tubes of a thousand gauss (Figure 1.7).

(iv) Many new details of solar flares have been discovered, the most important being the detections of γ-rays (Chupp, 1984), the imaging of hard X-rays (de Jager and Svestka, 1985) and the fragmentation of the energy release site as revealed by millisecond spikes in the radio (Benz and Aschwanden, 1992). Also, with coronographs huge erupting bubbles, called coronal mass ejections, have been seen propagating ahead of erupting prominences (Figure 1.8).

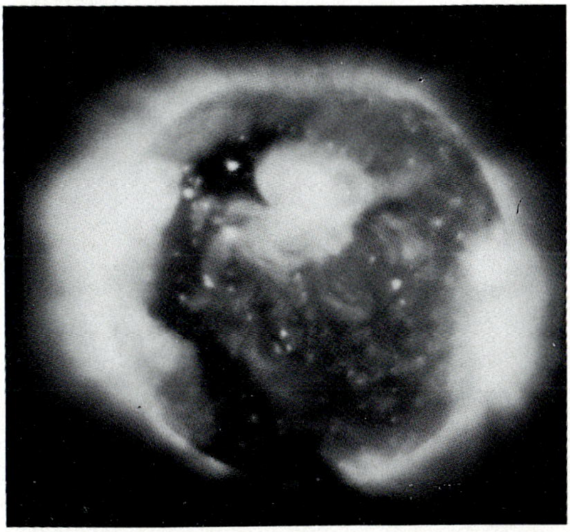

Fig. 1.9. A soft x-ray picture of the corona from Skylab (D.Webb)

(v) The corona has been revealed direct by soft x-ray telescopes (Figure 1.9). It is an intriguing new world with myriads of loops and possesses a three-fold structure of coronal holes (magnetically open regions from which the fast solar wind is escaping), coronal loops and x-ray bright points (where magnetic fields are interacting).

2 Magnetohydrodynamic Equations

Most of the structure and interesting behaviour on the Sun is produced by the magnetic field, so we need a theory for the interaction between the solar atmosphere and its magnetic field - magnetohydrodynamics (or MHD for short). The solar atmosphere is not a normal gas but a ***plasma***. We are all familiar on Earth with the three states (solid, liquid and gas) of matter. You change from one state to another (such as ice to water) by heating, and if you raise the temperature of a gas sufficiently it changes to the fourth state, plasma. In the plasma state the atoms have split into positive ions and negative electrons, which can flow around freely, so the gas becomes electrically conducting and a current can flow.

Plasma is very important since most of the universe is in this state. Indeed, on Earth we are in an extremely unusual part of the cosmos, a tiny island of solid, liquid and gas. But, as soon as we go up to the ionosphere, the plasma universe begins, including the whole of the region between Earth and Sun and the whole of the Sun itself.

MHD is then the study of the interaction between a magnetic field and a plasma, treated as a continuous medium (e.g. Priest, 1982; Priest and Hood, 1991), so that we are not concerned with individual particles. It builds partly on electromagnetism and partly on fluid mechanics. The assumption of a continuous medium is valid for length-scales much larger than the mean-free path

$$\lambda_{\mathrm{mfp}} \approx 300 \left(\frac{T}{10^6\mathrm{K}}\right)^2 \left(\frac{n}{10^{17}\mathrm{m}^{-3}}\right)^{-1} \mathrm{m} ,$$

which is typically 3 cm in the chromosphere and 30 km in the corona.

The magnetic field has several physical effects:
(i) it exerts a force, which may accelerate plasma or create structure;
(ii) it stores energy, which may later be released as an eruption or a solar flare;
(iii) it acts as a thermal blanket, which, when wrapped around a cool prominence, may protect it from the surrounding corona;
(iv) it channels fast particles and plasma;
(v) it drives instabilities and supports waves.

2.1 Flux Tubes

A *magnetic field line* is such that the tangent at any point is in the direction of the field. Its equation is given in two dimensions by

$$\frac{dy}{dx} = \frac{B_{\mathrm{y}}}{B_{\mathrm{x}}} , \tag{2.1}$$

or $r^{-1}dr/d\theta = B_{\mathrm{r}}/B_\theta$ in polars, or $dx/B_{\mathrm{x}} = dy/B_{\mathrm{y}} = dz/B_{\mathrm{z}}$ in three dimensions.

A *magnetic flux tube* is the surface generated by the set of field lines which intersect a simple closed curve. Flux tubes are the building blocks of a magnetic configuration, but they must not be thought of as independent isolated structures.

The *strength (F)* of a flux tube is the amount of magnetic flux crossing a section S, i.e.

$$F = \int_s \mathbf{B}.\mathbf{dS} . \tag{2.2}$$

Consider a finite segment of a flux tube bounded by sections S_1 and S_2 (Figure 2.1). There is no flux across the walls of the tube, and so physically, if no flux is created inside the tube, the flux (F_1) entering through section S_1 equals the flux (F_2) leaving through section S_2.

In other words, the strength (F) is constant along the tube. Mathematically, integrating over the whole surface (S_v) of the segment of the tube between S_1 and S_2 gives, since **B.dS** vanishes on the curved part,

$$\int_{S_v} \mathbf{B}.\mathbf{dS} = \int_{S_1} \mathbf{B}.\mathbf{dS} + \int_{S_2} \mathbf{B}.\mathbf{dS} = -F_1 + F_2$$

if F_1 is the flux from left to right and, on S_v, **dS** is in the direction of the outwards normal. However, by the Gauss divergence theorem,

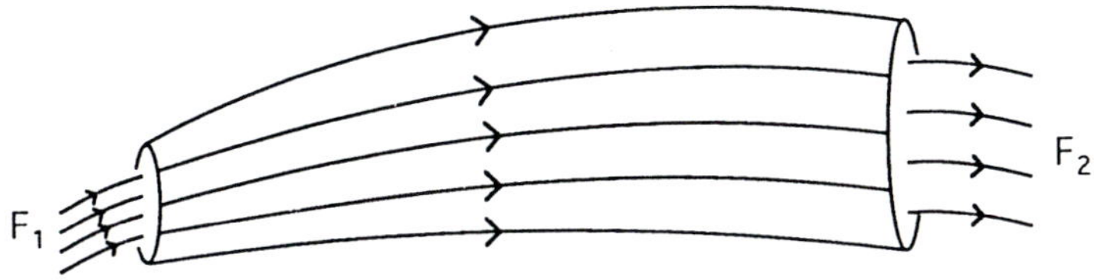

Fig. 2.1. Segment of a magnetic flux tube

$$\int_{S_v} \mathbf{B}.\mathbf{dS} = \int_V \nabla.\mathbf{B}\, dV = 0 \ ,$$

since $\nabla.\mathbf{B} = 0$. Thus $F_1 = F_2$, as required, and therefore $\nabla.\mathbf{B} = 0$ implies that no flux is indeed created and the strength is constant along the flux tube.

If the cross-sectional area (A) of a flux tube is small, then $F \approx BA$. Thus, as the magnetic field lines become closer together, so A becomes smaller and, since F is constant, B increases in value, and vice versa (Figure 2.2).

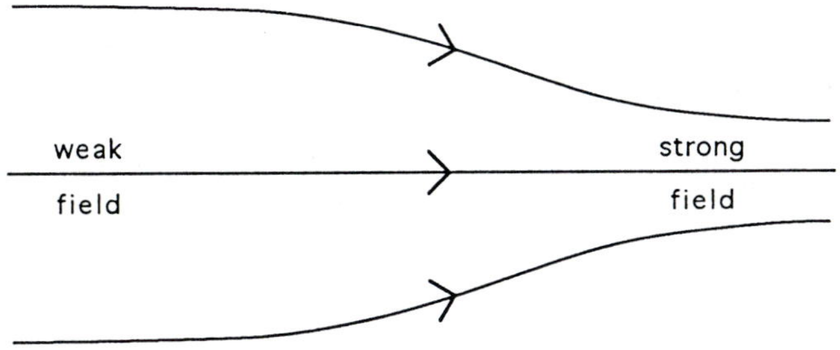

Fig. 2.2. Flux tube with varying section and therefore field strength

When sketching field lines from expressions for the field components there are three stages:
(i) evaluate the expressions for the field lines and sketch a typical one ;
(ii) decide the directions of the arrows;
(iii) decide the spacings of other field lines.
Thus, for example, if $B_x = y$ and $B_y = x$, the field lines are given by $dy/dx = x/y$ or $y^2 - x^2 =$ *constant*. When the *constant* is zero we obtain the two (separatrix) field lines $y = \pm x$ through the origin; when the constant is positive (negative) we have branches of a rectangular hyperbola intersecting the y-axis (x-axis). From the orientations of the axes in Figure 2.3 we have decided that the positive directions are to the right and upward; thus, for instance, on the positive x-axis

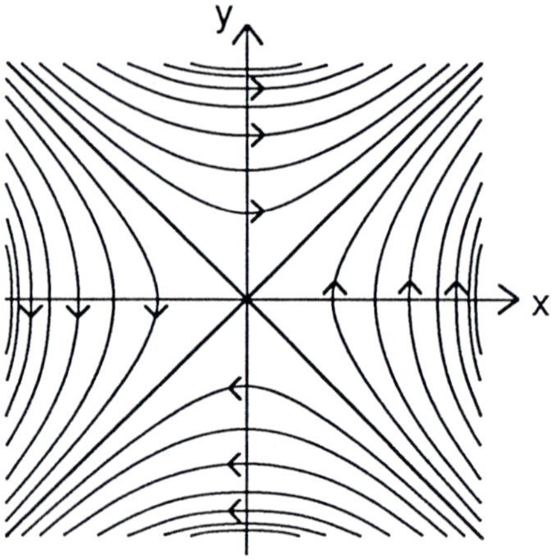

Fig. 2.3. Field lines for the magnetic field $(B_x, B_y) = (y, x)$

the field is simply $B_y = x$, which is positive and so the arrow is directed upwards. Also, as one moves out along the x-axis the magnitude of the field increases and so the field lines become more closely spaced. The origin is a special point, an X-type neutral point, where the field vanishes; as we shall see, it represents a weak spot in a configuration where magnetic energy can be released.

2.2 Fundamental Equations

The MHD equations are a unification of the equations of slow electromagnetism and fluid mechanics. Maxwell's equations are

$$\nabla \times \mathbf{H} = \mathbf{j} + \frac{\partial \mathbf{D}}{\partial t}, \quad \nabla . \mathbf{B} = 0 \ , \tag{2.3}$$

$$\nabla \times \mathbf{E} = -\frac{\partial \mathbf{B}}{\partial t}, \quad \nabla . \mathbf{D} = \rho_c \ , \tag{2.4}$$

where for a vacuum or a low-density plasma $\mathbf{B} = \mu \mathbf{H}$, $\mathbf{D} = \epsilon \mathbf{E}$.

Here $\mathbf{H}$ is the magnetic field, $\mathbf{B}$ the magnetic induction (although we shall loosely refer to it as the magnetic field), μ the magnetic permeability of free space, $\mathbf{E}$ the electric field, $\mathbf{D}$ the electric displacement, ϵ the electrical permittivity of free space, ρ_c the charge density, $\mathbf{j}$ the electric current density. These are supplemented by Ohm's Law

$$\mathbf{E} = \mathbf{j}/\sigma \ , \tag{2.5}$$

where σ is the electrical conductivity.

The equations of fluid mechanics, on the other hand, are

$$\rho \frac{d\mathbf{v}}{dt} = -\nabla p \ , \quad \frac{d\rho}{dt} + \rho \nabla . \mathbf{v} = 0 \ , \quad p = \mathcal{R} \rho T \tag{2.6}$$

and an energy equation, where ρ is the plasma density (the mass per unit volume), $\mathbf{v}$ the plasma velocity, p the plasma pressure, T the temperature, $\mathcal{R}$ the gas constant. The operator $d/dt = \partial/\partial t + \mathbf{v}.\nabla$ is the total (or material) derivative and represents the time rate of change moving with an element of plasma, in contrast to $\partial/\partial t$ which represents the time rate of change at a fixed point of space. Equation (2.6a) is the equation of motion, which says that the mass times acceleration of a moving element of plasma equals the sum of the forces acting on the element (here just the pressure gradient, $-\nabla p$). Equation (2.6b) is the equation of mass continuity and is simply a mathematical expression of the physical fact that no plasma is being created or destroyed: thus, if for example mass is flowing outwards away from an element (so that the flow is "diverging" and the divergence $\nabla.\mathbf{v}$ is positive), then the density of the element must be decreasing (so that from (2.6b) $d\rho/dt$ must be negative). Equation (2.6c) is the perfect gas law, stating that the pressure of a plasma is proportional to its density and temperature. In principle (2.6a) determines the velocity $\mathbf{v}$, (2.6b) the density ρ, (2.6c) the pressure p and the energy equation the temperature T.

In a gas the electromagnetic ((2.3)-(2.5)) and fluid dynamic (2.6) equations are decoupled and so the electromagnetic and fluid properties are independent. However, in MHD, we modify the above equations in three ways:

(i) A plasma feels an extra force, the so-called Lorentz force ($\mathbf{j} \times \mathbf{B}$), which is added on to the right-hand side of (2.6a). It is this term which couples the fluid equations to the electromagnetic equations. It is well known that an element $\mathbf{dl}$ of wire carrying a current J in a magnetic field feels a force $J\mathbf{dl}\times\mathbf{B}$ perpendicular to the wire and to the field, and so it is natural that a plasma element of volume dV carrying a current of density $\mathbf{j}$ per unit volume should feel a force $\mathbf{j}dV \times \mathbf{B}$.

(ii) Ohm's Law states that the electric field in a frame moving with the plasma is proportional to the current, but the total electric field on moving plasma is $\mathbf{E} + \mathbf{v} \times \mathbf{B}$, where $\mathbf{E}$ is the field acting on plasma at rest, so (2.5) is modified by adding $\mathbf{v} \times \mathbf{B}$ to the left-hand side. This too couples the electromagnetic equations to the fluid equations.

(iii) We consider processes with plasma speeds much slower than the speed of light ($v << c$), so that the displacement current ($\partial\mathbf{D}/\partial t$) in (2.3a) is negligible. This in turn means that we don't need to consider (2.4b), since it just determines ρ_c if needed.

2.3 Induction Equation

With the above assumptions, equations (2.3), (2.4) and (2.5) become

$$\mathbf{j} = \nabla \times \mathbf{B}/\mu, \quad \frac{\partial \mathbf{B}}{\partial t} = -\nabla \times \mathbf{E} \ , \tag{2.7}$$

$$\mathbf{E} = -\mathbf{v} \times \mathbf{B} + \mathbf{j}/\sigma, \quad \text{where } \nabla.\mathbf{B} = 0 \ . \tag{2.8}$$

We may therefore eliminate $\mathbf{j}$ and $\mathbf{E}$ by substituting for $\mathbf{j}$ from (2.7a) in (2.8a) and for $\mathbf{E}$ from (2.8a) in (2.7b), with the result that $\partial\mathbf{B}/\partial t = \nabla \times (\mathbf{v} \times \mathbf{B}) -$

$\eta\nabla \times (\nabla \times \mathbf{B})$, where $\eta = 1/(\mu\sigma)$ is the *magnetic diffusivity*. By expanding out the triple vector product in the last term and using (2.8b), we obtain finally

$$\frac{\partial \mathbf{B}}{\partial t} = \nabla \times (\mathbf{v} \times \mathbf{B}) + \eta \nabla^2 \mathbf{B} \ , \tag{2.9}$$

which is known as the *induction equation.*

This is the basic equation for magnetic behaviour in MHD: it determines **B** once **v** is known. In electromagnetism the electric current and electric field are primary variables, with the current driven by electric fields, and then the magnetic field is a secondary variable produced by the currents. However, in MHD the basic physics is quite different since the plasma velocity (**v**) and magnetic field (**B**) are the primary variables, determined by the induction equation and the equation of motion, while the resulting current (**j**) and electric field (**E**) are secondary and may be deduced from (2.7a) and (2.8a) if required.

If V_0, l_0 are typical velocity and length-scale, the ratio of the first to the second term on the right-hand side of (2.9) is, in order of magnitude, the *magnetic Reynolds number*

$$R_{\rm m} = \frac{l_0 V_0}{\eta} \ . \tag{2.10}$$

Thus, for example, in an active region where $\eta \approx 1\ m^2 \mathrm{s}^{-1}, l_0 \approx 10^5\, m, V_0 \approx 10^4\, m\,\mathrm{s}^{-1}$, we find $R_{\rm m} \approx 10^9$ and so the second term on the right of (2.9) is completely negligible. In turn, equation (2.8a) reduces to $\mathbf{E} = -\mathbf{v} \times \mathbf{B}$ to a very high degree of approximation. This is the case in almost all of the solar atmosphere, indeed in almost all of the universe - the only exception is in regions (such as current sheets) where the length-scale is extremely small, so small that $R_{\rm m} \lesssim 1$ and the second term on the right of (2.9) becomes important.

If $R_{\rm m} \leq 1$, the induction equation reduces to

$$\frac{\partial \mathbf{B}}{\partial t} = \eta \nabla^2 \mathbf{B} \ , \tag{2.11}$$

and so **B** is governed by a diffusion equation, which implies that field variations (irregularities) on a scale L_0 diffuse away on a time-scale of

$$\tau_{\rm d} = \frac{L_0}{\eta} \ , \tag{2.12}$$

which is obtained simply by equating the orders of magnitude of both sides of (2.11). The corresponding speed at which they diffuse is

$$v_{\rm d} = \frac{L_0}{\tau_{\rm d}} = \frac{\eta}{L_0} \ . \tag{2.13}$$

With $\eta \approx 1\, m^2\, \mathrm{s}^{-1}$, the decay time for a sunspot is (with $L_0 = 10^6 m$) 10^{12} sec = 30,000 years, so that the process whereby sunspots disappear in a few years cannot just be diffusion. Similarily, the diffusion time for a magnetic field pervading the Sun as a whole (with $L_0 = 7 \times 10^8\, m$) is 5×10^{17}sec = 10^{10}years.

This is of the order of the age of the universe, so a magnetic field in a star at its formation has not had time to diffuse much.

Consider, for example, a one-dimensional magnetic field $B(x,t)\hat{\mathbf{y}}$ satisfying

$$\frac{\partial B}{\partial t} = \eta \frac{\partial^2 B}{\partial x^2} , \tag{2.14}$$

and suppose that initially the field is $B(x,0) = B_0$ for $x > 0$ and $B(-x,0) = -B(x,0)$, as shown in Figure 2.4. Physically, what do we expect to happen? Since (2.14) has the form of a heat conduction equation and we know that heat

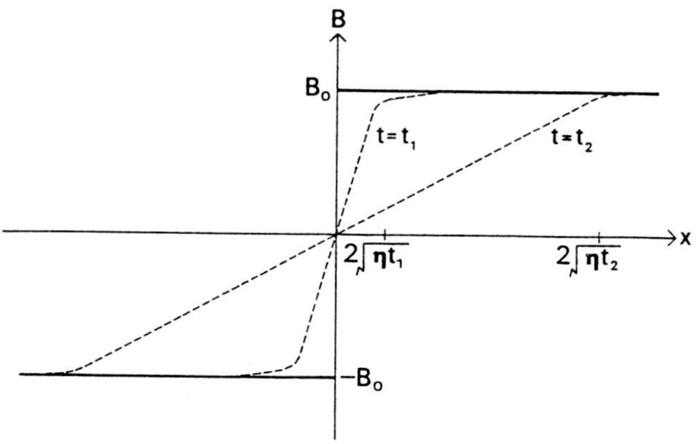

Fig. 2.4. B as a function of x for $t = 0, t_1, t_2$

tends to flow from a hot region to a cool one and smooth out a temperature gradient, we expect the same diffusive process to occur for our magnetic field and for the initially steep magnetic gradient at $x = 0$ to smooth out, as shown in Figure 2.4.

Mathematically, the required solution of (2.14) turns out to be, in terms of the error function ($\text{erf}(\xi)$)

$$B(x,t) = B_0 \operatorname{erf}\left(\frac{x}{\sqrt{(4\eta t)}}\right) = \frac{2B_0}{\sqrt{\pi}} \int_0^{x/\sqrt{(4\eta t)}} e^{-u^2} du .$$

This solution does indeed have the form shown in Figure 2.4 and it may be verified *a posteriori* by substituting back into (2.14). The resulting field lines at subsequent times are sketched in Figure 2.5, from which it can be seen that they are diffusing through the plasma and cancelling at $x = 0$. The total magnetic flux $\int_{-\infty}^{\infty} B\, dx$ remains constant (namely zero), but the energy $\int_{-\infty}^{\infty} B^2/(2\mu) dx$ decreases in time. Indeed, the rate at which it decreases is

$$\frac{\partial}{\partial t} \int_{-\infty}^{\infty} \frac{B^2}{2\mu} dx = \int_{-\infty}^{\infty} \frac{B}{\mu} \frac{\partial B}{\partial t} dx ,$$

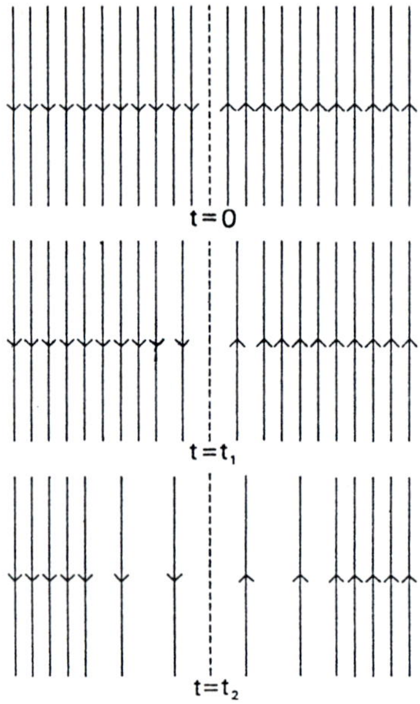

Fig. 2.5. Field lines at three times

or, substituting for $\partial B/\partial t$ from (2.20) and integrating by parts, this becomes

$$\int_{-\infty}^{\infty} \frac{B\eta}{\mu}\frac{\partial^2 B}{\partial x^2}dx = \frac{1}{\mu^2\sigma}\left\{\left[B\frac{\partial B}{\partial x}\right]_{-\infty}^{\infty} - \int_{-\infty}^{\infty}\left(\frac{\partial B}{\partial x}\right)^2 dx\right\} .$$

Now $\partial B/\partial x$ remains equal to zero at infinity so that the first term vanishes, and from (2.7a) the electric current is $j = \mu^{-1}\partial B/\partial x$ so that finally we have

$$\frac{\partial}{\partial t}\int_{-\infty}^{\infty}\frac{B^2}{2\mu}dx = -\int\frac{j^2}{\sigma}dx .$$

In other words the loss of magnetic energy has gone entirely into ohmic heating (j^2/σ per unit volume).

The main reason for variations in $R_{\rm m}$ from one phenomenon to another is variations in the appropriate length-scale L_0. If $R_{\rm m} \gg 1$, the induction equation becomes

$$\frac{\partial \mathbf{B}}{\partial t} = \nabla \times (\mathbf{v} \times \mathbf{B}) , \tag{2.15}$$

and Ohm's Law reduces to $\mathbf{E}+\mathbf{v}\times\mathbf{B} = \mathbf{0}$ so that the total electric field vanishes. Then, if we consider a curve C (bounding a surface S) which is moving with the plasma, in a time dt an element $\mathbf{ds}$ of C sweeps out an element of area $\mathbf{v}dt \times \mathbf{ds}$. The rate of change $d/dt \int \mathbf{B}.\mathbf{dS}$ of magnetic flux through C then consists of two parts, namely

$$\int_S \frac{\partial \mathbf{B}}{\partial t}.\mathbf{dS} + \int_C \mathbf{B}.\mathbf{v}\times\mathbf{ds} .$$

As C moves, so the flux changes partly because the magnetic field is changing with time (the first term) and partly because of the motion of the boundary (the second term). Then, by putting $\mathbf{B}.\mathbf{v} \times \mathbf{ds} = -\mathbf{v} \times \mathbf{B}.\mathbf{ds}$ and applying Stokes theorem to the second term, we obtain

$$\frac{d}{dt}\int \mathbf{B}.\mathbf{dS} = \int_S \left(\frac{\partial \mathbf{B}}{\partial t} - \nabla \times (\mathbf{v} \times \mathbf{B})\right).\mathbf{dS} \ ,$$

which vanishes in the present approximation.

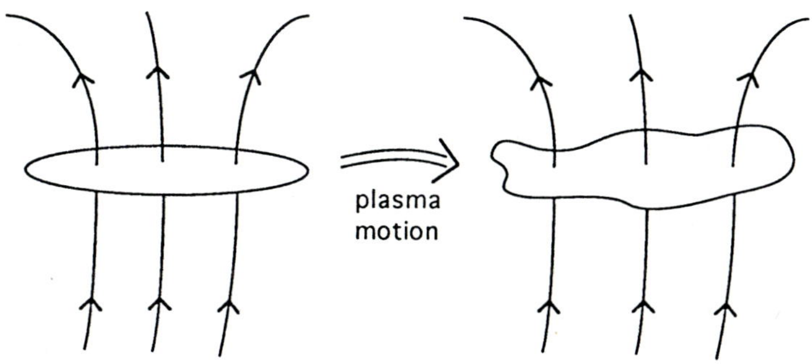

Fig. 2.6. Magnetic flux conservation - if a curve C_1 is distorted into C_2 by a plasma motion, the flux through C_1 at t_1 equals the flux through C_2 at t_2

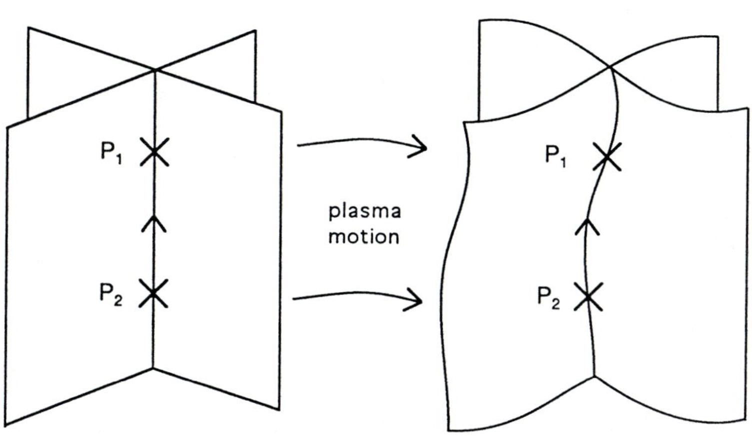

Fig. 2.7. Magnetic field line conservation - if plasma elements P_1 and P_2 lie on a field line at t_1, then they will lie on the same line at t_2

Thus the total magnetic flux through C remains constant as it moves with the plasma. In other words we have proved *magnetic flux conservation*, namely that plasma elements that form a flux tube initially do so at all later times (Figure 2.6). There is also *magnetic field line conservation*, namely that, if two plasma elements lie on a field line initially, they will always do so (Figure 2.7). At $t = t_1$, say, suppose elements P_1 and P_2 lie on a field line, which may be defined as the intersection of two flux tubes.

Then at some later time, $t = t_2$, by magnetic flux conservation P_1 and P_2 will still lie on both tubes, and so they will lie on the field line defined by their intersection. We interpret the above results by saying that the magnetic field lines move with the plasma - we say that they are frozen into the plasma - plasma can move freely along field lines, but in motion perpendicular to them they are dragged with the plasma or vice versa.

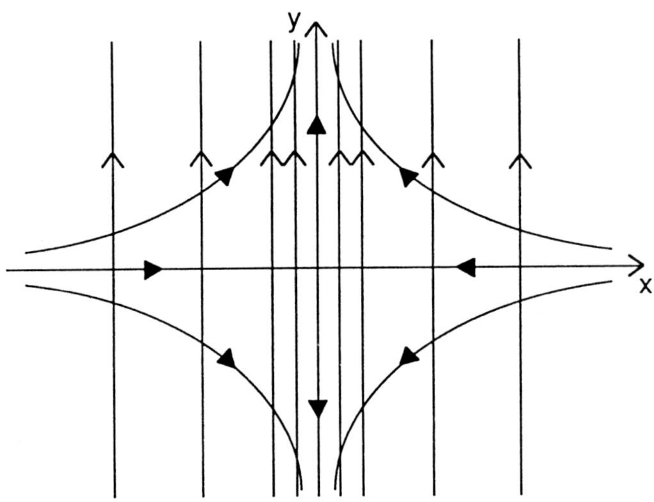

Fig. 2.8. A sketch of the magnetic lines at $t = 0$ for $|x| < \pi a/2$ and the streamlines (solid-headed arrows)

As an example, consider the effect of a flow

$$v_x = -\frac{V_0 x}{a} \; , \; v_y = \frac{V_0 y}{a} \tag{2.16}$$

on a field that is initially

$$\mathbf{B} = B_0 \cos\frac{x}{a}\hat{\mathbf{y}} \text{ at } t = 0 \; . \tag{2.17}$$

The equations of the streamlines are obtained from $dy/dx = v_y/v_x = -y/x$, or, after separating and integrating, xy =constant. These are rectangular hyperbolae (Figure 2.8) with inflow along the x-axis and outflow along the y-axis when $V_0 > 0$.

This represents a stagnation-point flow, so physically what would you expect the effect to be on the initially vertical field lines? Since the flow is carrying the field lines inwards from left and right they will accumulate near $x = 0$ where the field strength should grow. Also, since the component (v_x) of the velocity perpendicular to the field lines is constant along a particular field line ($x =$ constant), the field line will not be distorted but will remain straight as it comes in.

Now, the y-component of the induction equation (2.15) when $R_m \gg 1$ is $\partial B/\partial t = -\partial/\partial x\,(v_x B)$or

$$\frac{\partial B}{\partial t} - \frac{V_0 x}{a}\frac{\partial B}{\partial x} = \frac{V_0 B}{a} , \tag{2.18}$$

and this determines $B(x,t)$. In order to solve this partial differential equation, we consider characteristic curves in the $x - t$ plane which are defined to be such that

$$\frac{dx}{dt} = -\frac{V_0 x}{a} , \tag{2.19}$$

with solution

$$x = x^* e^{-V_0 t/a} , \tag{2.20}$$

where $x = x^*$, say, at $t = 0$.

We wish to determine $B(x,t)$ at every point of the $x - t$ plane and the elegance of considering characteristic curves, $x = x(t)$ given by (2.20), is that on such curves $B(x(t),t)$ has derivative

$$\frac{dB}{dt} = \frac{\partial B}{\partial t} + \frac{dx}{dt}\frac{\partial B}{\partial x} = \frac{\partial B}{\partial t} - \frac{V_0 x}{a}\frac{\partial B}{\partial x} ,$$

by (2.19), or, from (2.18), $dB/dt = V_0 B/a$.

In other words, on the characteristic curves we have a simple ordinary differential equation to solve in place of (2.18): the solution is $B = constant\ e^{V_0 t/a}$ or, since at $t = 0, x = x^*$ and $B = B_0 \cos x^*/a$, we have $B(x,t) = B_0 \cos x^*/a e^{V_0 t/a}$.

However, in this solution x^* is a constant we have introduced for our own benefit which was not present in the initial statement of the problem, so we should eliminate it by using (2.20), with the final result

$$B(x,t) = B_0 \cos\left(\frac{x}{a} e^{V_0 t/a}\right) e^{V_0 t/a} . \tag{2.21}$$

This solution is plotted in Figure 2.9 against x for several times and for the field initially between $x = \pm\pi a/2$. It can be seen that the field does indeed concentrate near $x = 0$ as time proceeds. The field strength at the origin is $B(0,t) = B_0 e^{V_0 t/a}$, and so grows exponentially in time (or decreases if the flow is reversed by taking $V_0 < 0$). The resulting field lines are shown in Figure 2.10.

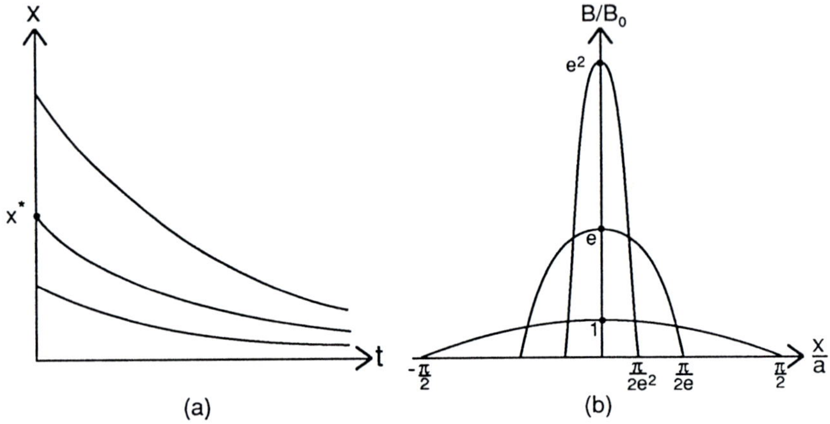

Fig. 2.9. (a) Characteristic curves $x = x^* \exp(-V_0 t/a)$; (b) the solution for B as a function of x for several times

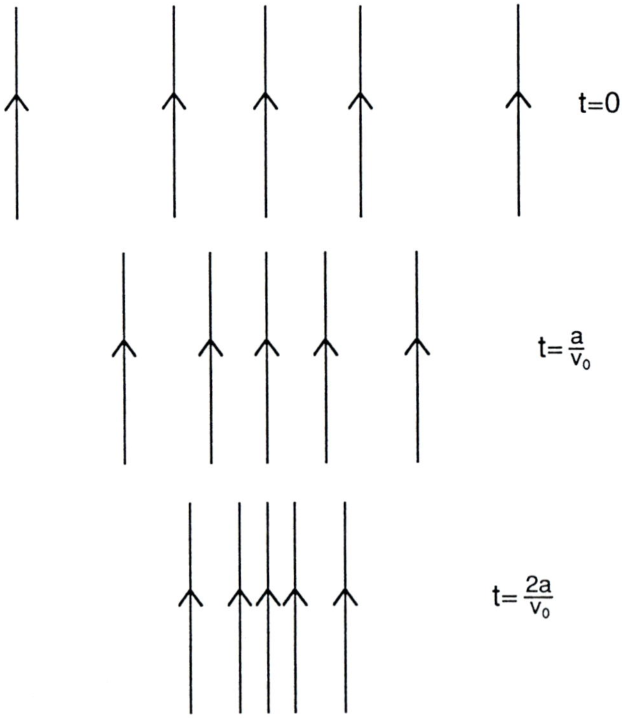

Fig. 2.10. Magnetic field lines initially and at two subsequent times

2.4 The Lorentz Force

One equation relating our fundamental variables $\mathbf{v}$ and $\mathbf{B}$ is the induction equation; the other is the equation of motion

$$\rho\frac{d\mathbf{v}}{dt} = -\nabla p + \mathbf{j}\times\mathbf{B} \ . \tag{2.22}$$

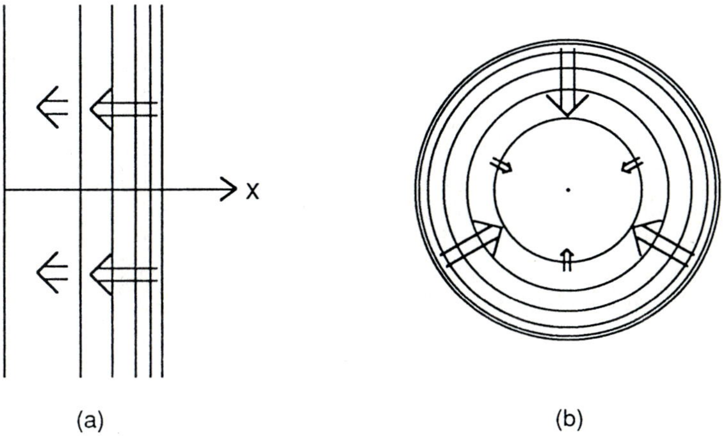

Fig. 2.11. Isobars and directions of pressure gradient for (a) $p = x^2$ and (b) $p = r^2$

The term $-\nabla p$ is the plasma pressure gradient: it acts from regions of high plasma pressure to low pressure and is perpendicular to the isobars (curves of constant pressure).

Consider, for example, the pressure $p = x^2$ for which the isobars are the straight lines $x =$ constant. For this pressure $-\nabla p = -\partial p/\partial x\,\hat{\mathbf{x}} = -2x\,\hat{\mathbf{x}}$ and so the pressure gradient is, for $x > 0$, in the negative x-direction and increases with x (Figure 2.11a). This is therefore perpendicular to the isobars and acts from high to low pressure.

Similarily, the pressure $p = r^2$ has circles $r =$ constant for isobars and the pressure gradient $-\nabla p = -\partial p/\partial r\,\hat{\mathbf{r}} = -2r\,\hat{\mathbf{r}}$ acts radially inwards, again perpendicular to the isobars (Figure 2.11b).

Now consider the magnetic force $(\mathbf{j}\times\mathbf{B})$. Since this is perpendicular to the magnetic field, any plasma acceleration parallel to the magnetic field must be caused by other forces. By substituting for $\mathbf{j}$ from Ampère's law (2.7a) and using an identity for the triple vector product, we can write it as

$$\mathbf{j}\times\mathbf{B} = (\nabla\times\mathbf{B})\times\mathbf{B}/\mu = (\mathbf{B}.\nabla)\frac{\mathbf{B}}{\mu} - \nabla\left(\frac{B^2}{2\mu}\right) \ . \tag{2.23}$$

The two terms on the right-hand side have important physical interpretations. Since the second term has the same form as $-\nabla p$, we can say that it represents the effect of a *magnetic pressure* of magnitude $B^2/(2\mu)$. It gives a force when B^2 varies with position, and the direction of the force is from regions of high magnetic pressure to low magnetic pressure.

The first term represents the effect of a *magnetic tension* parallel to the magnetic field of magnitude B^2/μ per unit area. It gives a force when the field lines are curved, just like an elastic band or rope. By writing $\mathbf{B} = B\hat{\mathbf{s}}$ in terms of the unit vector $\hat{\mathbf{s}}$ along the magnetic field, the tension term $(\mathbf{B}.\nabla)\mathbf{B}/\mu$ may be written

$$\frac{B}{\mu}\frac{d}{ds}(B\hat{\mathbf{s}}) = \frac{B^2}{\mu}\frac{d\hat{\mathbf{s}}}{ds} + \frac{B}{\mu}\frac{dB}{ds}\hat{\mathbf{s}} = \frac{B^2}{\mu}\frac{\hat{\mathbf{n}}}{\mathcal{R}} + \frac{d}{ds}\left(\frac{B^2}{2\mu}\right)\hat{\mathbf{s}} ,$$

where $\hat{\mathbf{n}}$ is the principal normal and $\mathcal{R}$ is the radius of curvature. The second term on the right of this equation is irrelevant, since it cancels with the component of $-\nabla(B^2/(2\mu))$ parallel to $\mathbf{B}$, as it must since $\mathbf{j} \times \mathbf{B}$ is perpendicular to $\mathbf{B}$. However, the first term on the right is the magnetic tension term, which shows that when the radius of curvature is small the tension force is large.

Thus we have shown that the magnetic force has two distinct effects: the magnetic pressure tries to compress the plasma and produces a net force if B varies with position; the magnetic tension tends to make the field lines shorten themselves like elastic bands and gives a net force if the field lines are curved.

As a first example consider $\mathbf{B} = x\hat{\mathbf{y}}$, which represents a current sheet with uniform current in the $z-$ direction:

$$\mathbf{j} = \nabla \times \mathbf{B}/\mu = \frac{1}{\mu}\frac{\partial B_y}{\partial x}\hat{\mathbf{z}} = \frac{\hat{\mathbf{z}}}{\mu} .$$

The field lines are straight lines $x =$ constant which reverse at $x = 0$ and become more closely spaced as $|x|$ increases (Figure 2.12a). Physically we expect the magnetic tension force to vanish (since field lines are straight) and the magnetic pressure force to act towards the origin away from the regions of high magnetic pressure. Mathematically, we find

$$(\mathbf{B}.\nabla)\frac{\mathbf{B}}{\mu} = \frac{B_y}{\mu}\frac{\partial B_y}{\partial y}\hat{\mathbf{y}} = 0$$

and

$$-\nabla\left(\frac{B^2}{2\mu}\right) = -\nabla\left(\frac{x^2}{2\mu}\right) = -\frac{x}{\mu}\hat{\mathbf{x}} ,$$

whose properties agree with what was expected physically.

Consider next a magnetic field $\mathbf{B} = \hat{\mathbf{x}} + x\hat{\mathbf{y}}$, for which the field lines are given by $dy/dx = x$ and so are the parabolae $y = \frac{1}{2}x^2 + c$. On the y-axis the field is uniform ($\mathbf{B} = \hat{\mathbf{x}}$) and so we expect no magnetic pressure force, but the

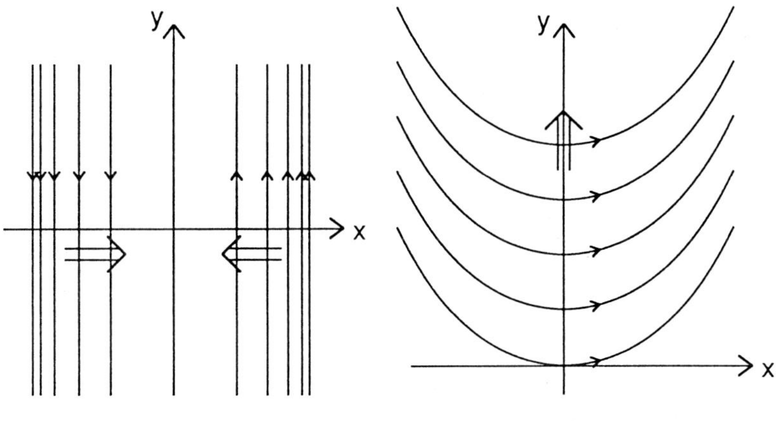

Fig. 2.12. Magnetic forces due to (a) $\mathbf{B} = x\hat{\mathbf{y}}$ and (b) $\mathbf{B} = \hat{\mathbf{x}} + x\hat{\mathbf{y}}$

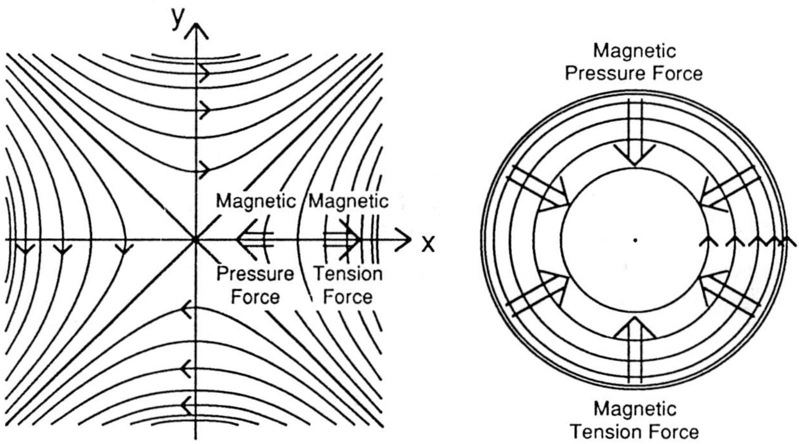

Fig. 2.13. Magnetic forces due to the fields (a) $\mathbf{B} = y\hat{\mathbf{x}} + x\hat{\mathbf{y}}$ and (b) $\mathbf{B} = r\hat{\theta}$

field lines are curved (Figure 2.12b) and so we expect an upwards tension force. The mathematics again supports this physical intuition, since

$$(\mathbf{B}.\nabla)\frac{\mathbf{B}}{\mu} = \left(\frac{\partial}{\partial x} + x\frac{\partial}{\partial y}\right)(\hat{\mathbf{x}} + x\hat{\mathbf{y}})/\mu = \frac{\hat{\mathbf{y}}}{\mu} \ ,$$

which is vertically upwards and

$$-\nabla\left(\frac{B^2}{2\mu}\right) = -\nabla\left(\frac{1+x^2}{2\mu}\right) = -\frac{x}{\mu}\hat{\mathbf{x}} \ ,$$

which vanishes on the x-axis.

Consider next the field with components $B_x = y$, $B_y = x$, with X-type field lines shown in Figure 2.13a. At a point on the positive x-axis, the curvature of the field lines suggests a magnetic tension force to the right, while the fact that the magnetic pressure is increasing (like x^2) away from the origin suggests a magnetic pressure force to the left. However, from the field lines alone, it is not evident which dominates. Now

$$(\mathbf{B}.\nabla)\frac{\mathbf{B}}{\mu} = \left(y\frac{\partial}{\partial x} + x\frac{\partial}{\partial y}\right)\left(\frac{y}{\mu}\hat{\mathbf{x}} + \frac{x}{\mu}\hat{\mathbf{y}}\right) = \frac{x}{\mu}\hat{\mathbf{x}} + \frac{y}{\mu}\hat{\mathbf{y}} \ ,$$

which becomes $x\hat{\mathbf{x}}/\mu$ on the x-axis and so does indeed act away from the origin. Also the magnetic pressure force is

$$-\nabla\left(\frac{B^2}{2\mu}\right) = -\nabla\left(\frac{x^2+y^2}{2\mu}\right) = -\frac{x}{\mu}\hat{\mathbf{x}} - \frac{y}{\mu}\hat{\mathbf{y}} \ ,$$

which becomes $-x\hat{\mathbf{x}}/\mu$ on the x-axis and so does act towards the origin. We find therefore that the tension and pressure forces are in precise balance, as they must be since the electric current is

$$\mathbf{j} = \frac{1}{\mu}\left(\frac{\partial B_y}{\partial x} - \frac{\partial B_x}{\partial y}\right)\hat{\mathbf{z}} = 0 \ ,$$

and so the Lorentz force $\mathbf{j} \times \mathbf{B}$ vanishes everywhere.

Finally, consider $\mathbf{B} = r\hat{\theta}$ for which the field lines are the circles r = constant, becoming more closely spaced as one moves outwards. Thus, physically the magnetic tension and pressure forces are expected both to act inwards. Mathematically, we find the magnetic pressure force is

$$-\nabla\left(\frac{B^2}{2\mu}\right) = -\nabla\left(\frac{r^2}{2\mu}\right) = -\frac{r}{\mu}\hat{\mathbf{r}} \ ,$$

which does indeed act inwards. Also the magnetic tension is

$$(\mathbf{B}.\nabla)\frac{\mathbf{B}}{\mu} = \left(\frac{B_\theta}{r}\frac{\partial}{\partial\theta}\right)\left(\frac{B_\theta\hat{\theta}}{\mu}\right) = \frac{\partial}{\partial\theta}\left(\frac{r\hat{\theta}}{\mu}\right) \ ,$$

and now r/μ does not vary with θ but $\hat{\theta}$ does! In fact $\partial\hat{\theta}/\partial\theta = -\hat{\mathbf{r}}$ and so $(\mathbf{B}.\nabla)\,\mathbf{B}/\mu = -r\hat{\mathbf{r}}/\mu$ which again acts inwards, as expected.

2.5 Concluding Comment

One feature I have been trying to stress by example is the importance of building both a physical and mathematical understanding of magnetic field behaviour, since they both complement one another and together give a deeper understanding than one alone.

3 Magnetohydrostatics

3.1 Introduction

The equation of motion is

$$\rho \frac{d\mathbf{v}}{dt} = -\nabla p + \mathbf{j} \times \mathbf{B} + \rho \mathbf{g} \ , \tag{3.1}$$

where

$$\mathbf{j} = \nabla \times \mathbf{B}/\mu \ , \tag{3.2}$$

and, if $L, v_0, L/v_0$ are typical values for the length-scale, plasma velocity and time-scale, respectively, the orders of magnitude of the current from (3.2) is $j_0 = B_0/(\mu L_0)$. Then, in terms of the typical density (ρ_0), pressure (p_0) and magnetic field (B_0), the sizes of each term in equation (3.1) are $\rho_0 v_0{}^2/L$, p_0/L, $B_0{}^2/(\mu L)$, $\rho_0 g$, respectively.

Now, provided the magnetic term is of the same order as the largest force term, we have *force balance* if the first term is much smaller than the third, namely

$$v_0{}^2 \ll \frac{B_0{}^2}{\mu \rho_0} \equiv v_A{}^2 \ , \tag{3.3}$$

where v_A is known as the Alfven speed. Then (3.1) reduces to the equation for *magnetohydrostatic force balance*

$$\mathbf{0} = -\nabla p + \mathbf{j} \times \mathbf{B} + \rho \mathbf{g}, \tag{3.4}$$

where $\mathbf{j} = \nabla \times \mathbf{B}/\mu$, $\nabla.\mathbf{B} = \mathbf{0}$ and $\rho = P/\mathcal{R}T$.

In this equation, when gravity is negligible we have *magnetostatic balance*

$$0 = -\nabla p + \mathbf{j} \times \mathbf{B} \tag{3.5}$$

if the fourth term in (3.1) is much smaller than the third, namely $L \ll B_0{}^2/(\mu \rho_0 g) = 2H/\beta$, where

$$H = \frac{p_0}{\rho_0 g} = \frac{\mathcal{R}T_0}{g} \tag{3.6}$$

is the *pressure scale height* and

$$\beta = \frac{p_0}{B_0{}^2/(2\mu)} \tag{3.7}$$

is the *plasma beta*, namely the ratio of plasma to magnetic pressure.

In turn, in (3.5) the magnetic force dominates if $2\beta \ll 1$, and then (3.5) reduces further to the equation

$$\mathbf{j} \times \mathbf{B} = \mathbf{0} \tag{3.8}$$

for a *force-free field*, in which the magnetic field is in equilibrium with itself under a balance between magnetic pressure and magnetic tension forces.

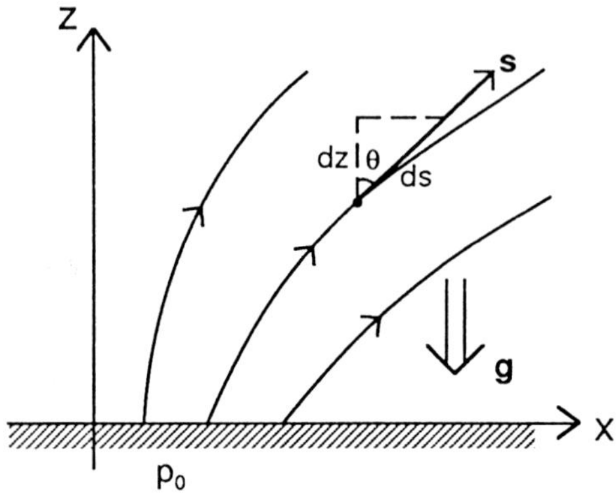

Fig. 3.1. Nomenclature for plasma in equilibrium along a field line

Suppose now that gravity is directed vertically downwards in the negative z-direction $\mathbf{g} = -g\hat{\mathbf{z}}$. Then the component of (3.4) parallel to a particular magnetic field line is $0 = -dp/ds - \rho g \cos\theta$, where s is measured along the field, or, since $ds \cos\theta = dz$, we have $0 = -dp/dz - \rho g$. After putting $\rho = p/(\mathcal{R}T)$ this becomes

$$\frac{dp}{dz} = -p\frac{g}{\mathcal{R}T} ,$$

which may be integrated to give

$$p = p_0 \exp\left(-\int_0^z \frac{g}{\mathcal{R}T}dz\right) , \tag{3.9}$$

where p_0 is the pressure at the base ($z = 0$) of the field line. If the variation $T(z)$ of the temperature with height is known, (3.9) determines the pressure and therefore the density.

If in particular the temperature is uniform ($T = T_0$) then

$$p = p_0 \, e^{-z/H} , \tag{3.10}$$

so that the pressure (and density) decrease exponentially with height, with H being the vertical distance over which the pressure falls off by a factor e.

Down in the photosphere where T_0= 5000K, the scale height is about 150 km or less and so, over for instance a vertical distance of 1.5 Mm, the pressure and density would fall off by a factor of $e^{10} \approx 20,000$. By contrast, in the corona where $T_0 = 2 \times 10^6$K, say, the scale height is about 100 Mm, and so the density falls off much more slowly. Indeed for many purposes in the corona we can neglect the effect of gravity - i.e. when the vertical scales of interest are 100 Mm or less.

3.2 Potential Fields

If the pressure gradients and gravitational force are negligible, (3.8) holds and a particular case of interest is when the current vanishes, so that

$$\nabla \times \mathbf{B} = 0 , \tag{3.11}$$

where

$$\nabla.\mathbf{B} = 0 \ . \tag{3.12}$$

Equation (3.11) may be satisfied identically by putting $\mathbf{B} = \nabla\Psi$ and then (3.12) gives Laplace's equation

$$\nabla^2\Psi = 0. \tag{3.13}$$

Thus many of the general results of potential theory may be applied. For example: *If the normal field component ($\mathrm{B_n}$) is imposed on the boundary S of a volume V, then the potential solution inside V is unique.*

Proof. Consider two potential solutions Ψ_1 and Ψ_2 such that

$$\nabla^2\Psi_1 = \nabla^2\Psi_2 = 0 \text{ in } V \ , \tag{3.14}$$

and

$$\frac{\partial\Psi_1}{\partial n} = \frac{\partial\Psi_2}{\partial n} \text{ on } S \ . \tag{3.15}$$

Put $\Psi^* = \Psi_1 - \Psi_2$, and consider

$$\int_V B^{*2} dV = \int_V \nabla\Psi^* \ .\nabla\Psi^* dV = \int_V \nabla.(\Psi^*\nabla\Psi^*) - \Psi^*\nabla^2\Psi^* dV \ .$$

But $\nabla^2\Psi^* = 0$ by (3.14) and so applying the divergence theorem to the first term gives

$$\int_V B^{*2} dV = \int_S \Psi^* \frac{\partial\Psi^*}{\partial n} dS \ ,$$

which vanishes since $\partial\Psi^*/\partial n = 0$ on S by (3.15). Thus B^* is identically zero so that $\mathbf{B_1} \equiv \mathbf{B_2}$ and the theorem is proved.

Another useful theorem is: *If B_n is prescribed on $\mathcal{S}$, then the field with the minimum magnetic energy is the potential field.*

Proof. Put

$$\mathbf{B} = \mathbf{B}_0 + \mathbf{b} \ ,$$

where $\nabla \times \mathbf{B}_0 = 0$ and $\nabla.\mathbf{B}_0 = 0$ so that $\mathbf{B}_0$ is a potential field, and suppose both $\mathbf{B}_0$ and $\mathbf{B}$ have the same value for their normal components on $\mathcal{S}$. Then

$$\nabla.\mathbf{b} = 0 \tag{3.16}$$

and $b_n = 0$ on $\mathcal{S}$.

Next, write $\mathbf{B}_0 = \nabla\Psi_0$ and consider

$$\int_V B^2\, dV = \int_V \mathbf{B}.\mathbf{B}\, dV = \int_V {B_0}^2 + 2\mathbf{B}_0.\mathbf{b} + b^2\, dV \ , \qquad (3.17)$$

where

$$\int_V \mathbf{B}_0.\mathbf{b}\, dV = \int_V \nabla\Psi.\mathbf{b}\, dV \ ,= \int_V \nabla.(\Psi\mathbf{b}) - \Psi\nabla.\mathbf{b}\, dV \ .$$

However, the second term vanishes by (3.16) and the first may be transformed by the divergence theorem so that the net result is

$$\int_V \mathbf{B}_0.\mathbf{b}\, dV = \int_S \Psi b_n\, dS \ ,$$

which vanishes since b_n is zero on S. Thus from (3.17)

$$\int_V B^2 dV > \int_V {B_0}^2 dV \ ,$$

and therefore $\mathbf{B}_0$ has the minimum magnetic energy.

For example, it is known that during a large solar flare the normal field component at the solar surface does not change appreciably, and so the magnetic energy source for such a flare must come from a sheared force-free field with energy in excess of potential.

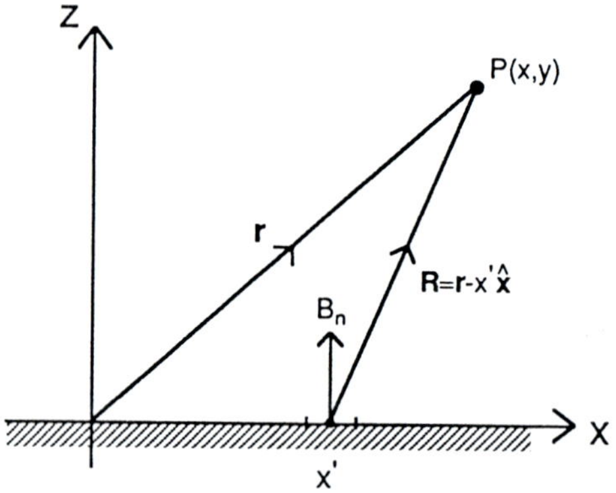

Fig. 3.2. Notation for calculating the potential field at P due to an imposed B_n on the x-axis

Several types of potential solution are of interest.

(a) We may calculate the field at a point P(x,y) due to an imposed $B_n(x)$ on the x-axis (Figure 3.2). Imagine a pole at the point $x'\hat{\mathbf{x}}$ on the x-axis producing a field $\mathbf{B} = (m/R)\hat{\mathbf{R}}$ at P, where $\mathbf{R} = \mathbf{r} - x'\hat{\mathbf{x}}$. Since this field has the same magnitude (m/R) at every point on a circle centred on the pole and having radius R, the magnetic flux crossing the upper semicircle of length πR is mR. But this flux is produced by an element of length dx' containing a field B_n and so equals $B_n dx'$. Thus $m = B_n dx'/\pi$.

Integrating along the x-axis we obtain the field at P produced by a continuous distribution of poles with a field $B_n(x')$ on the x-axis as

$$\mathbf{B} = \frac{1}{\pi}\int \frac{B_n(x')}{R}\hat{\mathbf{R}}\,dx' \ . \tag{3.18}$$

A similar technique may be used to calculate the field above a plane or the surface of a sphere rather than a line.

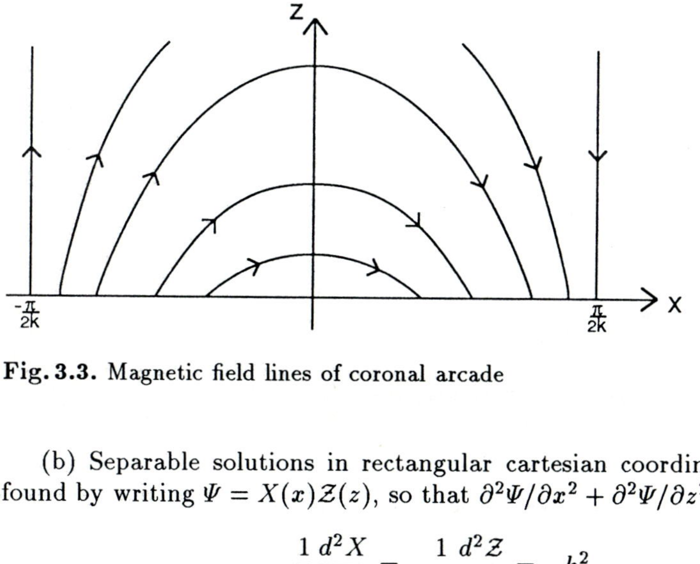

Fig. 3.3. Magnetic field lines of coronal arcade

(b) Separable solutions in rectangular cartesian coordinates (x,z) may be found by writing $\Psi = X(x)\mathcal{Z}(z)$, so that $\partial^2\Psi/\partial x^2 + \partial^2\Psi/\partial z^2 = 0$ implies

$$\frac{1}{x}\frac{d^2X}{dx^2} = -\frac{1}{\mathcal{Z}}\frac{d^2\mathcal{Z}}{dz^2} = -k^2 \ ,$$

say, where k is a constant. A useful solution is $\Psi = (B_0/k)\sin kx\, e^{-kz}$, so that

$$B_x = \frac{\partial\Psi}{\partial x} = B_0 \cos kx\, e^{-kz} \ ,$$

$$B_z = \frac{\partial\Psi}{\partial z} = B_0 \sin kx\, e^{-kz} \ . \tag{3.19}$$

The resulting field lines shown in Figure 3.3 for $|x| < \pi/(2k), z > 0$ are a reasonable model for a potential coronal arcade.

(c) Separable solutions in spherical polars (r, θ, ϕ) of Laplace's equation are

$$\Psi = \sum_{l=0}^{\infty}\sum_{m=-l}^{l}\left(a_{lm}r^{l} + b_{lm}r^{-l-1}\right)P_l^m(\cos\theta)\,e^{im\phi} \ ,$$

where P_l^m is the associated Legendre polynomial. If Ψ is independent of ϕ, this reduces to

$$\Psi = \sum_{l=0}^{\infty} \left(a_l r^r + b_l r^{-l-1}\right) P_l(\cos\theta) \ ,$$

where P_l is the Legendre polynomial. These may be used in a variety of problems in a spherical geometry.

(d) Similarly, problems in a cylindrical geometry may be treated in terms of cylindrical polars, for which the general solution is

$$\Psi = \sum_{n=-\infty}^{\infty} \left(c_n J_n(kR) + d_n Y_n(kR)\right) e^{in\phi \pm kz} \ ,$$

or, with no z-dependence,

$$\Psi = c \log R + \sum_{n=-\infty}^{\infty} \left(c_n R^n + d_n R^{-n}\right) e^{in\phi} \ ,$$

where J_n and Y_n are Bessel functions.

3.3 Force-Free Fields

If pressure gradients and gravity are negligible we have

$$\mathbf{j} \times \mathbf{B} = \mathbf{0} \ , \tag{3.20}$$

so that $\mathbf{j} = \nabla \times \mathbf{B}/\mu$ is parallel to $\mathbf{B}$. In other words

$$\nabla \times \mathbf{B} = \alpha \mathbf{B} \ , \tag{3.21}$$

where α is a function of position. Such a magnetic field is called *force-free*. Equation (3.20) looks disarmingly simple, but very little has been done so far in understanding the nature of its solutions in general.

Taking the divergence of (3.21) we find $\nabla.\nabla \times \mathbf{B} = \nabla.(\alpha\mathbf{B})$, or

$$0 = \alpha \nabla.\mathbf{B} + \mathbf{B}.\nabla\alpha \ .$$

Thus, since $\nabla.\mathbf{B}$ vanishes, $\mathbf{B}.\nabla\alpha = 0$, which implies that, although in general α varies in space, it has a constant value along each field line.

If instead we take the curl of (3.21)

$$\nabla \times (\nabla \times \mathbf{B}) = \nabla \times (\alpha\mathbf{B}) \ ,$$

or

$$\nabla(\nabla.\mathbf{B}) - \nabla^2\mathbf{B} = \alpha\nabla \times \mathbf{B} + \mathbf{B} \times \nabla\alpha \ .$$

Thus, after putting $\nabla.\mathbf{B} = \mathbf{0}$ and substituting for $\nabla \times \mathbf{B}$ from (3.22),

$$-\nabla^2\mathbf{B} = \alpha^2\mathbf{B} + \mathbf{B} \times \nabla\alpha. \tag{3.22}$$

However, in general this is no simpler than the original equation (3.21) since it represents three coupled equations for the three coefficients of **B**.

In the particular case when α is uniform and so possesses the same constant value on all field lines, (3.22) reduces to the simple equation

$$\left(\nabla^2 + \alpha^2\right)\mathbf{B} = 0. \tag{3.23}$$

Its solutions give so-called "constant- α" or linear force-free fields and may be found by a variety of techniques, generalising many of the techniques for solving Laplace's equation. The beauty of (3.23) is that it is a linear equation so that the solutions may be superposed.

There are several general theorems which lend an air of mystery to the structure of force-free fields, since very little is known of their general properties:

(i) If the flux and topological connections on the surface S of a simply connected volume V are given and the field possesses a minimum energy, then it is force-free. However, the converse is not true, so that a force-free field does not necessarily produce a minimum energy.

(ii) If $\mathbf{j}\times\mathbf{B}$ vanishes everywhere within V and on S, then the magnetic field **B** is identically zero. Thus, a nontrivial (i.e. non-zero) field that is force-free inside V must be stressed somewhere on S. In other words, force-free fields are possible but they must be anchored down somewhere on the boundary. You cannot build a force-free field from currents enclosed entirely within a volume.

(iii) No magnetic field having a finite energy can be force-free everywhere. Thus a nontrivial field that is force-free everywhere must possess a singularity somewhere: this is also true in particular for a potential field where one is familiar with fields produced by, for instance, point monopoles or dipoles.

(iv) An axisymmetric, force-free, poloidal magnetic field must be current-free. Thus one cannot construct force-free fields in polars that are too simple.

(v) For a perfectly conducting plasma in a closed volume V, the *magnetic helicity* is

$$K = \int_{V} \mathbf{A}.\mathbf{B}\, dV \ , \tag{3.24}$$

where $\mathbf{B} = \nabla \times A$ is invariant, and the state of minimum energy is a linear force-free field. K is a measure of the sum of the twist and the linkage of flux tubes. Thus for two linked tubes

$$K = \Theta_1 {F_1}^2 + \Theta_2 {F_2}^2 + 2LF_1F_2, \tag{3.25}$$

where Θ_1 and Θ_1 are the twists of the two tubes, F_1, F_2 their fluxes and L the linking number (Figure 3.4). Reconnection can convert linkage helicity to twist helicity, but it preserves the total helicity.

Simple linear force-free fields may be set up by generalising the various separable solutions for a potential field. The simplest solution is for a one-dimensional field

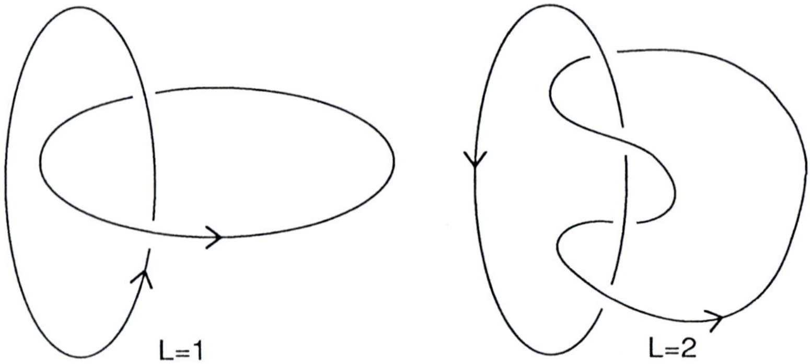

Fig. 3.4. Linked tubes with linkage (L) of 1 and 2

$$\mathbf{B} = (0, B_y(x), B_z(x)) \ ,$$

with y- and z- components that vary with x. The magnetic tension force vanishes because the field lines are straight and so $\mathbf{j}\times\mathbf{B} = 0$ reduces to $d/dx({B_y}^2+{B_z}^2) = 0$, with solution

$$B_z(x) = ({B_0}^2 - {B_y}^2)^{\frac{1}{2}}, \tag{3.26}$$

so that the field lines lie in planes parallel to the $y-z$ plane and simply rotate with constant magnitude as one moves along the x-axis. In particular, for a constant-α field, the z-component of (3.21) is $dB_y/dx = \alpha B_z$, or from (3.26)

$$\frac{dB_y}{dx} = \alpha({B_0}^2 - {B_y}^2)^{\frac{1}{2}} \ ,$$

with solution (such that $B_y(0) = 0$)

$$B_y = B_0 \sin \alpha x, \ B_z = B_0 \cos \alpha x \ .$$

A model for a two-dimensional force-free coronal arcade may be built by seeking solutions of (3.21) and (3.23) in the form $B_z = X(x)\mathcal{Z}(z)$, for which the z-component of (3.23) gives

$$\frac{1}{X}\frac{d^2X}{dx^2} = -\frac{1}{\mathcal{Z}}\frac{d^2\mathcal{Z}}{dz^2} - \alpha^2 = -k^2, \quad \text{say} \ .$$

A solution of interest is

$$B_z = -B_0 \sin kx . e^{-lz} \ , \tag{3.27}$$

where

$$l^2 = k^2 - \alpha^2 \ . \tag{3.28}$$

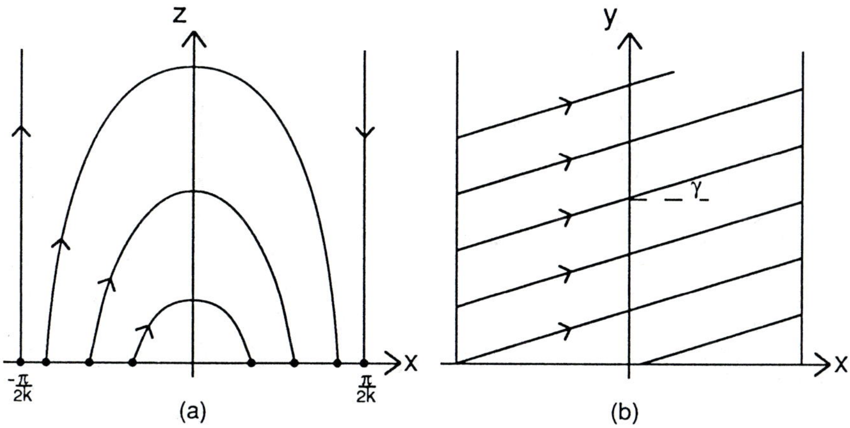

Fig. 3.5. Field lines in a vertical (x-z) and horizontal (x-y) plane for a force-free coronal arcade with z measured vertically upwards from the photosphere

Then the z- and x- components of (3.21) are $\alpha B_z = \partial B_y/\partial x$ and $\alpha B_x = -\partial B_y/\partial z$, which have as solutions

$$B_y = \frac{\alpha B_0}{k} \cos kx\, e^{-lz} , \tag{3.29}$$

$$B_x = \frac{l B_0}{k} \cos kx\, e^{-lz} . \tag{3.30}$$

The resulting field lines in vertical and horizontal planes are shown in Figure 3.5. As α increases from 0 (the potential case with vanishing current) to k, so according to (3.28) l decreases from l to 0. The field lines in the x-z plane expand upwards keeping the same footpoints, while in the x-y plane the shear angle $\gamma = \tan^{-1} \alpha/(k^2 - \alpha^2)^{\frac{1}{2}}$ increases from 0 to $\pi/2$.

Non-constant-α fields in two dimensions ($\mathbf{B}(x, z)$) may be considered by writing the components in terms of a flux function (A) as

$$B_x = \frac{\partial A}{\partial z}, \quad B_y(x, z), \quad B_z = -\frac{\partial A}{\partial x} , \tag{3.31}$$

so that $\nabla.\mathbf{B} = 0$ is satisfied automatically. We are interested in the field above the surface ($z = 0$, say) of a star or other object with the footpoints of the field anchored in the surface (Figure 3.6). The equations of projections of the field lines in the x-z plane are given by $dx/B_x = dz/B_z$, or $\partial A/\partial x\, dx + \partial A/\partial z\, dz = 0$, or $dA = 0$ so that $A=$ constant.

The resulting current components are

$$j_x = -\frac{1}{\mu}\frac{\partial B_y}{\partial z}, \quad j_y = \frac{1}{\mu}\left(\frac{\partial B_x}{\partial z} - \frac{\partial B_z}{\partial x}\right), \quad j_z = \frac{1}{\mu}\frac{\partial B_y}{\partial x} ,$$

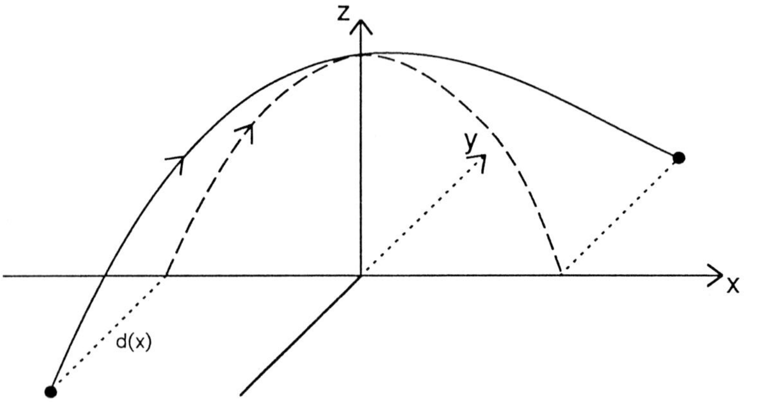

Fig. 3.6. A magnetic field line (solid) sheared by a displacement $d(x)\hat{\mathbf{y}}$ of its footpoints. The dashed curve indicates the projection of the field line onto the x-z plane

so that the components of $\mathbf{j} \times \mathbf{B} = \mathbf{0}$ become

$$\nabla^2 A \frac{\partial A}{\partial x} + b_y \frac{\partial b_y}{\partial x} = 0 \ , \tag{3.32}$$

$$\frac{\partial b_y}{\partial z}\frac{\partial A}{\partial x} - \frac{\partial b_y}{\partial x}\frac{\partial A}{\partial z} = 0 \ , \tag{3.33}$$

$$\nabla^2 A \frac{\partial A}{\partial z} + b_y \frac{\partial b_y}{\partial z} = 0 \ . \tag{3.34}$$

Now (3.33) is simply $\nabla B_y \times \nabla A = 0$, so that the vectors ∇B_y and ∇A are parallel. But they are perpendicular to the surfaces B_y = constant and A= constant, respectively, and so these surfaces must coincide and

$$B_y = B_y(A). \tag{3.35}$$

Then in (3.34), $\partial B_y/\partial z = (dB_y/dA)(\partial A/\partial z)$, so that (3.34) becomes

$$\nabla^2 A + B_y \frac{dB_y}{dA} = 0 \ . \tag{3.36}$$

This is the basic equation for determining the flux function (A) and therefore from (3.32) the x- and z- field components: it is known as the *Grad-Shafranov equation*. In general it is a nonlinear equation for A, but analytical solutions are known for particular forms of the function $B_y(A)$, namely B_y = constant (potential), $B_y = cA^{\frac{1}{2}}$ (constant-current), $B_y = cA$ (constant-alpha), $B_y = e^{-2A}$, and $B_y = A^{-1}$. Otherwise, usually one needs to resort to a numerical treatment and then deduce the footpoint displacement afterwards.

The more difficult problem of imposing the footpoint displacement $d(x)\hat{\mathbf{y}}$ needs to be solved numerically, for then $B_y(A)$ is determined by $dy/B_y = dx/B_x$, or, after integrating from one end of a field line to the summit,

$$d = B_\mathrm{y} \int_{A=\mathrm{const}} \frac{dx}{B_\mathrm{x}} \; .$$

If a pressure gradient $-\nabla p(x,z)$ is present too, one simply adds a term dp/dA to (3.36).

3.4 Magnetic Flux Tubes

Straight tubes with cylindrical symmetry have field lines lying on cylindrical surfaces (Russell et al, 1990) and magnetic and current components in cylindrical polars of

$$\mathbf{B} = (0, B_\phi(R), B_\mathrm{z}(R)) \; , \tag{3.37}$$

$$\mathbf{j} = \left(0, -\frac{dB_\mathrm{z}}{dR}, \frac{1}{R}\frac{d}{dR}(RB_\phi)\right) . \tag{3.38}$$

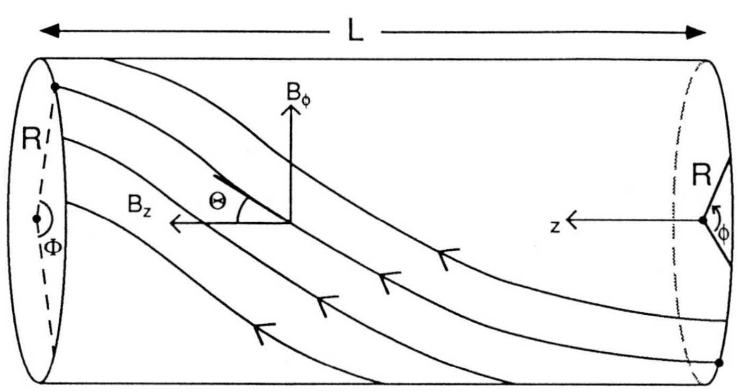

Fig. 3.7. Notation for a straight magnetic flux tube

The amount Φ (the *twist*) by which a field line is twisted about the z-axis in going from one end of the tube to another is

$$\Phi = \frac{LB_\phi}{RB_\mathrm{z}} \; . \tag{3.39}$$

In general Φ varies with R and so field lines on different surfaces have different inclinations. The radial component of the magnetostatic force balance ($\nabla p = \mathbf{j} \times \mathbf{B}$) is

$$\frac{dp}{dR} + \frac{d}{dR}\left(\frac{{B_\phi}^2 + {B_\mathrm{z}}^2}{2\mu}\right) + \frac{{B_\phi}^2}{\mu R} = 0, \tag{3.40}$$

where the terms simply represent the plasma pressure gradient, the magnetic pressure gradient and the magnetic tension forces, respectively. We are allowed to prescribe two of the four functions p, B_ϕ, B_z, Φ of R and deduce the other two from (3.39) and (3.40): which ones are prescribed depends on the extra physics of the particular problem being considered.

Example (i) If $p=$ constant and $\Phi(R) = \Phi_0$ (a "uniform-twist" field), then the solution of (3.39) and (3.40) is

$$B_z = \frac{B_0}{1+\tilde{R}^2}, \qquad B_\phi = \frac{B_0 \tilde{R}}{1+\tilde{R}^2},$$

where $\tilde{R} = \Phi_0 R/L$. Thus B_z decreases with radius, while B_ϕ increases to a maximum and then decreases. As the twist increases, the ϕ-component of the field increases, so the inwards tension force increases, leading to a compensating rise in the outwards magnetic pressure force.

Example(ii) If $B_z = 1$ and the twist is imposed to be $\Phi = \Phi_0/(1+R^2)$, so that it declines with radial distance, the resulting solutions are

$$B_\phi = \frac{R\Phi_0}{L(1+R^2)}, \qquad p = p_0 + \frac{{\Phi_0}^2}{2\mu L^2(1+R^2)}.$$

In this case, as the twist and therefore tension force increase, so the outwards pressure gradient rises to balance it.

Many more extensions of the theory of flux tubes are described in the review by Priest (1990). For example:

(i) the mean-square axial field of a force-free flux tube confined by an external plasma pressure is not affected by twisting (Parker's theorem);

(ii) if an untwisted tube is compressed then the field strength and plasma density increase by the same proportion;

(iii) if a twisted tube is compressed, it tends to become less twisted;

(iv) tubes with a slowly varying cross-section become more twisted in the regions where they are wider;

(v) a general theory for the motion or equilibrium of slender tubes has been developed;

(vi) analytical and numerical results for thick tubes show that the effect of twisting is to make the core contract and become straight over most of its length, while the outer part of the tube expands; too much twist leads to a loss of equilibrium;

(vii) toroidal effects have been included for curved tubes, notably in tokamak and reversed-field pinch orderings;

(viii) the effect of gravity is to make a horizontal tube rise by magnetic buoyancy and a vertical tube to spread out with height;

(ix) a wide variety of instabilities are possible, including kink modes, sausage modes and resistive modes.

4 Magnetohydrodynamic Waves

In a gas there are sound waves which propagate equally in all directions at the sound speed. In a plasma there are also waves, but they are of several types. Waves are very important in the solar atmosphere and throughout the cosmos. For example, they may be seen propagating out of sunspots or away from large solar flares. They are also a prime candidate for heating the solar atmosphere.

In talking about waves, it is fair enough just to wave your hands! But how do we describe them mathematically? For an elastic string one may derive the equations for a small displacement from equilibrium and show that they are of the form of a wave equation with a wave speed equal to the square root of the tension divided by the mass density. The approach is similar for a plasma, where one sets up the equations for small disturbances of the plasma and finds that they too satisfy a wave equation.

4.1 Sound Waves

Consider a uniform plasma at rest with no magnetic field, a pressure p_0 and density ρ_0. Suppose there is a small disturbance with velocity $\mathbf{v}'$, pressure p_0+p', density $\rho_0 + \rho'$, satisfying the equations

$$\rho\frac{d\mathbf{v}}{dt} = -\nabla p \ , \quad \frac{\partial \rho}{\partial t} + \nabla.(\rho\mathbf{v}) = 0 \ , \quad \frac{p}{\rho^\gamma} = K = \frac{p_0}{{\rho_0}^\gamma} \ ,$$

where motions are so rapid that there is no heat exchange and γ is the ratio of specific heats (5/3 for a fully ionised hydrogen plasma). These may be linearised, neglecting squares and products of small quantities, to give

$$\rho_0\frac{\partial \mathbf{v}'}{\partial t} = -\nabla p' \ , \quad \frac{\partial \rho'}{\partial t} + \rho_0\nabla.\mathbf{v}' = 0 \ , \quad p' = {c_s}^2\rho' \ , \tag{4.1}$$

where ${c_s}^2 = \gamma p_0/\rho_0$.

Suppose first of all that $\rho' = \rho'(x,t)$ and $\mathbf{v}' = v'(x,t)\hat{\mathbf{x}}$, so that (4.1) reduce to

$$\rho_0\frac{\partial v'}{\partial t} = -\frac{\partial p'}{\partial x}, \quad \frac{\partial \rho'}{\partial t} + \rho_0\frac{\partial v'}{\partial x} = 0, \quad p' = {c_s}^2\rho' \ .$$

If $\mathbf{v}'$ and p' are eliminated between these three equations, we obtain

$$\frac{\partial^2 \rho'}{\partial t^2} = c_s^2\frac{\partial^2 \rho'}{\partial x^2} \ , \tag{4.2}$$

so that ρ' satisfies a wave equation and disturbances propagate along the x-axis with speed c_s (the *sound speed*). Formally, we may Fourier analyse an arbitrary disturbance into components, since this is a linear equation, and consider each component separately, namely

$$\rho' = \text{const} \times e^{i(kx-\omega t)} \ , \tag{4.3}$$

where ω is the wave frequency, k the wavenumber, $2\pi/k$ the wavelength and $2\pi/\omega$ the period of the wave. Then $\partial\rho'/\partial t = -i\omega\rho'$ and $\partial\rho'/\partial x = ik\rho'$, so that (4.2) gives

$$\omega^2 = k^2 {c_s}^2 \ . \tag{4.4}$$

This is the *dispersion relation* for sound waves, relating ω and k. The solution (4.6) represents a wave moving without change of shape at constant speed $\omega/k = c_s$, which is known as the *phase speed* of the wave.

More generally, for an arbitrary linear disturbance $\rho'(x,y,z,t)$, Equation (4.1) implies, after eliminating $\mathbf{v}'$ and p', that

$$\frac{\partial^2\rho'}{\partial t^2} = {c_s}^2\,\nabla^2\rho' \ . \tag{4.5}$$

Instead of (4.3) we now consider a general Fourier component

$$\rho' = \text{constant} \times e^{i(k_x x + k_y y + k_z z - \omega t)} = \text{constant} \times e^{i(\mathbf{k}.\mathbf{r} - \omega t)} \ . \tag{4.6}$$

Then $\partial\rho'/\partial x = ik_x\rho'$, $\partial\rho'/\partial y = ik_y\rho'$, $\partial\rho'/\partial z = ik_z\rho'$ or, in compact notation, $\nabla\rho' = i\mathbf{k}\rho'$, and, as before, $\partial\rho'/\partial t = -i\omega\rho'$.

Equation (4.5) therefore implies the dispersion relation

$$\omega^2 = k^2 {c_s}^2 \ , \tag{4.7}$$

so that the waves propagate equally in all directions with phase speed c_s. In addition, Equation (4.1a) becomes $\rho_0\omega\mathbf{v}' = -\mathbf{k}p'$, so that $\mathbf{v}'$ is parallel to the direction $\mathbf{k}$ of propagation. In other words we have a longitudinal wave, with *phase velocity* $(\omega/k)\hat{\mathbf{k}} = c_s\hat{\mathbf{k}}$.

The overall procedure therefore for considering simple linear waves is to start with the MHD equations for variations from equilibrium of a uniform plasma. Then, by neglecting squares and products of small quantities, these are reduced to a set of linear partial differential equations. By looking for plane wave solutions of the form (4.6), these are in turn reduced to a set of homogeneous algebraic equations for the amplitudes of the variables, which leads to the dispersion relation.

4.2 Alfven Waves

On disturbing a uniform magnetic field, one would expect, by analogy with an elastic band, the magnetic tension to make a wave propagate along the field lines with speed

$$v_A = \sqrt{\left(\frac{\text{tension}}{\rho}\right)} = \frac{B}{\sqrt{(\mu\rho)}} \ , \tag{4.8}$$

known as the *Alfven speed*, since the tension is B^2/μ. Consider, therefore, an ideal plasma (with negligible dissipation), initially at rest with a uniform field $\mathbf{B}_0 = B_0\hat{\mathbf{z}}$ and density ρ_0. The effect of a disturbance is to introduce a velocity $\mathbf{v}'$ and to make the other variables $\mathbf{B}_0 + \mathbf{B}'$, $\rho_0 + \rho'$. Suppose it is so small that squares and products of v', B', ρ' can be neglected. Then, the pressureless MHD equations become

$$\frac{\partial \mathbf{B}'}{\partial t} = \nabla \times (\mathbf{v}' \times \mathbf{B}_0) \ , \tag{4.9}$$

$$\rho_0 \frac{\partial \mathbf{v}'}{\partial t} = (\nabla \times \mathbf{B}') \times \frac{\mathbf{B}_0}{\mu} \ , \tag{4.10}$$

where $\nabla.\mathbf{B}' = 0$ and ρ' is given by

$$\frac{\partial \rho'}{\partial t} + \nabla.(\rho_0 \mathbf{v}') = 0 \ , \tag{4.11}$$

and p' by $p' = c_s^2 \rho'$.

Look for wave-like solutions by supposing the perturbation quantities behave like $\exp[i(\mathbf{k}.\mathbf{r} - \omega t)]$ so that (4.9) and (4.10) reduce to

$$-\omega \mathbf{B}' = \mathbf{k} \times (\mathbf{v}' \times \mathbf{B}_0) = (\mathbf{B}_0.\mathbf{k})\mathbf{v}' - \mathbf{B}_0(\mathbf{k}.\mathbf{v}') \ , \tag{4.12}$$

$$-\mu\rho_0\omega \mathbf{v}' = (\mathbf{k} \times \mathbf{B}') \times \mathbf{B}_0 = \mathbf{B}'(\mathbf{B}_0.\mathbf{k}) - \mathbf{k}(\mathbf{B}'.\mathbf{B}_0) \ , \tag{4.13}$$

where $\mathbf{k}.\mathbf{B}' = 0$, $p' = c_s^2 \rho'$ and

$$-\omega\rho' + \rho_0 \mathbf{k}.\mathbf{v}' = 0 \ . \tag{4.14}$$

For *Alfven waves* propagating along the magnetic field $\mathbf{B}_0$, we make $\mathbf{k}$ parallel to $\mathbf{B}_0$ and assume $\mathbf{v}'$ is normal to $\mathbf{k}$. Then (4.12) and (4.13) reduce to

$$-\omega \mathbf{B}' = B_0 k \, \mathbf{v}', \tag{4.15}$$

$$-\mu\rho_0\omega \, \mathbf{v}' = B_0 k \, \mathbf{B} - \mathbf{k}(\mathbf{B}'.\mathbf{B}_0), \tag{4.16}$$

where the last term vanishes, since by (4.15) $\mathbf{B}'$ is parallel to $\mathbf{v}'$, which is normal to k by assumption, and $\mathbf{B}_0$ is parallel to k so that $\mathbf{B}'.\mathbf{B}_0 = 0$. Thus (4.15) and (4.16) imply that $\omega^2 = k^2 v_A{}^2$.

More generally, the Alfven waves may also propagate at some angle θ to $\mathbf{B}_0$, in which case we still assume $\mathbf{k}.\mathbf{v}' = 0$. Then (4.12) and (4.13) this time become

$$-\omega \mathbf{B}' = (\mathbf{B}_0.\mathbf{k})\mathbf{v}' \ ,$$

$$-\mu\rho\omega \mathbf{v}' = (\mathbf{B}_0.\mathbf{k})\mathbf{B}' - \mathbf{k}(\mathbf{B}'.\mathbf{B}_0) \ ,$$

where again $\mathbf{B}'.\mathbf{B}_0 = 0$ (which may be seen most easily by taking $\mathbf{k}$. the second equation and remembering that $\mathbf{k}.\mathbf{v}' = \mathbf{k}.\mathbf{B}' = 0$). Thus these two equations give the dispersion relation for Alfven waves as

$$\omega^2 = k^2 v_A{}^2 \cos^2\theta, \tag{4.17}$$

since $\mathbf{B}_0.\mathbf{k} = B_0 k \cos\theta$ and $v_A{}^2 = B_0{}^2/(\mu\rho_0)$.

By comparison with sound waves, which are longitudinal and have non-zero density perturbation (ρ'), Alfven waves are transverse in the sense that $\mathbf{v}'$ is perpendicular to the direction $\mathbf{k}$ of propagation and so by (4.14) ρ' (and therefore

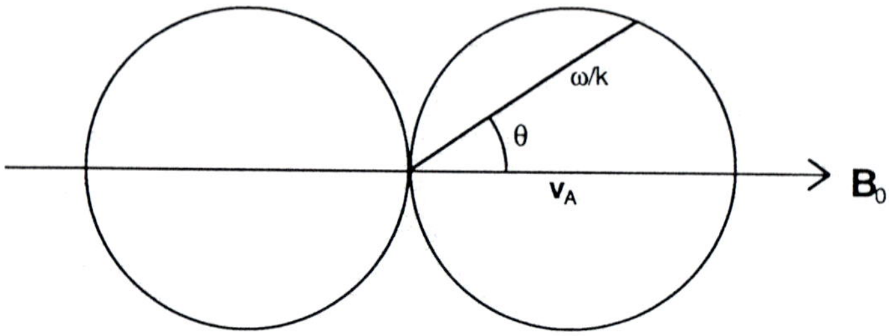

Fig. 4.1. Phase diagram for Alfven waves

also p') vanishes. The phase velocity is $\omega/k = \pm v_{\rm A}\cos\theta$ in the direction of $\mathbf{k}$, so that the speed of propagation depends on the direction and there is no propagation perpendicular to $\mathbf{B}_0$ ($\theta = \frac{1}{2}\pi$). The variation of phase speed with θ may be represented on a phase diagram (Figure 4.1) in which the distance of a point on the curve from the origin is proportioned to ω/k. In this case the curves are circles.

4.3 Compressional Alfven Waves

Consider now the case when $\mathbf{k}.\mathbf{v}' \neq 0$. Substitute for $\mathbf{B}'$ from (4.12) in (4.13), so that

$$\mu\rho_0\omega^2\mathbf{v}' = \left[(\mathbf{B}_0.\mathbf{k})\mathbf{v}' - \mathbf{B}_0(\mathbf{k}.\mathbf{v}')\right](\mathbf{B}_0.\mathbf{k}) - \mathbf{k}\left[(\mathbf{B}_0.\mathbf{k})(\mathbf{v}'.\mathbf{B}_0) - B_0^2(\mathbf{k}.\mathbf{v}')\right] . \tag{4.18}$$

This represents three linear homogeneous equations for three unknowns (v_x', v_y', v_z') and so in principle the determinant of coefficients would give a relation between the coefficients, namely the dispersion relation. But since $\mathbf{v}'$ only appears in the forms $\mathbf{v}', \mathbf{k}.\mathbf{v}'$ and $\mathbf{B}_0.\mathbf{v}'$, we make take in turn $\mathbf{B}_0$. and $\mathbf{k}$. this equation to obtain two equations for $\mathbf{k}.\mathbf{v}'$ and $\mathbf{B}_0.\mathbf{v}'$, namely

$$\mu\rho_0\omega^2(\mathbf{B}_0.\mathbf{v}') = 0 , \tag{4.19}$$

and

$$\mu\rho_0\omega^2(\mathbf{k}.\mathbf{v}') = k^2B_0^2(\mathbf{k}.\mathbf{v}') . \tag{4.20}$$

Thus, from (4.20), either $\mathbf{k}.\mathbf{v}' = 0$ or

$$\omega^2 = k^2{v_{\rm A}}^2 , \tag{4.21}$$

which is the dispersion relation for compressional Alfven waves. These waves propagate equally in all directions, like sound waves, and, since $\mathbf{k}.\mathbf{v}' \neq 0$, (4.14) implies that ρ' and p' are in general non-zero. For propagation across the field ($\mathbf{k}.\mathbf{B}_0 = 0$) it can easily be seen from (4.18) that $\mathbf{v}'$ is parallel to $\mathbf{k}$ and therefore the mode is longitudinal.

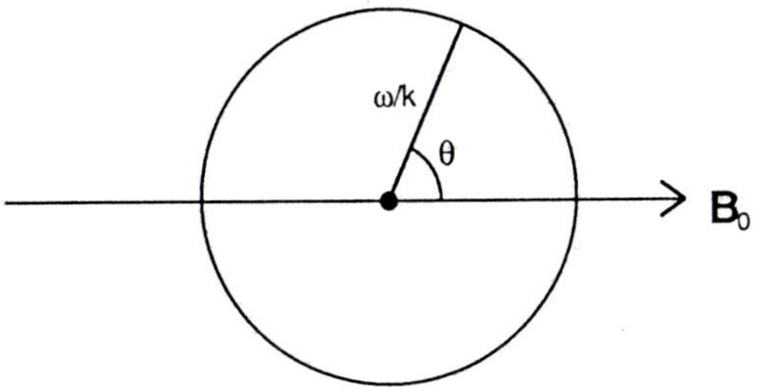

Fig. 4.2. Phase diagram for compressional Alfven waves

4.4 Magnetoacoustic Waves

We have found that there are two waves when the pressure vanishes, namely the Alfven and compressional Alfven waves, and one wave when the magnetic field vanishes, namely the sound wave. If pressure fluctuations are included in the MHD equations by adding a term $-\nabla p'$ to the right of (4.10), the effect is to add a term $-\mu \mathbf{k} p'$ to the right of (4.13) and $\mathbf{k} c_s^2 \mu \rho_0 (\mathbf{k} . \mathbf{v}')$ to the right of (4.18). The Alfven waves (for which $\mathbf{k}.\mathbf{v}'$ vanishes) are unaltered since ρ' and p' vanish. However, the sound and compressional Alfven waves are coupled together to give two *magneto-acoustic waves.*

Equations (4.19) and (4.20) become

$$-\omega^2 (\mathbf{B}_0 . \mathbf{v}') = -(\mathbf{B}_0 . \mathbf{k}) c_s^2 (\mathbf{k} . \mathbf{v}') \ , \tag{4.22}$$

$$\left(\mu \rho_0 \omega^2 - k^2 c_s^2 \mu \rho_0 - k^2 B_0^2\right) (\mathbf{k} . \mathbf{v}') = -k^2 (\mathbf{B}_0 . \mathbf{k}) (\mathbf{B}_0 . \mathbf{v}') \ . \tag{4.23}$$

Thus, either $\mathbf{k}.\mathbf{v}'$ and $\mathbf{B}_0.\mathbf{v}'$ both vanish, when (4.18) gives the dispersion relation (4.17) for Alfven waves; or (4.22) and (4.23) imply

$$\omega^4 - \omega^2 k^2 (c_s^2 + v_A^2) + c_s^2 v_A^2 k^4 \cos^2 \theta = 0 \ . \tag{4.24}$$

This is the dispersion relation for slow and fast magnetoacoustic waves. The smallest root for ω^2/k^2 gives the slow mode and the largest the fast mode. The particular cases $p_0 = 0$ (i.e. $c_s^2 = 0$) and $B_0 = 0$ (i.e. $v_A^2 = 0$) reduce to the dispersion relations ($\omega^2 = k^2 v_A^2$ and $\omega^2 = k^2 c_s^2$) for compressional Alfven and sound waves, respectively, as expected. The resulting phase diagrams are sketched in Figure 4.3.

4.5 Effect of Magnetic Diffusion on Alfven Waves

Assume $\nabla . \mathbf{v}' = 0$. Then the linearised continuity equation implies $\rho' = 0$ as usual for Alfven waves, while the equations of induction (with diffusion) and motion become

$$\frac{\partial \mathbf{B}'}{\partial t} = \nabla \times (\mathbf{v}' \times \mathbf{B}_0) + \eta \nabla^2 \mathbf{B}', \quad \mu \rho_0 \frac{\partial \mathbf{v}'}{\partial t} = (\nabla \times \mathbf{B}') \times \mathbf{B}_0 \ ,$$

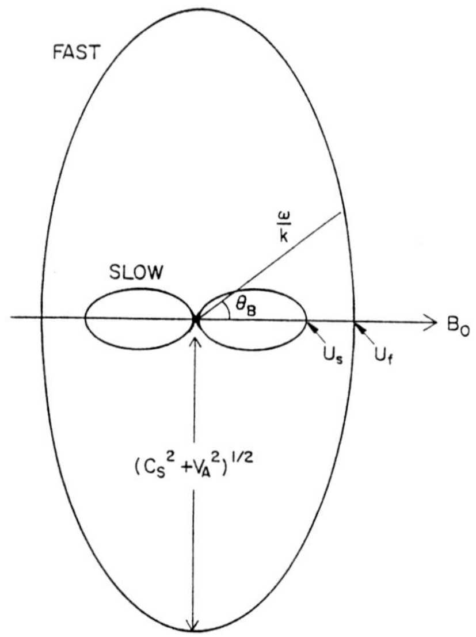

Fig. 4.3. Phase diagram for magnetoacoustic waves

where $\nabla.\mathbf{B}' = 0$. After making the usual wave assumption, they reduce to

$$-i\omega\mathbf{B}' = i(\mathbf{B}_0.\mathbf{k})\mathbf{v}' - k^2\zeta\mathbf{B}', \quad -\mu\rho_0\omega\mathbf{v}' = \mathbf{B}'(\mathbf{B}_0.\mathbf{k}) - \mathbf{k}(\mathbf{B}'.\mathbf{B}_0) \ ,$$

where, as usual, $\mathbf{B}'.\mathbf{B}_0 = 0$.

These two equations may be combined to yield the dispersion relation

$$\omega^2 = k^2{v_A}^2 - i\omega\frac{kv_A}{R_m} \ , \tag{4.25}$$

where

$$R_m = \frac{v_A}{k\eta} \ .$$

By splitting ω into a real (ω_r) and imaginary (ω_i) part, (4.25) gives

$$\omega_i = -k\frac{v_A}{2R_m}, \quad {\omega_r}^2 = k^2{v_A}^2\left(1 - \frac{1}{4{R_m}^2}\right) \ . \tag{4.26}$$

Thus if $R_m \gg 1$, the effect of diffusion is to produce a slow decay of the wave and a small reduction in the frequency ω_r.

4.6 Shock Waves

(i) Hydrodynamic Shock

Small-amplitude sound waves propagate without change of shape, but when the amplitude is finite the crest can move faster than its trough, causing a progressive steepening. Ultimately, the gradients become so large that dissipation becomes important, and a steady shock wave shape may be attained with a balance between the steepening effect of the nonlinear convective term and the broadening effect of dissipation. The dissipation inside the shock front converts the energy being carried by the wave gradually into heat. The effect of the passage of the shock is to compress and heat the gas.

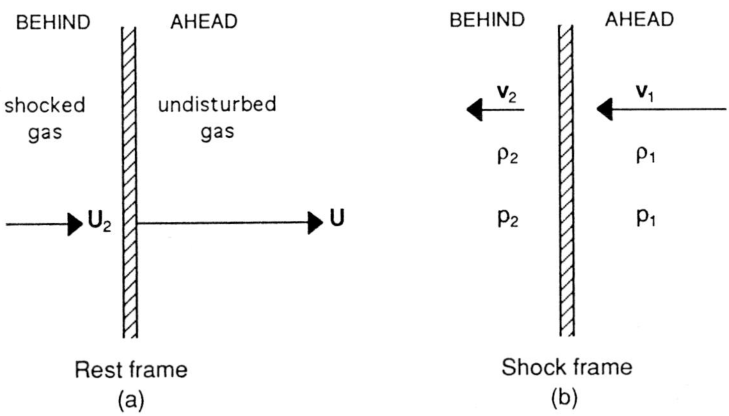

Fig. 4.4. Notation for a hydrodynamic shock

We model a shock as a plane discontinuity, although in reality it is a very thin transition region (Figure 4.4). The shock travels at speed U, say, into a gas at rest and accelerates the shocked gas to a speed U_2. In a frame of reference moving with the shock the fluid ahead has speed $v_1 = U$, density ρ' and pressure p', while the corresponding variables behind the shock are $v_2 = U - U_2, \rho_2$ and p_2, say.

Conservation of mass, momentum and energy then gives

$$\rho_2 v_2 = \rho_1 v_1 \ , \tag{4.27}$$

$$p_2 + \rho_2 v_2^2 = p_1 + \rho_1 v_1^2 \ , \tag{4.28}$$

$$p_2 + \left(\rho_2 e_2 + \frac{1}{2}\rho_2 v_2^2\right) v_2 = p_1 v_1 + (\rho_1 e_1 + \frac{1}{2}\rho_1 v_1^2) v_1 \ , \tag{4.29}$$

where $e = p/[(\gamma - 1)\rho]$ is the internal energy. These may be solved to give the *Rankine-Hugoniot jump relations*:

$$\frac{\rho_2}{\rho_1} = \frac{(\gamma+1)M_1^2}{2+(\gamma-1)M_1^2}, \ \frac{v_2}{v_1} = \frac{2+(\gamma-1)M_1^2}{(\gamma+1)M_1^2}, \ \frac{p_2}{p_1} = \frac{2\gamma M_1^2 - (\gamma-1)}{\gamma+1}, \tag{4.30}$$

which are supplemented by the condition for an isolated system from the second law of thermodynamics that the entropy $s = c_v \log p/\rho^\gamma$ must increase, namely,

$s_2 \geq s_1$, where c_v is the specific heat at constant volume and $M_1 = v_1/c_{s1}$ is the *Mach number*, namely the ratio of the shock speed to the sound speed ($c_{s1} = (\gamma p_1/\rho_1)^{1/2}$).

Consequences of these equations are:

(i) $M_1 \geq 1$, so that the shock speed (v_1) exceeds the sound speed ahead of the shock;

(ii) $v_2 \leq c_{s2}$, so that in the shock frame the flow is subsonic behind the shock (and supersonic ahead of it);

(iii) $p_2 \geq p_1$ and $\rho_2 \geq \rho_1$, so that the shock is compressive;

(iv) $v_2 \leq v_1$ and $T_2 \geq T_1$, so that the shock slows down the gas and heats it up;

(v) $1 \leq \frac{\rho_2}{\rho_1} < \frac{\gamma+1}{\gamma-1}$, so that the maximum density ratio is $(\gamma+1)/(\gamma-1)$, whereas the pressure ratio increases with M_1, like $M_1{}^2$.

(ii) Perpendicular Magnetic Shock

In the presence of a magnetic field, we now have three wave modes, and when their amplitudes are large the Alfven wave can propagate without steepening, whereas the slow and fast magnetoacoustic modes steepen to form shocks. Derivation of the jump relations is more complicated, since: there is an extra variable (**B**); **B** and **v** may be inclined away from the shock normal; and the entropy condition is replaced by an "evolutionary condition".

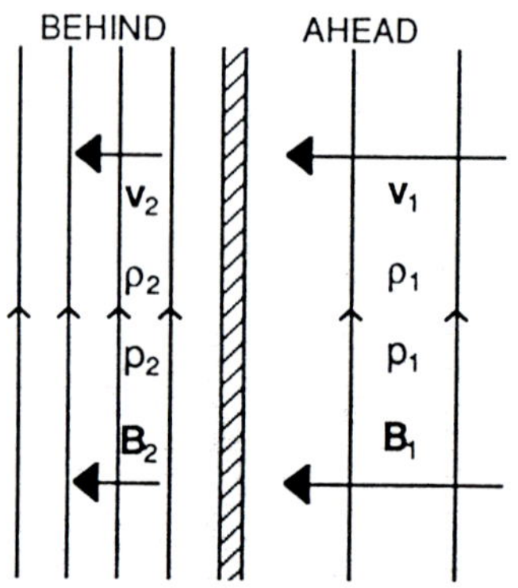

Fig. 4.5. Notation for a perpendicular shock

The jump relations for mass, momentum, energy and magnetic flux are now

$$\rho_2 v_2 = \rho_1 v_1 \ ,$$

$$p_2 + \rho_2 v_2^2 + \frac{B_2^2}{2\mu} = p_1 + \rho_1 v_1^2 + \frac{B_1^2}{2\mu} \ ,$$

$$\left(p_2 + \frac{B_2^2}{2\mu}\right) v_2 + \left(\rho_2 e_2 + \frac{1}{2}\rho_2 v_2^2 + \frac{B_2^2}{2\mu}\right) v_2 =$$

$$\left(p_1 + \frac{B_1{}^2}{2\mu}\right) v_1 + \left(\rho_1 e_1 + \frac{1}{2}\rho_1 v_1{}^2 + \frac{B_1{}^2}{2\mu}\right) v_1 \ ,$$

$$B_2 v_2 = B_1 v_1 \ , \tag{4.31}$$

which imply that

$$\frac{v_2}{v_1} = \frac{1}{X}, \quad \frac{B_2}{B_1} = X, \quad \frac{p_2}{p_1} = \gamma M_1{}^2 \left(1 - \frac{1}{X}\right) - \frac{1 - X^2}{\beta_1} \ ,$$

where $\beta_1 = 2\mu p_1/B_1{}^2$ and $X = \rho_2/\rho_1$ is the density ratio, which is the positive solution of

$$2(2 - X)X^2 + [2\beta_1 + (\gamma - 1)\beta_1 M_1^2 + 2]\gamma X - \gamma(\gamma + 1)\beta_1 M_1^2 = 0 \ . \tag{4.32}$$

Consequences of these jump relations together with the evolutionary condition (that the perturbation caused by a small disturbance be small and unique) are:

(i) Equation (4.32) has only one root since $1 < \gamma < 2$;

(ii) the effect of the magnetic field is to reduce X below its hydrodynamic value;

(iii) the shock is compressive with $X \geq 1$;

(iv) the shock speed (v_1) must exceed the fast magnetoacoustic speed $\surd(c_{s1}{}^2 + v_{A1}^2)$ ahead of the shock;

(v) magnetic compression is limited to the range $1 < B_2/B_1 < (\gamma+1)/(\gamma-1)$, where for $\gamma = 5/3$ the upper limit is 4.

(iii) Oblique Magnetic Shocks

Set up axes in a frame moving with the shock, as before, and assume **v** and **B** lie in the xy plane (Figure 4.6). Then the jump relations for mass, $x-$ and $y-$momentum and magnetic flux may be written

$$\rho_2 v_{2x} = \rho_1 v_{1x} \ ,$$
$$p_2 + B_2^2/(2\mu) - B_{2x}{}^2/\mu + \rho_2 v_{2x}{}^2 = p_1 + B_1^2/(2\mu) - B_{1x}{}^2 + \rho_1 v_{1x}{}^2 \ ,$$
$$\rho_2 v_{2x} v_{2y} - B_{2x} B_{2y}/\mu = \rho_1 v_{1x} v_{1y} - B_{1x} B_{1y}/\mu \ ,$$
$$\left(p_2 + B_2^2/(2\mu)\right) v_{2x} - B_{2x}(\mathbf{B}_2.\mathbf{v}_2)/\mu + \left(\rho_2 e_2 + \tfrac{1}{2}\rho_2 v_2^2 + B_2^2/(2\mu)\right) v_{2x} =$$
$$\left(p_1 + B_1^2/(2\mu)\right) v_{1x} - B_{1x}(\mathbf{B}_1.\mathbf{v}_1)/\mu + \left(\rho_1 e_1 + \tfrac{1}{2}\rho_2 v_1^2 + B_1^2/(2\mu)\right) v_{1x} \ ,$$
$$B_{2x} = B_{1x} \ ,$$
$$v_{2x} B_{2y} - v_{2y} B_{2x} = v_{1x} B_{1y} - v_{1y} B_{1x} \ .$$

Now if we choose axes moving parallel to the shock at such a speed that

$$v_{1y} = v_{1x} \frac{B_{1y}}{B_{1x}} \ , \tag{4.33}$$

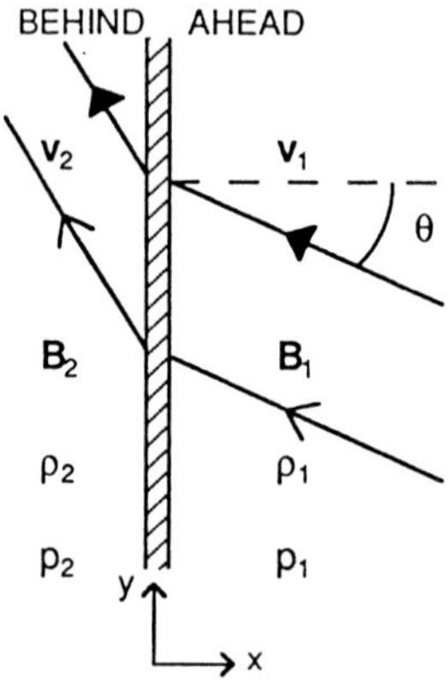

Fig. 4.6. Notation for an oblique shock

so that **v** is parallel to **B**, the equations simplify greatly and may be solved to give

$$\frac{v_{2x}}{v_{1x}} = \frac{1}{X}, \quad \frac{v_{2y}}{v_{1y}} = \frac{v_1{}^2 - v_{A1}^2}{v_1{}^2 - X v_{A1}}, \quad \frac{B_{2x}}{B_{1x}} = 1 \,,$$
$$\frac{B_{2y}}{B_{1y}} = \frac{(v_1{}^2 - v_{A1}{}^2)X}{v_1{}^2 - X v_{A1}{}^2}, \quad \frac{p_2}{p_1} = X + \frac{(\gamma - 1) X v_1{}^2}{2 c_{s1}{}^2}\left(1 - \frac{v_2{}^2}{v_1{}^2}\right), \tag{4.34}$$

where the compression ratio is

$$X = \frac{\rho_2}{\rho_1}, \quad c_{s1}{}^2 = \frac{\gamma p_1}{\rho_1}, \quad v_{A1}{}^2 = \frac{B_1{}^2}{\mu \rho_1}, \quad \cos\theta = \frac{v_{1x}}{v_1}\,, \tag{4.35}$$

and X is a solution of

$$(v_1{}^2 - X\, v_{A1}{}^2)^2 \left\{X c_{s1}{}^2 + \tfrac{1}{2} v_1{}^2 \cos^2\theta\, [X(\gamma - 1) - (\gamma + 1)]\right\} \tag{4.36}$$
$$+\tfrac{1}{2} v_{A1}{}^2 v_1{}^2 \sin^2\theta\, X \left\{[\gamma + X(2 - \gamma)]\, v_1{}^2 - X v_{A1}{}^2\, [(\gamma + 1) - X(\gamma - 1)]\right\} = 0 \,.$$

Equation (4.32) has three solutions, which give the slow shock, Alfven wave and fast shock, the forms of the resulting field lines being shown in Figure 4.7.

The slow and fast shocks have the following properties:

(i) they are compressive with $X > 1$ and $p_2 > p_1$;

(ii) they conserve the sign of B_y so that $B_{2y}/B_{1y} > 0$;

(iii) for the slow shock $B_2 < B_1$, so that **B** refracts towards the shock normal and B decreases as the shock passes;

(iv) for the fast shock $B_2 > B_1$, so that the shock makes **B** refract away from the normal and B increase;

(v) the flow speed (v_{1x}) ahead of the shock exceeds the appropriate wave speed, while the speed (v_{2x}) behind it is smaller than the wave speed;

(vi) the flow normal to the shock is slowed down ($v_{2x} < v_{1x}$);

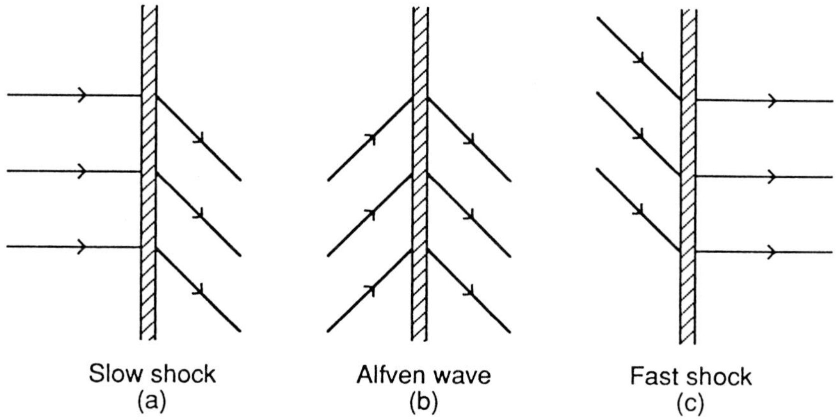

Fig. 4.7. Magnetic field lines for special oblique waves

(vii) in the limit as $B_x \to 0$, the fast shock becomes a perpendicular shock, while the slow shock becomes a *tangential discontinuity*, for which v_x and B_x vanish and there are arbitrary jumps in v_y and B_y, subject only to total pressure balance $\left(p_2 + B_2^2/(2\mu) = p_1 + B_1^2/(2\mu)\right)$.

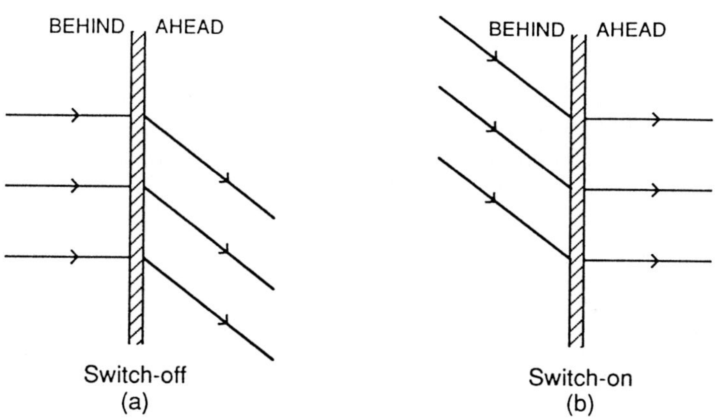

Fig. 4.8. Notation for switch-off and switch-on shocks

In the limit when $v_1 = v_{A1}$ and $X \neq 1$, (4.34) implies that $B_{2y} = 0$ and we have a *switch-off shock*. Since $\mathbf{v}_1$ and $\mathbf{B}_1$ are parallel, this implies that

$$v_{1x} = \frac{B_{1x}}{\sqrt{(\mu\rho)}} , \tag{4.37}$$

so that the shock propagates at the Alfven speed based on the normal component of the magnetic field.

For a shock propagating along $\mathbf{B}_1$, the fast-shock solution is $X = {v_1}^2/{v_{A1}}^2$ for a *switch-on shock*.

When $v_1 = v_{A1}$, the conservation of flux implies that $v_{2y}/v_{1y} = B_{2y}/B_{1y}$, while the equations for conservation of x-momentum and energy reduce to $p_2 = p_1$ and ${B_{2y}}^2 = {B_{1y}}^2$. Thus, in addition to the trivial solution ($\mathbf{B}_2 = \mathbf{B}_1, \mathbf{v}_2 = \mathbf{v}_1$), we also have

$$B_{2y} = -B_{1y} \quad , \quad B_{2x} = B_{1x} \ ,$$

$$v_{2y} = -v_{1y} \quad , \quad v_{2x} = v_{1x} \ ,$$

for a finite-amplitude *Alfven wave*, sometimes called an *intermediate wave* or *rotational discontinuity*. The tangential component of magnetic field is reversed by the wave, but there is no change in pressure or density.

5 The Solar Wind

5.1 Introduction

In the nineteenth century, variations in the Earth's magnetic field were first observed, with a sudden increase (by about $10^{-3}G$) followed by a slow decrease - a geomagnetic storm. Such storms were related to the Sun since they often occur one or two days after a large solar flare and the eleven-year period in sunspot numbers is echoed by one in geomagnetic activity.

Chapman (1929) suggested geomagnetic storms are caused by streams of plasma from solar flares, travelling at 1000 km s^{-1} through a vacuum - a simple cause and effect - although now we believe there is a much more complex interaction. Biermann (1951) observed comets and suggested that continuously emitted "solar corpuscles" rather than radiation pressure cause comet tails to point away from the Sun. However, this was not taken seriously since MHD was not yet well developed and the interaction between magnetic field and plasma not well understood.

Chapman (1957) considered a *static* corona with energy transfer by conduction and asked the question: how far does the corona reach? For a steady, spherically symmetric conduction, the total heat flux across successive spherical surfaces is constant and so

$$4\pi r^2 \kappa \frac{dT}{dr} = \text{constant} \ , \tag{5.1}$$

where $4\pi r^2$ is the surface area of a sphere of radius $r, \kappa dT/dr$ is the heat flux per unit area, and $\kappa = \kappa_0 T^{5/2}$ is the coefficient of heat conduction. Integrating (5.1) and assuming T vanishes at infinity and equals T_0 at $r = r_0$, he found

$$T = T_0 \left(\frac{r_0}{r}\right)^{2/7} \ . \tag{5.2}$$

If $T_0 = 10^6$ K at the Sun (say, at $r_0 = 10^3$ Mm), then at the Earth's orbit ($r = 150 \times 10^3$ Mm), the temperature would be 2×10^5 K. Thus the corona doesn't terminate near the Sun but extends to the Earth and beyond. This prediction that the Earth is enveloped by a hot plasma was staggering, since previously it had been thought it is sitting in a vacuum.

For hydrostatic equilibrium

$$\frac{dp}{dr} = -\frac{M_0 G\rho}{r^2} \ , \tag{5.3}$$

where $p = \mathcal{R}\rho T$ and M_0 is the mass of the sun, while G is the gravitational constant. In other words, (5.3) represents a balance between an outwards pressure gradient and the inwards gravitational attraction, with gravity falling off as r^{-2}. Thus, from (5.2) - (5.3)

$$\frac{dp}{dr} = -Ar_0^{5/7}\frac{p}{r^{12/7}} \ , \tag{5.4}$$

where $A = M_0 G/(r_0 RT_0)$ is approximately 15. Integrating (5.4) we find that, if $p = p_0$ at $r = r_0$,

$$p = p_0 \exp\frac{7A}{5}\left[\left(\frac{r_0}{r}\right)^{5/7} - 1\right] \ . \tag{5.5}$$

One consequence of this model is that from (5.5), as $r \to \infty$, so $p \to p_0 e^{-21} \approx p_0/10^9$, and, although this is a very small pressure, it far exceeds the interstellar pressure of $p_0/10^{16}$. Another fault with the model is that the resulting density is

$$\rho = \rho_0\left(\frac{r}{r_0}\right)^{2/7}\exp\frac{7A}{5}\left[\left(\frac{r_0}{r}\right)^{5/7} - 1\right] \ ,$$

which tends to infinity as r increases.

5.2 Parker Solution

Parker (1958) resolved the inconsistency by suggesting that the corona is not in static equilibrium and is instead expanding - i.e., in the absence of a "lid" to hold it in (p_∞), such an extended corona cannot remain in equilibrium but must stream out steadily, and he christened it "the solar wind". He found solutions with the pressure tending to zero at infinity.

He modelled the corona very simply by assuming radial symmetry and steady flow. His set of equations was

$$4\pi r^2\rho v = \text{constant} \ , \tag{5.6}$$

$$\rho v\frac{dv}{dr} = -\frac{dp}{dr} - \frac{GM\rho}{r^2} \ , \tag{5.7}$$

$$\frac{p}{\rho^\gamma} = K \ . \tag{5.8}$$

Since ρv represents the rate of flow of mass per unit area and $4\pi r^2$ is the surface area of a sphere, (5.7) is the equation of mass continuity which implies that the rate of flow of mass across all spherical surfaces is the same. Equation (5.8) is the equation of motion and (5.9) is an adiabatic law.

By eliminating p and ρ, (5.7) becomes

$$\left(v - \frac{c_s^2}{v}\right)\frac{dv}{dr} = \frac{2c_s^2}{r} - \frac{GM}{r^2} , \tag{5.9}$$

which may be integrated analytically to give

$$\frac{1}{2}v^2 + \frac{c_s^2}{\gamma - 1} - \frac{GM}{r} = C , \tag{5.10}$$

where $c_s = (\gamma p/\rho)^{1/2}$ is the sound speed, which in general varies with r, and C is a constant of integration.

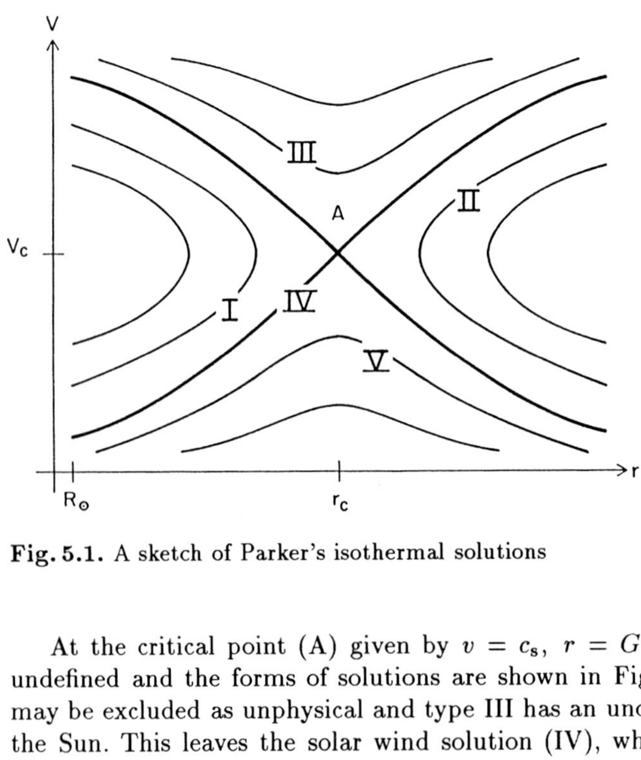

Fig. 5.1. A sketch of Parker's isothermal solutions

At the critical point (A) given by $v = c_s$, $r = GM/(2c_s^2) \equiv r_c$, dv/dr is undefined and the forms of solutions are shown in Figure 5.1. Types I and II may be excluded as unphysical and type III has an unobserved high speed near the Sun. This leaves the solar wind solution (IV), which becomes supersonic, and the "solar breeze" solutions (V), which remain subsonic.

In the particular case when $\gamma = 1$ so that $p/\rho = \mathcal{R}T$ is constant, (5.10) is not valid and (5.9) may be integrated instead to give

$$\tfrac{1}{2}v^2 - c_s^2 \log v = 2c_s^2 \log r + \frac{GM}{r} + \text{constant}$$

or

$$\frac{v^2}{c_s^2} - \log\frac{v^2}{c_s^2} = 4\log\frac{r}{r_c} + \frac{2GM}{rc_s^2} + \text{constant} . \tag{5.11}$$

Thus, for the wind solution when $r \gg r_c$ and $v \gg c_s$, the first and third terms in (5.11) balance so that

$$v \approx 2c_s\sqrt{(\log(r/r_c))}$$

and (5.6) implies that

$$p \sim \rho \sim \frac{1}{r^2\sqrt{(\log r)}} ,$$

which tends to zero at large distances, as required. In contrast, for the breeze solutions when $r \gg r_c$ and $v \gg c_s$, the second and third terms balance in (5.11) so that $v/c_s \approx r_c^2/r^2$ and ρ and p tend to constant values at large distances. Thus the solar wind solution is the more relevant.

Fig. 5.2. A composite solar eclipse showing the outer corona in white light and the inner corona in soft X-rays. (Courtesy High Altitude Observatory, NCAR and American Science and Engineering)

But Parker's solution was not widely accepted. The great controversy was resolved in 1959 when the solar wind was observed directly by Lunik III and Venus I. At 1 AU for the quiet solar wind one finds roughly a speed of 320 km s^{-1}, a density of 8 particles per cc, a magnetic field of 5×10^{-5} G, and electron and proton temperatures of 1.5×10^5K and 0.4×10^5K, respectively.

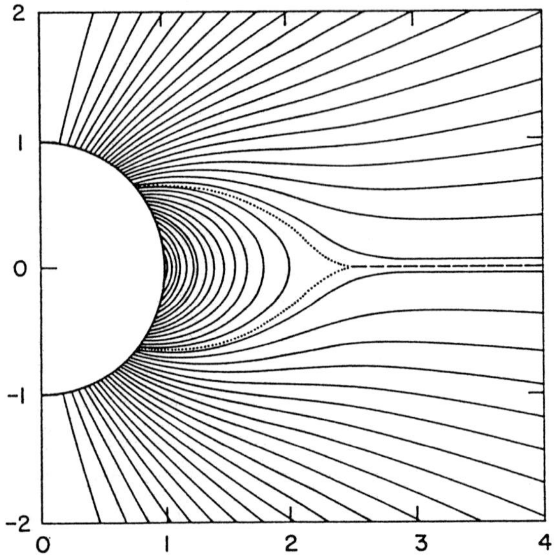

Fig. 5.3. Magnetic field lines for Pneuman-Kopp model

In research there is very occasionally a brilliant new idea, such as Parker's solar wind, which is subsequently extended and modified in many ways. In this case, an energy equation has been incorporated to give the temperature $T(r)$, or two energy equations for a two-fluid plasma give $T_i(r)$ and $T_e(r)$. Also the effect of the magnetic field was added. Eclipse and soft x-ray photographs show coronal holes (Figure 5.2), which were modelled most simply by replacing $4\pi r^2$ in (5.7) by an imposed area function $A(r)$. A two-dimensional MHD model was set up by Pneuman and Kopp (1971), as shown in Figure (5.3), and more recently separable solutions have been constructed by Tsinganos (1991) and Lima and Priest (1993). Schematically, the three-dimensional structure expected for Figure 5.3 is shown in Figure 5.4.

5.3 Effect of the Magnetic Field

Suppose first that the magnetic field just acts as a tracer and does not affect the flow velocity. The rotation of the Sun makes the magnetic field twist up into a spiral, like a rotating gramophone record with the grooves corresponding to magnetic field lines and the motion of the needle to the solar wind outflow. Thus, if there is no magnetic force, the plasma just moves straight out in an inertial frame (Figure 5.5).

If plasma and magnetic field are frozen together and the field line rotates at a speed $\Omega(R - R_0)$, the component of $\mathbf{v}$ perpendicular to $\mathbf{B}$ equals the speed of the field line in that direction, so that $v \sin\psi = \Omega(R - R_0)\cos\psi$ or $\tan\psi = \Omega(R - R_0)/v$, where ψ is the inclination of the field line to the radial direction. Thus ψ vanishes at the solar surface so that the field line is perpendicular to the

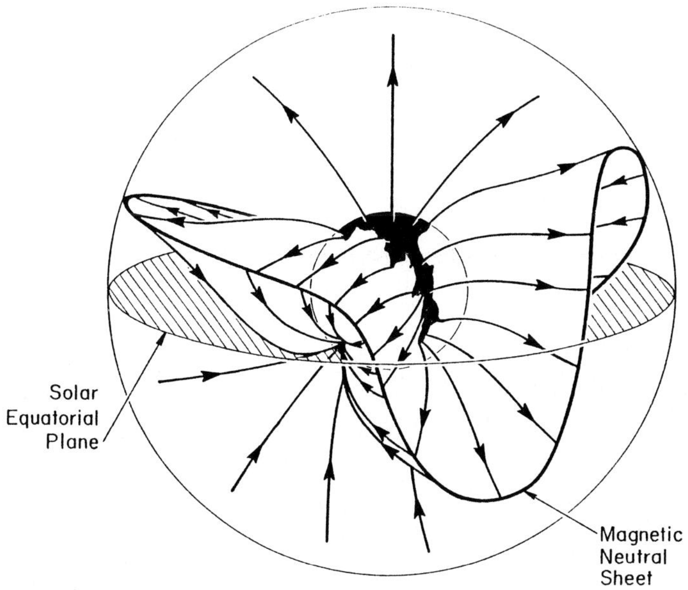

Fig. 5.4. Schematic drawing of warped neutral sheet

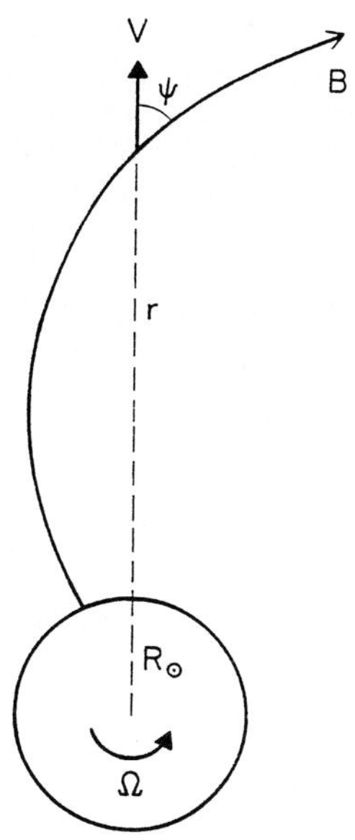

Fig. 5.5. A spiral magnetic field line attached to the Sun and rotating with angular speed Ω viewed from above the north pole

solar surface, and at 1 AU ψ is about 45°. If v is constant the equation of the field line is given by $\phi = \Omega t = \Omega(R_0 - R)/v$ so that

$$R = R_0 - \frac{v\phi}{\Omega} ,$$

which represents an Archimedes spiral.

Suppose next that $\mathbf{B}$ affects $\mathbf{v}$ and consider their interaction in the equatorial plane (Weber and Davis, 1967), where

$$\mathbf{B} = (B_r(r), 0, B_\phi(r)) , \quad \mathbf{v} = (v_r(r), 0, v_\phi(r)) ,$$

and the equation $\nabla.\mathbf{B} = 0$ implies that $r^{-2}\partial/\partial r(r^2 B_r) = 0$ or

$$B_r = \frac{B_0 R_0^2}{r^2} . \tag{5.12}$$

The other basic equations are

$$\rho(\mathbf{v}.\nabla)\mathbf{v} = -\nabla\left(p + \frac{B^2}{2\mu}\right) + \mathbf{B}.\nabla\frac{\mathbf{B}}{\mu} - \frac{GM_0\rho}{r^2}\hat{\mathbf{r}} , \tag{5.13}$$

$$\nabla.(\rho\mathbf{v}) = 0 , \tag{5.14}$$

$$\frac{p}{\rho^\gamma} = \text{constant} , \tag{5.15}$$

$$p = R\rho T , \tag{5.16}$$

$$\nabla \times (\mathbf{v} \times \mathbf{B}) = 0 . \tag{5.17}$$

In this geometry (5.14) may be integrated to give

$$\rho\, v_r\, r^2 = \text{constant} \tag{5.18}$$

and (5.17) implies

$$r(v_r B_\phi - v_\phi B_r) = c , \tag{5.19}$$

where the boundary conditions $B_\phi = 0$ and $v_\phi = \Omega R_0$ at $r = R_0$ imply that $c = -\Omega R_0^2 B_0$, or from (5.12)

$$c = -\Omega r^2 B_r .$$

Thus (5.19) determines B_ϕ as

$$B_\phi = \frac{v_\phi - r\Omega}{v_r} B_r . \tag{5.20}$$

Next, the ϕ-component of (5.13) is

$$\rho(\mathbf{v}.\nabla)v_\phi = (\mathbf{B}.\nabla)\frac{B_\phi}{\mu}$$

or

$$\mu\rho v_r \frac{d}{dr}(r v_\phi) = B r \frac{d}{dr}(r B_\phi) .$$

But, from (5.12) and (5.14), $\rho v_r/B_r$ is constant, and so this may be integrated to give

$$rv_\phi - \frac{B_r}{\mu\rho v_r} rB_\phi = L \ , \tag{5.21}$$

where L is constant and represents the total angular momentum per unit mass carried in the plasma and magnetic stresses. The solar wind carries angular momentum away from the Sun, and the $\mathbf{j} \times \mathbf{B}$ force transmits a torque to the solar surface which gradually slows it down.

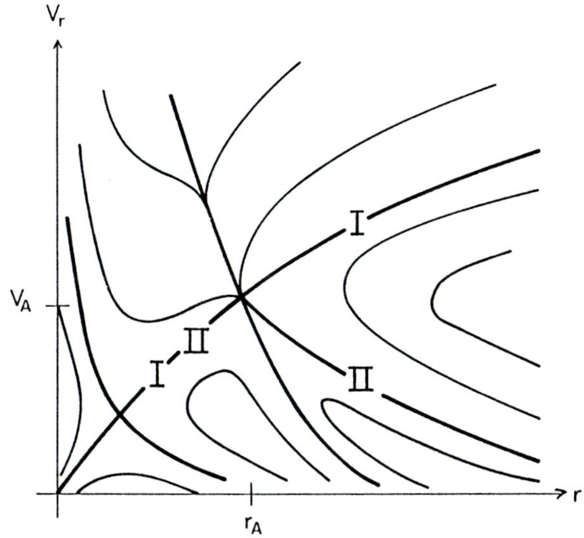

Fig. 5.6. A sketch of the solution topology for a magnetised solar wind

After substituting for B_ϕ from (5.21), this gives

$$v_\phi = \Omega r \frac{M_A{}^2 L/(r^2\Omega) - 1}{M_A{}^2 - 1} \ , \tag{5.22}$$

where $M_A = v_r\sqrt{(\mu\rho)}/B_r$ is the Alfven Mach number. The Alfven critical point ($r = r_A$), where $M_A = 1$, is typically at about $12R_0$. For v_ϕ to remain finite at $r = r_A$, (5.22) implies that

$$L = \Omega r_A{}^2 \ ,$$

so that L is what would be given by solid-body rotation out to r_A. Finally, the radial component of (5.13) is

$$\rho v_r \frac{dv_r}{dr} - \frac{\rho v_\phi{}^2}{r} = -\frac{dp}{dr} - \frac{B_\phi}{\mu r}\frac{d}{dr}(rB_\phi) - \frac{GM_0\rho}{r^2} \ ,$$

which determines v_r. The resulting solutions are sketched in Figure 5.6 and have now not one but three critical points, the outer two being very close together. Again, the required solar wind solution (I) passes through the critical points.

6 Magnetic Reconnection

6.1 Introduction

The induction equation

$$\frac{\partial \mathbf{B}}{\partial t} = \nabla \times (\mathbf{v} \times \mathbf{B}) + \eta \nabla^2 \mathbf{B} \tag{6.1}$$

shows that the magnetic field changes due to advection and diffusion. The time-scale for diffusion is $\tau_d = L^2/\eta$, which, as we have seen, is very long for typical global length-scales, and the speed of diffusion is $v_\mathrm{d} = \eta/L$.

For example, we found in Section 2.3 that a current sheet diffuses away and converts magnetic energy into heat ohmically (Figure 6.1). The field lines diffuse in through the plasma and cancel, so that the region of diffused field spreads out at v_d, as shown by the solid-headed arrows in Figure 6.1. Therefore a steady state may be produced if magnetic flux (and plasma) are carried in at the same rate as it is trying to diffuse. However, in order to do so, we need to create an extremely small length-scale L (and therefore large magnetic gradient $\nabla \mathbf{B}$ and current $\mathbf{j}$). Furthermore, although the magnetic field may be destroyed by cancellation as it comes in, the plasma itself cannot be destroyed and needs to flow out sideways, as illustrated in the following model.

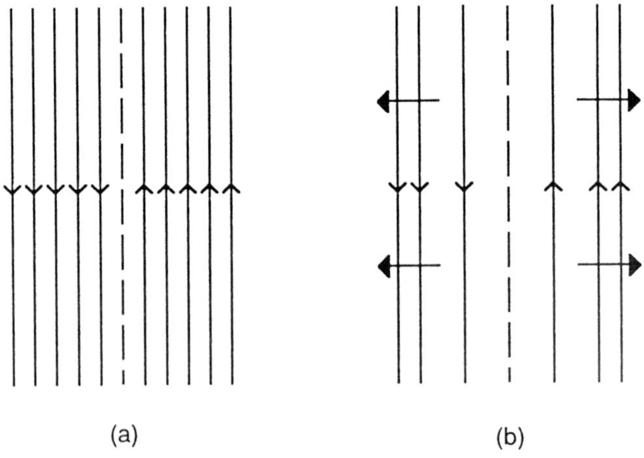

Fig. 6.1. Diffusion of oppositely directed field lines

6.2 Magnetic Annihilation (Sonnerup and Priest, 1975)

Suppose we have a steady-state flow

$$v_x = -\frac{V_0 x}{a}, \quad v_y = \frac{V_0 y}{a} \ , \tag{6.2}$$

so that the streamlines are the rectangular hyperbolae ($xy =$ constant). A prop-

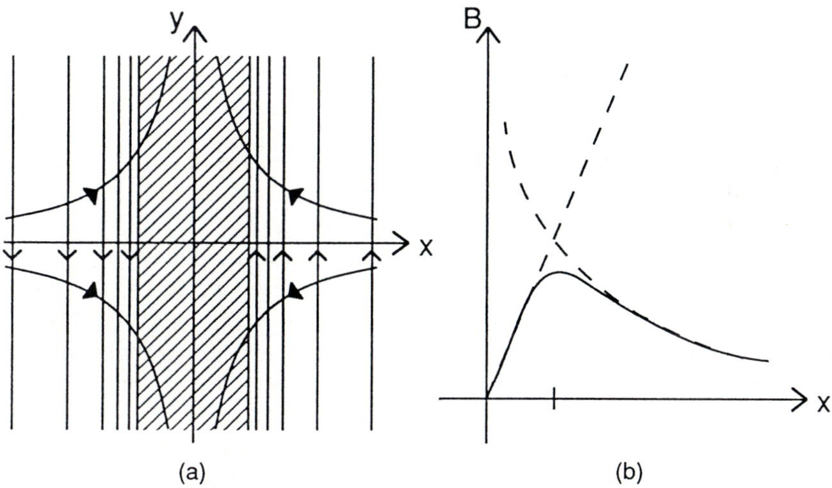

Fig. 6.2. (a) Stagnation-point flow creating a steady current sheet (shaded) (b) Magnetic field profile

erty of (6.2) is that $\nabla.\mathbf{v} = \mathbf{0}$, so that the steady-state continuity equation $(\mathbf{v}.\nabla)\rho + \rho(\nabla.\mathbf{v}) = \mathbf{0}$ reduces to $(\mathbf{v}.\nabla)\rho = \mathbf{0}$ which implies that the density (ρ) is uniform (if it is constant at the edge). The flow vanishes at the origin and therefore represents an incompressible, *stagnation-point flow.*

Suppose now that the magnetic field lines are straight with $\mathbf{B} = B(x)\hat{\mathbf{y}}$ and that they reverse sign at $x = 0$. Then in Ohm's Law,

$$\mathbf{E} + \mathbf{v} \times \mathbf{B} = \eta \nabla \times \mathbf{B} \ , \tag{6.3}$$

both $\mathbf{v} \times \mathbf{B}, \nabla \times \mathbf{B}$ and therefore $\mathbf{E}$ are directed purely in the z-direction. Thus for a steady state with $\mathbf{E} = E(x, y)\hat{\mathbf{z}}$, the equation $\nabla \times \mathbf{E} = \mathbf{0}$ implies that $\partial E/\partial y = \partial E/\partial x = 0$, so that

$$E = \text{constant} \ . \tag{6.4}$$

Indeed (6.4) is essentially an integral of the induction equation (6.1), and in the present case it reduces to

$$E - \frac{V_o x}{a} B = \eta \frac{dB}{dx} \ . \tag{6.5}$$

Now, when x is sufficiently large, the right-hand side of (6.5) is negligible and $B \approx E/(V_0 x)$, whereas when x is very small the second term is negligible and $B \approx Ex/\eta$. These approximate solutions are indicated by dashed curves in Figure 6.2b. When x is large the magnetic field lines are frozen to the plasma and are carried inwards, whereas when x is small the magnetic field diffuses through the plasma. The division between these two extremes, i.e. the half-width of the resulting current sheet, occurs (by equating the two approximations for B) when $x = (a\eta/V_0)^{1/2}$. The full solution of (6.5) is shown in Figure 6.2b. The steady-state equation of motion, however, is also satisfied and so the above solution represents an exact solution of the nonlinear MHD equations - one of the very few that exists.

The steady equation of motion for straight field lines (6.4) becomes

$$\rho(\mathbf{v}.\nabla)\mathbf{v} = -\nabla\left(p + \frac{B^2}{2\mu}\right) \tag{6.6}$$

or

$$\rho\left(-\mathbf{v}\times(\nabla\times\mathbf{v}) + \nabla(\frac{1}{2}v^2)\right) = -\nabla\left(p + \frac{B^2}{2\mu}\right) . \tag{6.7}$$

But the flow (6.2) has zero vorticity ($\nabla \times \mathbf{v} = \mathbf{0}$) and ρ is constant, so the first term in (6.7) vanishes and the remainder imply that

$$\nabla\left(p + \frac{B^2}{2\mu} + \frac{1}{2}\rho v^2\right) = 0 ,$$

so that

$$p = \text{constant } - \frac{1}{2}\rho v^2 - \frac{B^2}{2\mu} , \tag{6.8}$$

which just determines the plasma pressure. This model has been generalised to include a three-dimensional stagnation-point flow with a rotating field ($\mathbf{B}(\mathbf{x})$) and also to include time-dependent effects.

6.3 Qualitative Effects of Reconnection

In most of the universe the magnetic Reynolds number (R_m) is very much larger than unity and so the magnetic field is frozen to the plasma, but in very small regions it can slip through the plasma. Thus for example, a field line initially joining a plasma element at A to one at B (Figure 6.3a) may be carried towards another oppositely directed one and an extremely narrow region of very strong magnetic gradient (and containing an X-type neutral point) may be formed between them (Figure 6.3b). Then the field lines may diffuse, break and reconnect, so that the element A becomes linked instead to the element C (Figure 6.3c).

There are several important effects of this local process:

(i) Changes of global topology and connectivity of field lines, which affect the paths of fast particles and heat, since these travel mainly along field lines;

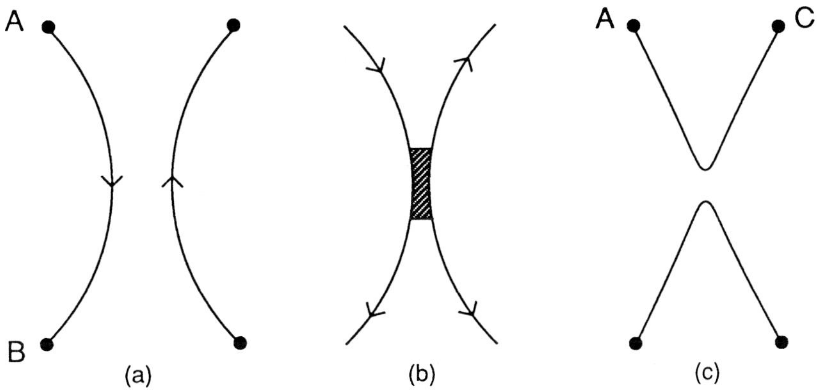

Fig. 6.3. Breaking and reconnection of magnetic field lines

(ii) Conversion of magnetic energy to heat, kinetic energy and fast particle energy;

(iii) Creation of large electric currents, large electric fields, shock waves and filamentation, all of which may help to accelerate fast particles.

6.4 Formation of a Current Sheet

(a) X-type Collapse. There are several ways of forming current sheets. One is through the collapse of the field near an X-type neutral point such as

$$B_{\mathrm{x}} = y, \quad B_{\mathrm{y}} = x \ , \tag{6.9}$$

which has field lines $y^2 - x^2 = $ constant. The field is in equilibrium since the electric current $\mu^{-1}\,(\partial B_{\mathrm{y}}/\partial x - \partial B_{\mathrm{x}}/\partial y)$ vanishes and so there is a balance everywhere between the magnetic pressure force (P) acting inwards and the magnetic force (T) acting outwards (Figure 6.4a).

Suppose now the field is distorted to $B_{\mathrm{x}} = y, B_{\mathrm{y}} = \alpha^2 x$, where $\alpha^2 > 1$, with field lines $y^2 - \alpha^2 x^2 = $ constant, as sketched in Figure 6.4b, and electric current $j = (\alpha^2 - 1)/\mu$.

Physically, we expect an inwards force on the x-axis since the tension force is smaller and the magnetic pressure force larger, whereas along the y-axis we expect an outwards force since the tension force is increased by the larger curvature. Mathematically, the magnetic force has components

$$\mathbf{j} \times \mathbf{B} = -\frac{(\alpha^2 - 1)\alpha^2 x}{\mu}\hat{\mathbf{x}} + \frac{(\alpha^2 - 1)y}{\mu}\hat{\mathbf{y}} \ , \tag{6.10}$$

which are indeed in the expected directions. These act in such a sense as to increase the perturbation and so the initial equilibrium is unstable.

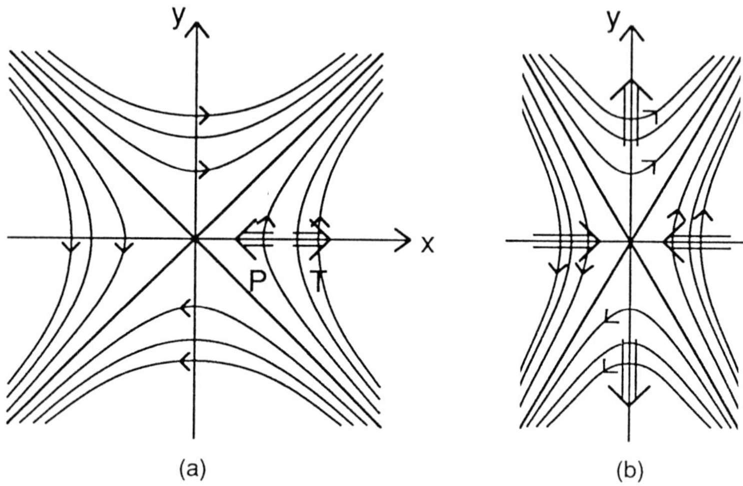

Fig. 6.4. Collapse of the field near an X-point

Instability may be demonstrated formally by showing that the linearised form of the ideal MHD equations

$$\frac{\partial \mathbf{B}}{\partial t} = \nabla \times (\mathbf{v} \times \mathbf{B}), \quad \rho \frac{d\mathbf{v}}{dt} = \mathbf{j} \times \mathbf{B}, \quad \frac{d\rho}{dt} = -\rho \nabla . \mathbf{v} \ , \tag{6.11}$$

possesses solutions of the form

$$B_x = B_0(1 - \epsilon e^{\omega t})\frac{y}{L} \ , \quad B_y = B_0(1 + \epsilon e^{\omega t})\frac{x}{L} \ , \tag{6.12}$$

$$v_x = -\epsilon v_0 e^{\omega t}\frac{x}{t} \ , \quad v_y = \epsilon v_0 e^{\omega t}\frac{y}{L}, \quad \rho = \rho_0 \ , \tag{6.13}$$

where $B_0, L, \rho_0, v_0 = B_0/\sqrt{(\mu\rho)}, \epsilon(\ll 1)$ are constants and $\omega = 2v_0/L$.

(b) Driving Source Motion. Suppose the sources of the magnetic field move slowly together and drive the formation of a series of equilibria containing a current sheet (Figure 6.5). Initially, as before we suppose a field of the form (6.9). Then the question is how to describe the resulting equilibrium containing a current sheet and having the appearance of Figure 6.5b. Outside the sheet, if the current vanishes, the magnetic field satisfies $\nabla \times \mathbf{B} = 0$ and $\nabla . \mathbf{B} = 0$ or, in two dimensions,

$$\frac{\partial B_y}{\partial x} - \frac{\partial B_x}{\partial y} = 0 \quad \text{and} \quad \frac{\partial B_x}{\partial x} + \frac{\partial B_y}{\partial y} = 0 \ . \tag{6.14}$$

Now, it may be shown as follows that if the combination

$$B_y + iB_x = f(z) \tag{6.15}$$

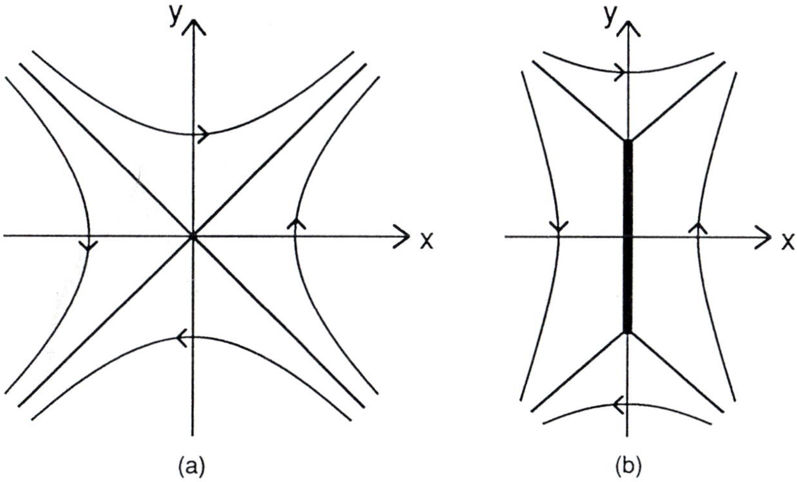

Fig. 6.5. Creation of a current sheet from an X-point

is any analytic (i.e. differentiable) function of the complex variable $z = x + iy$, then (6.14) is automatically satisfied. We are familiar with the fact that the derivative $f'(x)$ of a function of x exists if the gradient at x has the same value whether x is approached from the left or the right. In a similar way if $f(z)$ is analytic the gradient has the same value when z is approached from any direction, in particular keeping y constant (so that $z = x$) and keeping x constant (so that $z = iy$). In other words

$$\frac{\partial}{\partial x}(B_y + iB_x) = \frac{1}{i\partial y}(B_y + iB_x) , \tag{6.16}$$

or, by equating real and imaginary parts, we obtain (6.14) as required. Thus, we can treat the current sheet as a cut in the complex plane and the object is to find a function $f(z)$ which has such a cut.

Now the initial state (6.15) has $B_y + iB_x = z$ and when a sheet stretches from $z = -iL$ to $z = iL$ we may use

$$B_y + iB_x = (z^2 + L^2)^{1/2} , \tag{6.17}$$

which behaves like z when $z \gg L$ and reduces to z when $L = 0$. Thus the evolution through a series of equilibria with a slowly growing sheet may simply be modelled by letting L slowly increase in value in (6.17). The field has as limiting field lines (separatrices) through the ends of the sheet, field lines which are inclined to one another at the ends of the sheet by $2\pi/3$. This may be shown by noting that near the upper end of the sheet at $z = iL$, (6.17) becomes approximately $B_y + iB_x = \surd(2iL)\mathcal{Z}^{1/2}$ where $\mathcal{Z} = z - iL$. This may be written as

$$B_y + iB_x = \frac{da}{d\mathcal{Z}} , \tag{6.18}$$

where

$$a = \sqrt{(2iL)}\frac{2}{3}\mathcal{Z}^{3/2} = \sqrt{(2L)}\frac{2}{3}e^{i\pi/4}R^{3/2}e^{3i\Theta/z} \ , \tag{6.19}$$

and the complex number $\mathcal{Z}$ has been written in polar form as $\mathcal{Z} = Re^{i\Theta}$.

However, if A is the real part of a, then (6.18) implies that $B_y = \partial A/\partial X,\ B_x = -\partial A/\partial y$, where X and Y are the real and imaginary parts of $\mathcal{Z}$ and the magnetic field lines are given by

$$\frac{dY}{dX} = \frac{B_y}{B_x} = -\frac{\partial A/\partial X}{\partial A/\partial Y} \quad \text{or} \quad \frac{\partial A}{\partial X}dX + \frac{\partial A}{\partial Y}dY = 0 \ . \tag{6.20}$$

In other words $dA = 0$ and so $A =$ constant.

By taking the real part of (6.19) we can see that

$$A = \sqrt{(2L)}\frac{2}{3}R^{3/2}\cos\left(\frac{3\Theta}{2} + \frac{\pi}{4}\right) \ . \tag{6.21}$$

Thus the particular field lines $A = 0$ are given by

$$\frac{3\Theta}{2} + \frac{\pi}{4} = -\frac{\pi}{2}, \frac{\pi}{2}, \frac{3\pi}{2}, \frac{5\pi}{2} \ , \tag{6.22}$$

and so $\Theta = -\pi/2, \pi/6, 5\pi/6$ or $3\pi/2$. In other words the current sheet (at $-\pi/2$ or $3\pi/2$) is inclined to the separatrices ($\pi/6$ and $5\pi/6$) by $2\pi/3$, as required.

This technique for obtaining current sheets has been applied to a wide variety of different configurations (see Priest, 1985) and also it has been generalised to allow for a dynamic time-dependent formation of a current sheet (Priest and Titov, 1994) rather than just a passive evolution through a series of equilibria.

6.5 Linear Reconnection

Is slow steady, 2D, incompressible reconnection possible? We need to solve the equations

$$\mathbf{E} + \mathbf{v} \times \mathbf{B} = \frac{\mathbf{j}}{\sigma} \ , \tag{6.23}$$

$$\rho(\mathbf{v}.\nabla)\mathbf{v} = -\nabla \mathbf{p} + \mathbf{j} \times \mathbf{B} \ , \tag{6.24}$$

where $\nabla.\mathbf{v} = 0,\ \nabla.\mathbf{B} = 0,\ \mathbf{j} = \nabla \times \mathbf{B}/\mu$.

For steady 2D flows and magnetic fields (6.23) implies that $\mathbf{E} = E_z(x,y)\hat{\mathbf{z}}$ and $\nabla \times \mathbf{E} = \mathbf{0}$ implies that $\partial E_z/\partial x = \partial E_z/\partial y = 0$ and so E_z is constant. Also if the flow speeds are much smaller than the Alfven speed ($v \ll v_A = B\sqrt{(\mu\rho)}$) (6.24) reduces to

$$0 = -\nabla p + \mathbf{j} \times \mathbf{B} \ . \tag{6.25}$$

The answer to the above question has been given by Priest et al (1994) in terms of an *Anti-reconnection Theorem*: Steady 2D reconnection (with plasma flow across separatrices) is impossible if $v \ll v_A$ everywhere (and viscosity is negligible and the magnetic diffusivity η is uniform).

Proof.
Take the curl of (6.25), then $\nabla \times (\mathbf{j} \times \mathbf{B}) = 0$ or

$$\mathbf{j}(\nabla \times \mathbf{B}) + (\mathbf{B}.\nabla)\mathbf{j} - (\nabla.\mathbf{j})\mathbf{B} - (\mathbf{j}.\nabla)\mathbf{B} = 0 \ .$$

But $\nabla.\mathbf{B}$ vanishes and so does $\nabla.\mathbf{j}$ since $\mathbf{j} = \nabla \times \mathbf{B}/\mu$. Also, since $\mathbf{j}$ is in the z-direction for $\mathbf{B} = \mathbf{B}(\mathbf{x}, \mathbf{y})$, the latter term vanishes as well, and we are left with $(\mathbf{B}.\nabla)\mathbf{j} = \mathbf{0}$, so that $\mathbf{j}$ is constant along magnetic field lines. However, (6.23) implies that $\mathbf{E} = \mathbf{j}/\sigma$ at the X-point and so, since $\mathbf{E}$ is constant everywhere and $\mathbf{j}$ is constant along fieldlines, $\mathbf{E} = \mathbf{j}/\sigma$ along the separatrices (the field lines which pass through the X-point). Thus, from (6.23) $\mathbf{v} \times \mathbf{B} = \mathbf{0}$ along a separatrix and so there is no flow across it.

The main way to avoid this theorem and allow reconnection is to allow the flow speed to become Alfvenic somewhere so that the process becomes nonlinear, as in the following section.

6.6 Fast Steady-State Reconnection

(a) Sweet - Parker Model (1958). The Sweet - Parker model consists of a simple diffusion region of length 2L and width $2l$ between oppositely directed fields, for which an order-of-magnitude analysis is conducted. First of all, suppose the input flow speed and magnetic field are v_i, B_i, respectively (Figure 6.6) and ask: what is the outflow speed?

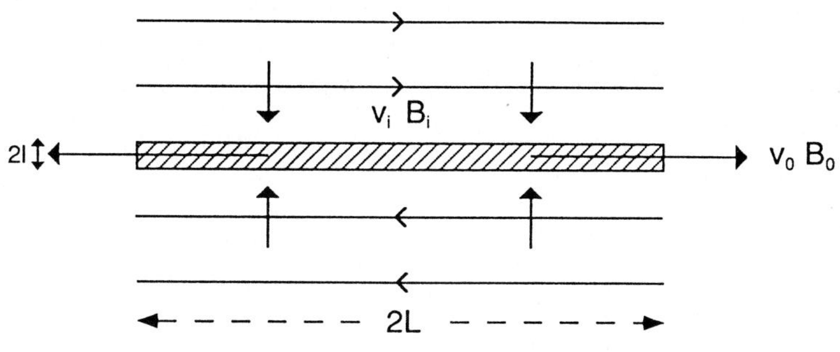

Fig. 6.6. Sweet - Parker Reconnection

The electric current in order of magnitude is $j \approx B_i/(\mu l)$ and so the Lorentz force along the sheet is $(\mathbf{j} \times \mathbf{B})_x \approx jB_0 = B_i B_0/(\mu l)$. This force accelerates the plasma from rest at the neutral point to v_0 over a distance L and so, by equating the magnitude of $\rho(\mathbf{v}.\nabla)v_x$ to the above Lorentz force, we have

$$\rho \frac{v_0^2}{L} \approx \frac{B_i B_0}{\mu l} \ . \tag{6.26}$$

However, from $\nabla.\mathbf{B} = \mathbf{0}$,

$$\frac{B_0}{l} \approx \frac{B_i}{L} \ , \tag{6.27}$$

and so the right-hand side of (6.26) may be rewritten as $B_{\rm i}^2/(\mu L)$ and we have

$$v_{\rm O}^2 = \frac{B_{\rm i}^2}{\mu\rho} \equiv v_{\rm Ai}^2 \ , \tag{6.28}$$

where $v_{\rm Ai}$ is the Alfven speed at the inflow. Not surprisingly, therefore, the magnetic force accelerates the plasma to the Alfven speed.

The next question is: how fast can field lines and plasma enter the diffusion region (at v_i)? First of all, note that for a steady state the plasma must carry the field lines in at the same speed that they are trying to diffuse outward, so that

$$v_i = \frac{\eta}{l} \ . \tag{6.29}$$

Also conservation of mass implies that the rate $(4\rho L v_{\rm i})$ at which mass is entering the sheet must equal the rate $(4\rho l v_{\rm O})$ at which it is leaving, so that

$$L v_{\rm i} = l v_{\rm Ai} \ . \tag{6.30}$$

The width l may be eliminated between these two equations to give $v_{\rm i}^2 = \eta v_{\rm Ai}/L$, or in dimensionless form

$$M_{\rm i} = \frac{1}{R_{\rm mi}{}^{1/2}} \tag{6.31}$$

in terms of the Alfven Mach number

$$M = \frac{v}{v_{\rm A}} \tag{6.32}$$

and the magnetic Reynolds number

$$R_m = \frac{L v_{\rm A}}{\eta} \tag{6.33}$$

based on the Alfven speed.

It is interesting to consider the energetics of the reconnection process in a Sweet–Parker diffusion layer which is long and thin ($l \ll L$). One implication of (6.30) is that $v_{\rm i} \ll v_{\rm Ai}$, so that the inflow speed is much smaller than the Alfven speed. Now the rate of inflow of electromagnetic energy is the Poynting flux ($\mathbf{E} \times \mathbf{H}$ per unit area), or, since $E = v_{\rm i} B_{\rm i}$ in magnitude,

$$E \frac{B_{\rm i}}{\mu} L = v_{\rm i} \frac{B_{\rm i}^2}{\mu} L \ . \tag{6.34}$$

Therefore by (6.32) the ratio of the inflow of kinetic to electromagnetic energy is

$$\frac{\text{Inflow } K.E.}{\text{Inflow } e.m.} = \frac{\frac{1}{2}\rho v_{\rm i}^2}{B_{\rm i}^2/\mu} = \frac{v_{\rm i}^2}{2 v_{\rm Ai}^2} \ll 1 \ . \tag{6.35}$$

In other words, most of the inflowing energy is magnetic.

Next consider the energy outflow. By conservation of flux

$$v_0 B_0 = v_i B_i \ , \tag{6.36}$$

(which is consistent with (6.27) and (6.30)) and so $B_0 \ll B_i$. Also the outflow of electromagnetic energy is $EB_0 l/\mu$, which is much less than the inflow of electromagnetic energy (6.35) since both $B_0 \ll B_i$ and $l \ll L$. So what has happened to the inflowing magnetic energy? Well, the ratio of outflowing kinetic to inflowing magnetic energy is

$$\frac{\text{Outflow}\, K.E.}{\text{Inflow}\, e.m.} = \frac{\frac{1}{2}\rho v_0^2 (v_0 l)}{v_i B_i^2 L/\mu} = \frac{\frac{1}{2} v_0^2}{v_{Ai}^2} = \frac{1}{2} \ .$$

Thus half of the inflowing magnetic energy is converted to kinetic energy, while the remaining half is converted to thermal energy. In other words, the effect of the reconnection is to create hot fast streams of plasma.

In this connection it is useful to remember, that by substituting for $\nabla \times \mathbf{H}$ from Ampere's law (2.11) and for $\nabla \times \mathbf{E}$ from Faraday's law (2.12), we may write

$$-\nabla.(\mathbf{E} \times \mathbf{H}) = \mathbf{E}.\nabla \times \mathbf{H} - \mathbf{H}.\nabla \times \mathbf{E}$$

or

$$-\nabla.(\mathbf{E} \times \mathbf{H}) = \mathbf{E}.\mathbf{j} + \frac{\partial}{\partial t}\left(\frac{B^2}{2\mu}\right) \ ,$$

which implies that an inflow of electromagnetic energy can produce electrical energy ($\mathbf{E}.\mathbf{j}$) for the plasma and a rise in the magnetic energy. Furthermore, by taking the scalar product of $\mathbf{j}$ with Ohm's law $\mathbf{E} = \mathbf{j}/\sigma - \mathbf{v} \times \mathbf{B}$, we obtain

$$\mathbf{E}.\mathbf{j} = \frac{j^2}{\sigma} + \mathbf{v}.\mathbf{j} \times \mathbf{B} \ ,$$

so that the electrical energy appears partly as ohmic heat and partly as the work done by the Lorentz force (in, say, accelerating the plasma). In our case the inflow of electromagnetic energy goes into electrical energy, half of which appears as heat and half as kinetic energy.

The fast regimes of reconnection that we shall consider next contain a tiny Sweet–Parker diffusion region around the X-point, and the flow speed and magnetic field at large distances L_e from the X–point are denoted by v_e and B_e. The properties of reconnection models depend on two dimensionless parameter, namely the reconnection rate ($M_e = v_e/v_{Ae}$) and the global magnetic Reynolds number ($R_{me} = Lv_{Ae}/\eta$).

Reconnection is said to be "fast" when the reconnection rate (M_e) is much larger than the Sweet–Parker rate (6.31). Properties at the inflow to the diffusion region (denoted by subscript 'i') may be related to the "external" values at large distances (denoted by subscript 'e'). Thus flux conservation ($v_i B_i = v_e B_e$) may be written in dimensionless form as

$$\frac{M_i}{M_e} = \frac{B_e^2}{B_i^2} \ . \tag{6.37}$$

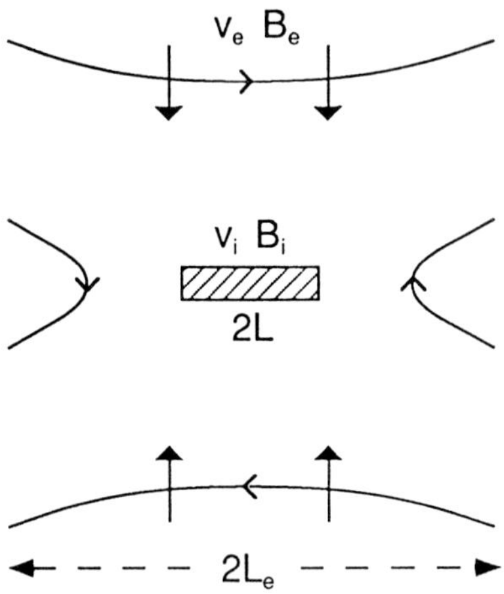

Fig. 6.7. The notation for fast regimes

Furthermore, the Sweet–Parker relations (6.29) and (6.30) may be rewritten in dimensionless form as

$$\frac{L}{L_e} = \frac{1}{R_{me}} \frac{1}{M_e^{3/2}} \frac{1}{M_i^{1/2}} , \tag{6.38}$$

$$\frac{l}{L_e} = \frac{1}{R_{me}} \frac{1}{M_e^{1/2}} \frac{1}{M_i^{1/2}} . \tag{6.39}$$

Thus, once B_i/B_e has been determined from a model of the external region outside the diffusion region, (6.37) determines M_i/M_e and (6.38) and (6.39) give the dimensions of the diffusion region in terms of M_e and R_{me} .

(b) Petschek Model (1964). Petschek's regime is "almost–uniform" in the sense that the field in the inflow region is a small perturbation to a uniform field B_e. It is also potential in the sense that there is no current in the inflow region. Most of the energy conversion takes place at standing slow-mode shocks, which are almost switch-off in nature (Figure 6.8a). These shock waves accelerate and heat the plasma, with $\frac{2}{5}$ of the inflowing magnetic energy being changed to heat and $\frac{3}{5}$ to kinetic energy.

The Petschek analysis is straightforward. The inflow region consists of slightly curved field lines and the magnetic field is a uniform horizontal field ($B_e\hat{\mathbf{x}}$), plus a solution of Laplace's equation which vanishes at large-distances and which has a normal component of, say, B_N at the shock waves and zero at the diffusion region. To lowest order, the inclination of the shocks may be neglected, and so the problem is to find a solution in the upper half-plane which vanishes at infinity and which equals $2B_N$ between L and L_e on the x-axis and, by symmetry $-2B_N$ between $-L_e$ and $-L$. Now, we may regard the normal component on the x-axis as being produced by a continuous series of poles. If each pole produces a field

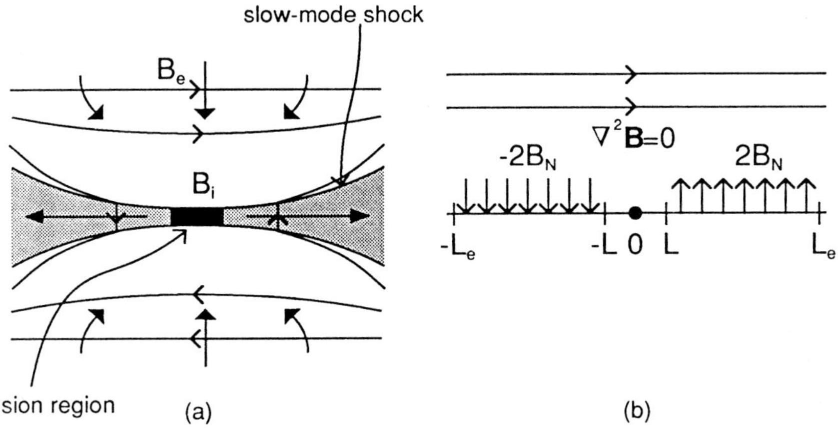

Fig. 6.8. (a) Petschek's model (b) Notation for the inflow region

m/r at a distance r, then the flux produced in the upper half plane by that pole will be πm: but, if the pole occupies a distance dx of the x-axis, the flux is also $2B_N dx$, so that $m = 2B_N/\pi$ and integrating along the x-axis gives the field at the origin produced by the poles as

$$\frac{1}{\pi}\int_{-L_e}^{L} \frac{2B_N}{x} dx - \frac{1}{\pi}\int_{L}^{L_e} \frac{2B_N}{x} dx \ . \tag{6.40}$$

Adding this to the field (B_e) at infinity gives

$$B_i = B_e - \frac{4B_N}{\pi} \log \frac{L_e}{L} \ . \tag{6.41}$$

But at the shock waves, remembering from (4.49) that slow shocks travel at the Alfven speed based on the normal field, $B_N/\sqrt{(\mu\rho)} = v_e$, so that (6.41) becomes

$$B_i = B_e \left(1 - \frac{4M_e}{\pi} \log \frac{L_e}{L}\right) \ , \tag{6.42}$$

which is the expression for B_i that we have been seeking.

Since $M_e \ll 1$ and $B_i \approx B_e$, the scalings (6.38) and (6.39) become

$$\frac{L}{L_e} \approx \frac{1}{R_{me} M_e^2}, \ \frac{l}{L_e} \approx \frac{1}{R_{me} M_e} \ , \tag{6.43}$$

which show that the dimensions of the central region decrease as the magnetic Reynolds number (R_{me}) or reconnection rate (M_e) increase. Petschek suggests that the mechanism chokes itself off when B_i becomes too small, and so he estimates a *maximum reconnection rate* (M_e^*) by putting $B_i = \frac{1}{2}B_e$ in (6.42) to give

$$M_e^* \approx \frac{\pi}{8 \log R_{me}} \ . \tag{6.44}$$

In value this is typically 0.01 since $\log R_{\rm me}$ is slowly varying, and so we see that for typical $R_{\rm me}$ values this is much faster than the Sweet–Parker rate.

Thus, for twenty years the problem of fast reconnection was thought to have been solved completely (by Petschek), until in 1986 a new generation of reconnection models was proposed (with Petschek's mechanism as a special case) and high-resolution numerical experiments were undertaken.

(c) Unified Theory of Fast Almost-Uniform Reconnection (Priest and Forbes, 1986). I was puzzled to find that often numerical experiments have quite different properties from Petschek's theory, such as diverging rather than slightly converging inflow and different scalings of the diffusion region dimensions with reconnection rate. The object is to analyse the inflow region and find $B_{\rm i}/B_{\rm e}$ and therefore how $M_{\rm i}$ varies with the reconnection rate ($M_{\rm e}$). The analysis is to take the steady ideal, 2D, MHD equations

$$\rho(\mathbf{v}.\nabla)\mathbf{v} = -\nabla\mathbf{p} + \mathbf{j}\times\mathbf{B} \ , \tag{6.45}$$

$$\mathbf{E} + \mathbf{v}\times\mathbf{B} = 0 \ , \tag{6.46}$$

where $\nabla.\mathbf{v} = 0, \nabla.\mathbf{B} = 0$, $\mathbf{j} = \nabla\times\mathbf{B}/\mu$ and $\mathbf{E}$ is constant, and then to seek solutions

$$\mathbf{B} = \mathbf{B}_{\rm e} + M_{\rm e}\mathbf{B}_1 + \ldots, \ \mathbf{v} = M_{\rm e}\mathbf{v}_1 + \ldots$$

that are a small perturbation about a uniform field $\mathbf{B}_{\rm e} = \mathbf{B}_{\rm e}\hat{\mathbf{x}}$. In these expansions $M_{\rm e} \ll 1$ is the expansion parameter.

Neglecting the pressure gradient, to lowest order (6.45) becomes

$$j_1 B_{\rm e} = 0 \ , \tag{6.47}$$

or, if $(B_{1\rm x}, B_{1\rm y}) = (\partial A_1/\partial y, -\partial A_1/\partial x)$,

$$\nabla^2 A_1 = 0 \ . \tag{6.48}$$

The boundary conditions are to impose $B_{1\rm x} = 0$ on the top boundary, $\partial B_{1\rm y}/\partial x = 0$ on the side boundaries and

$$B_{1\rm y} = f(x) = \begin{cases} 2B_{\rm N}, & L < x < L_{\rm e} \ , \\ 2B_{\rm N}x/L, & -L < x < L \ , \\ -2B_{\rm N} & -L_{\rm e} < x < -L \ . \end{cases}$$

The resulting separable solutions are

$$B_{1\rm x} = -\sum_0^\infty a_{\rm n} \cos\left[\left(n+\frac{1}{2}\right)\pi x\right] \sinh\left[\left(n+\frac{1}{2}\right)\pi(1-y)\right] \ ,$$

$$B_{1\rm y} = \sum_0^\infty a_{\rm n} \sin\left[\left(n+\frac{1}{2}\right)\pi x\right] \cosh\left[\left(n+\frac{1}{2}\right)\pi(1-y)\right] \ , \tag{6.49}$$

where

$$a_{\rm n} = \frac{4B_{\rm N} sinh\left[\left(n+\frac{1}{2}\right)\pi L(L_{\rm e}]\right.}{L/L_{\rm e}\left(n+\frac{1}{2}\right)^2 \pi^2 cosh\left[\left(n+\frac{1}{2}\right)\pi\right]} \ .$$

From these expressions we can calculate $B_{\rm i}/B_{\rm e}$ and substitute in (6.37) in place

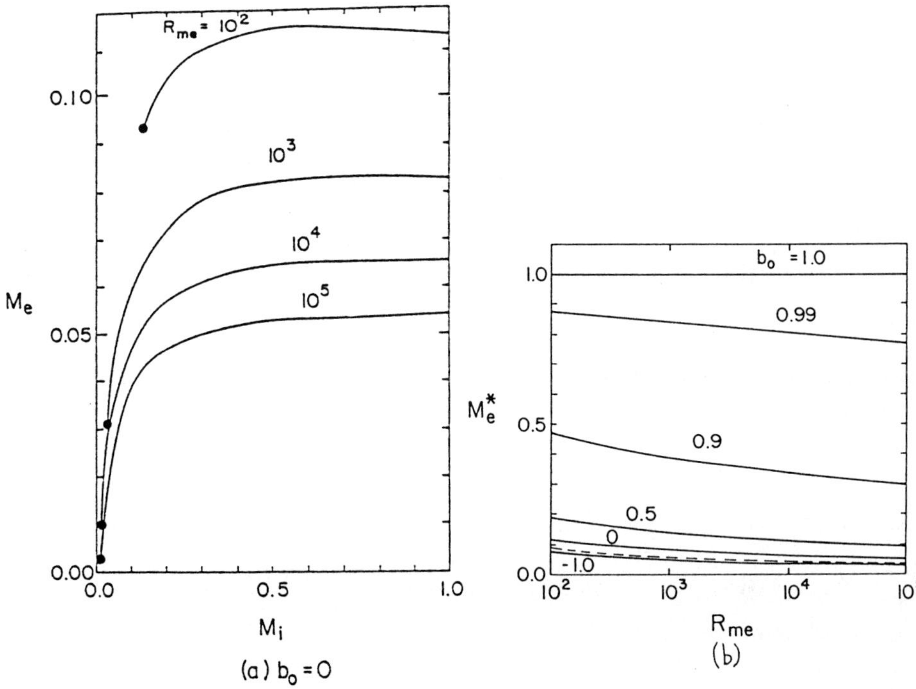

Fig. 6.9. (a) Reconnection rate $M_{\rm e}$ as a function of $M_{\rm i}$ for $b = 0$
(b) Maximum reconnection rate $M_{\rm e}^*$

of (6.42). The resulting graphs of $M_{\rm e}$ against $M_{\rm i}$ for given $R_{\rm me}$ are shown in Figure 6.10, from which it can be seen that for a given $R_{\rm me}$ there is indeed a maximum reconnection rate ($M_{\rm e}^*$), as Petschek had surmised. Furthermore, the variation of $M_{\rm e}^*$ with $R_{\rm me}$, shown as the curve marked $b = 0$ in Figure 6.10, is very close to Petschek's estimate, shown dashed.

It is always good to derive some else's result in your own way, to make it real for yourself and understand in depth their assumptions. But now, how can we generalise this analysis by relaxing one of the assumptions? What we decided to do is include pressure gradients, so that terms dp_1/dy and $-(\mu/B_{\rm e})dp_1/dy$ are added to the right-hand sides of (6.47) and (6.48). But the effect on the solution (6.49) is simply to add a constant $a_{\rm n}b$ to each term in the sum.

The new parameter b produces a whole range of different regimes: $b = 0$ gives Petschek's regime (a weak fast-mode expansion); $b = 1$ gives a Sonnerup–

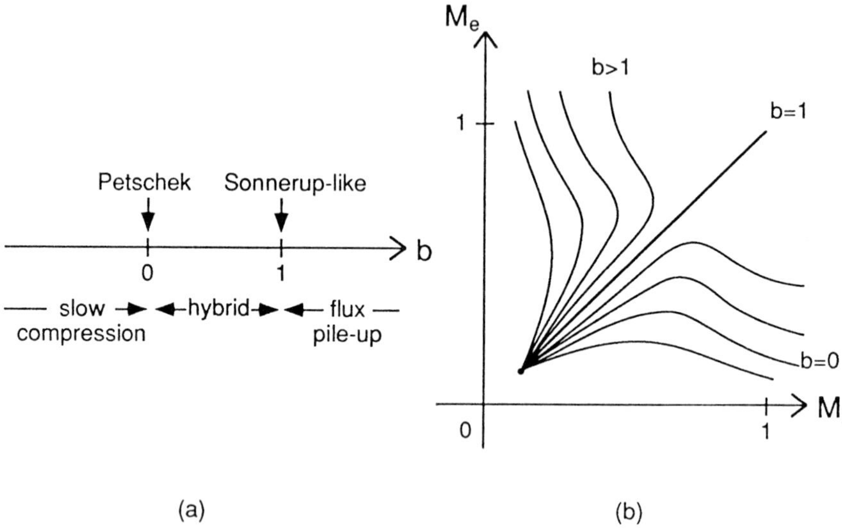

Fig. 6.10. (a) Different regimes of reconnection (b) Reconnection rate ($M_e = v_e/v_{Ae}$) as a function of $M_i = v_i/v_{Ai}$ for different b

like regime (a weak slow-mode expansion); $b < 0$ gives a family of slow-mode compressions; $0 < b < 1$ gives a hybrid family of slow- and fast- mode expansions; $b > 1$ gives a flux pile-up family of slow-mode expansions (Figure 6.11). For a *compression* the pressure increases as the plasma moves in towards the diffusion region, whereas for an *expansion* the pressure decreases. *Slow-mode* behaviour has the pressure and magnetic field behaving differently, so that for example a slow-mode compression makes the pressure increase and the magnetic field decrease. *Fast-mode* behaviour has the plasma and magnetic field both increasing or decreasing together.

The way the resulting reconnection rate (M_e) varies with M_i for a given R_{me} and several values of b is sketched in Figure 6.11b for M_e and M_i up to 1, although of course the analysis is only valid for $M \ll 1$. For $b = 1$, M_e increases linearly with M_i; for $b = 0$ the Petschek maximum can be seen; all other regimes with $b < 1$ also possess a maximum reconnection rate, although when $0 < b < 1$ the maximum rate is faster than Petschek's; when $b > 1$, there is no maximum rate, within the limitations of the theory.

Thus the solutions depend on the parameter b, which is in turn determined by the nature of the flow on the inflow boundary since v_x at the corner ($x = L_e, y = L_e$) is proportional to $b - 2/\pi$. The field lines and streamlines for several cases are shown in Figure 6.12. When $b < 0$ the streamlines near the axis $x = 0$ are converging and so tend to compress the plasma and hence produce slow-mode compressions. When $b > 1$ the streamlines diverge and so tend to expand the plasma, producing slow-mode expansions: we refer to this as a "flux pile–

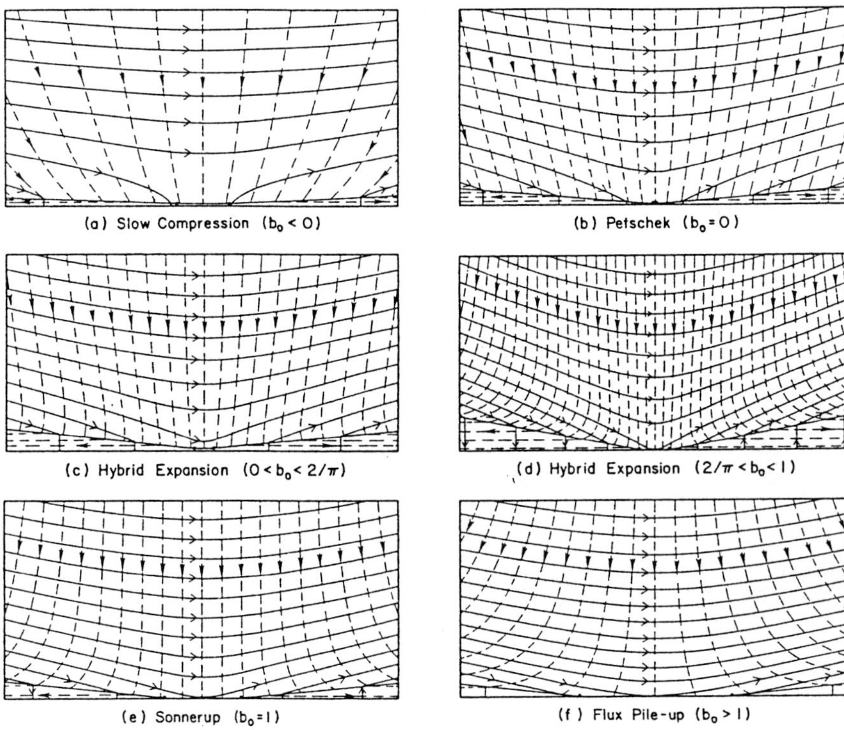

Fig. 6.11. Magnetic field lines (solid) and streamlines (dashed) in the upper half-plane for several different regimes of almost-uniform reconnection

up regime", since the magnetic field lines come closer together and the field strength increases as they are carried in. Another feature is that the central diffusion regions are much larger for the flux pile-up regime than the Petschek regime.

(d) Numerical Experiments. Biskamp has conducted a most impressive numerical experiment on fast steady reconnection. Figure 6.12 shows a series of runs for $M_e = 0.042$ and $R_{me} = 1746$ in the first column, $R_{me} = 3492$ in the second column and $R_{me} = 6984$ in the third column. He found that the length and width of the diffusion region increase as R_{me} (or M_e) increase, and, since this is the opposite behaviour to Petschek, he concluded that Petschek's mechanism and therefore fast reconnection does not exist for large R_{me}. However, Priest and Forbes (1992) have shown by using more recent models that Biskamp's particular scaling may be produced by adopting his boundary conditions and that fast reconnection does indeed exist with a wide range of different regimes, depending on the boundary conditions .

Biskamp's experiment contains several interesting features:

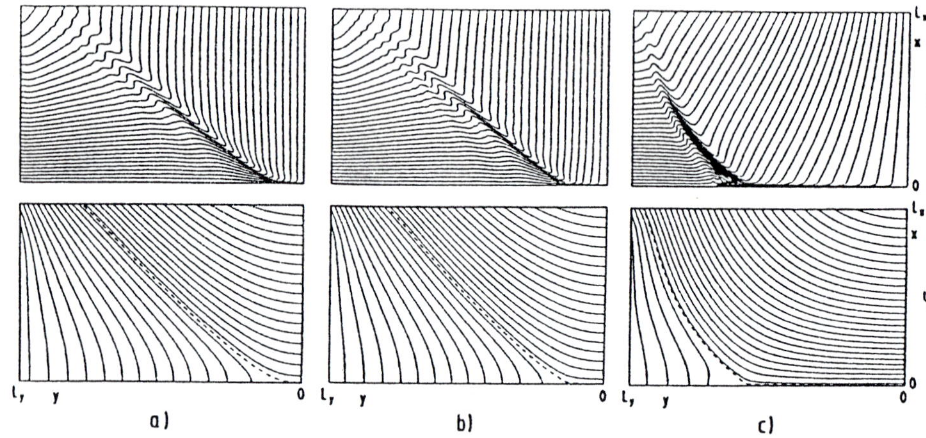

Fig. 6.12. Numerical experiment showing streamlines (top) and magnetic field lines (bottom) in the second quadrant (Biskamp, 1986)

(i) the nature of the inflow varies from converging to diverging streamlines;
(ii) the inflow field lines may be highly curved and the shock angle large;
(iii) strong jets of plasma flowing out along the separatrices;
(iv) spikes of reversed current at the ends of the diffusion region.

However, there is a new generation of theoretical models which generalizes Petschek's mechanism in several ways and gives explanations of most of these properties. The above family of Almost-Uniform Models (Priest and Forbes, 1986) explains feature (i), and another family of Non-Uniform Models (see (f) below) are able to reproduce features (ii) – (iv).

A recent numerical experiment by Yan, Lee and Priest (1992) has verified the configurations and scalings of the Almost-Uniform Regimes, provided the resistivity is enhanced in the diffusion region. As the outflow boundary condition the normal velocity component from the Almost-Uniform theory is imposed. 100 x 200 grid points are taken in one quadrant and R_{me} is taken to be 2000; the resulting field lines and streamlines for $b = 0$ and $M_e = 0.042$ (Petschek-like reconnection) are shown in Figure 6.13a. In contrast the case $b = 1$ (Sonnerup-like reconnection) is shown in Figure 6.13b for $M_e = 0.052$. The profiles along the y-axis show that the magnetic field is uniform and the pressure declines as the plasma comes in (a weak slow-mode expansion). Figure 6.13c shows a flux pile-up case ($b = 2$) for $M_e = 0.1$ with the pressure decreasing and magnetic field increasing as plasma flows in (a slow-mode expansion). Finally, Figure 6.13d shows a slow-mode compression case ($b = -3$) for $M_e = 0.035$ in which the magnetic field decreases and pressure increases inwards.

(e) Non-Uniform Reconnection (Priest and Lee, 1990). In order to avoid the assumption of small curvature inherent in the Almost-Uniform Theory, Priest

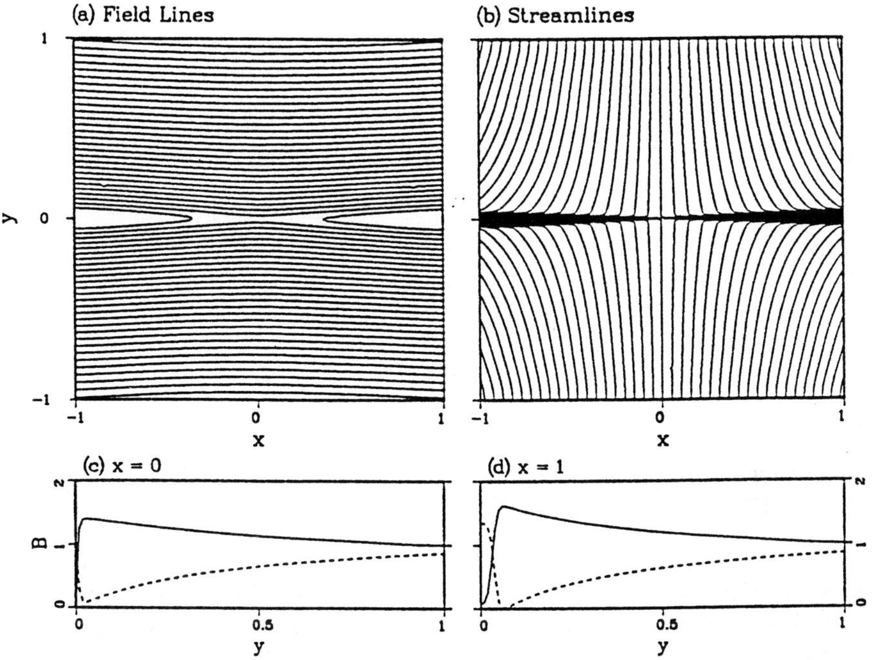

Fig. 6.13. Numerical experiment on Almost–Uniform Reconnection for $b = 2$

and Lee (1990) considered a strongly curved potential inflow. They wrote the plasma velocity and magnetic field in terms of a stream function (ψ) and flux function (A). First of all in the upstream region they assumed the Alfven speed is much smaller than the flow speed and sound speed, so that the field is potential and chose $B_y + iB_x = B_i \left(z^2/L^2 - 1\right)^{1/2}$. Then they deduced the velocity from $\mathbf{E}+\mathbf{v}\times\mathbf{B} = \mathbf{0}$. Next they calculated the position of the shock as the characteristic $\psi + A =$ constant passing through the end $z = L$ of the current sheet.

Finally they deduced the downstream flow by solving numerically the ideal MHD equations together with appropriate boundary conditions on the shock and the outflow boundary. Some of the results are shown in Figure 6.15. The field lines are highly curved because of the form of the inflow boundary conditions. A separatrix jet occurs for long diffusion regions because the inflow to the diffusion region is diverted along the separatrix. Also reversed current spikes at the ends of the diffusion region are revealed as regions of reversed magnetic curvature which slow down the plasma and divert it along the separatrices.

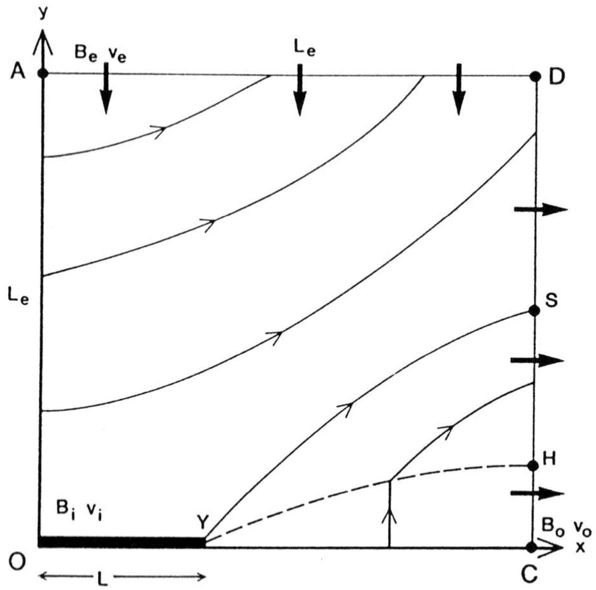

Fig. 6.14. Notation for Non-Uniform Theory

7 Coronal Heating

7.1 Introduction

The solar corona is a wonderfully beautiful and complex environment which is heated and structured by the magnetic field. It has a three-fold structure of loop, hole and bright point, in which different heating mechanisms may well be at work. Eclipse photographs such as those in Figure 1.3 show the corona indirectly by photospheric light going outwards and then being Thomson-scattered towards you, so that the brighter areas are more dense than the darker areas. One sees bright magnetically closed regions, where the magnetic field is strong enough to contain the plasma and hold it close to the Sun, and also dark magnetically open regions where the plasma is streaming outwards from the Sun as the solar wind. In both cases the structures are outlining the magnetic field.

I remember being captivated when I first saw pictures from Skylab, such as those in Figure 1.9, which revealed a new world - the corona directly in soft X-rays. Here there are myriads of coronal loops, including active regions which show up as rather fuzzy but very bright structures. In addition, the magnetically open regions are dark "coronal holes", and a new feature is the presence of many tiny bright dots, called X-ray bright points.

More recently, Leon Golub and collaborators have increased our sense of wonder with observations from a new Normal Incidence X-ray Telescope (NIXT) with a greatly improved resolution by a factor of five to only three-quarters of an arc-second (Golub, 1993). For instance in Fe XVI at $2 - 3 \times 10^6$K one sees

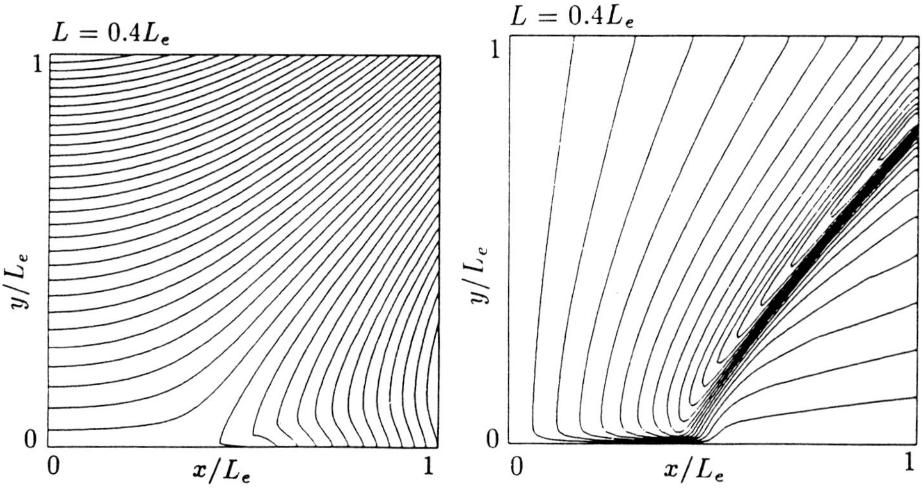

Fig. 6.15. (a) Magnetic field lines and (b) stream lines for one quadrant of Non-Uniform Reconnection

extremely fine-scale structure (Figure 7.1) with the active regions being very much better resolved than before.

Another dramatic improvement in coronal observations has come from the Japanese YOHKOH satellite, YOHKOH meaning "sunbeam", in particular from the soft X-ray telescope, which has much less scattering than Skylab and very good time-resolution . It again shows up much extremely fine-scale structure and reveals that the coronal magnetic field is continually evolving and reconnecting.

Coronal loops have a wide variation in their properties. Active region loops have, according to old Skylab measurements, typical lengths (L) of 10 - 200 Mm, temperatures (T) of $2.2 - 2.8 \times 10^6$K , densities (n) of $5 \times 10^{14} - 5 \times 10^{15}\,\mathrm{m}^{-3}$, while quiet region loops have $L \sim 20-700\mathrm{Mm}, n \sim 2\times10^{14}-10^{15}\mathrm{m}^{-3}$ and interconnecting loops have $L \leq 700\mathrm{Mm}, T \sim 2 \times 10^6\mathrm{K}, n \sim 7 \times 10^{14}\mathrm{m}^{-3}$. However, we now know from SMM and from YOHKOH that normal active region loops, for instance, can be very much hotter with temperatures up to $4-6\times10^6$K. In coronal holes the plasma is cooler (with T$\sim 1.4 - 1.8 \times 10^6$K) and rarer, with densities about one-third of quiet coronal values (n_c). Bright points are between 4 and 22 Mm in size and have temperatures $1.3 - 1.7 \times 10^6$K and densities between 2 and 4 times greater than n_c.

In the corona usually the plasma beta is very small ($\beta \ll 1$) and we have a force-free field. Coronal magnetic field lines are anchored in the dense photosphere where $\beta \gtrsim 1$, and their feet move about due to a variety of motions. Granulation has typical horizontal velocities of 0.25 - 2 km s^{-1} and lifetimes for individual granules of about 8 min., while supergranules involve horizontal motions of about 0.3 km s^{-1} for typically a day, and also there are motions associated with general active region evolution.

The way the corona responds to such motions on a timescale τ depends on

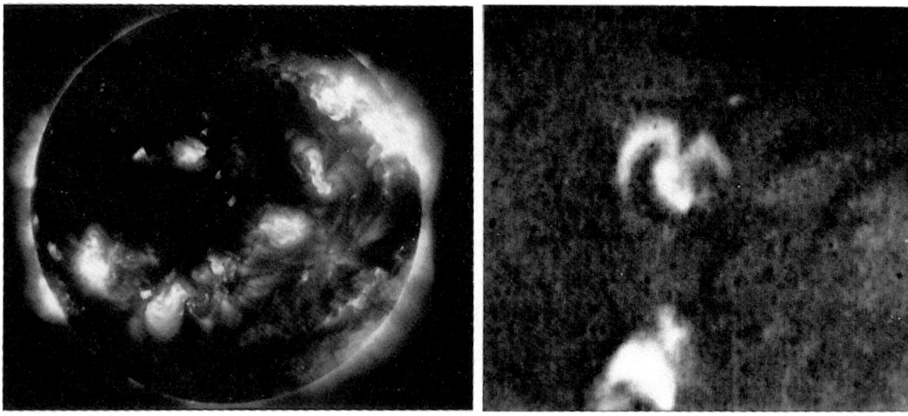

Fig. 7.1. (a) Soft X-ray picture from NIXT (b) a close-up of one of the bright points (Courtesy, L Golub)

the value of τ relative to the time $(10L/v_{\mathrm{A}})$ it takes an Alfven wave to propagate about ten times at the Alfven speed $(v_{\mathrm{A}} = B/(\mu\rho)^{\frac{1}{2}})$ along the length (L) of a coronal structure. If $\tau < 10L/v_{\mathrm{A}}$, which is typically 2 min for $B = 100\mathrm{G}, n = 10^{15}\mathrm{m}^{-3}, L = 100\mathrm{Mm}$, then the effect is to produce wave motions. Whereas if $\tau > 10L/v_{\mathrm{A}}$ there is a slow evolution of the coronal field through a series of equilibria.

Now, the energy is injected upwards from the photosphere as a Poynting Flux $\mathbf{S} = \mathbf{E} \times \mathbf{B}/\mu = -(v \times \mathbf{B}) \times \mathbf{B}/\mu$, but what happens to the energy? The answer is provided by Poynting's theorem

$$\int \mathbf{E} \times \mathbf{H}.\mathbf{dS} = \frac{\partial}{\partial t}\int \frac{B^2}{2\mu} dV + \int \frac{J^2}{\sigma} dV + \int \mathbf{v}.\mathbf{j} \times \mathbf{B} dV \ , \tag{7.1}$$

which shows that the inflow of electromagnetic energy (the term on the left) produces several effects (on the right), namely, in turn a rise of magnetic energy, ohmic heating and work done by the magnetic force. Thus some energy is stored, and may be eventually released as for example an erupting prominence or a solar flare, while some energy is continuously dissipated and heats the corona, and a final part of the energy accelerates the plasma in a wide variety of ways. The latter may either escape or be dissipated by viscous or resistive damping.

Twenty years ago the corona was thought to be quiet, almost-uniform and heated by acoustic waves. Now it is known to be a highly turbulent environment with a general unresolved "turbulent" velocity of 10- 40 km s^{-1} in coronal lines. Transition region lines from Skylab show an increase from 16 to 24 km s^{-1} as T rises from 5×10^4 to 5×10^5K . At 10^6K UV coronal lines give about 25 km s^{-1} and at 3×10^6K recent SMM observations in Mg XI give 40 km s^{-1} . Sometimes one finds turbulent events at 250 km s^{-1} in regions smaller than 1.5 Mm . Also,

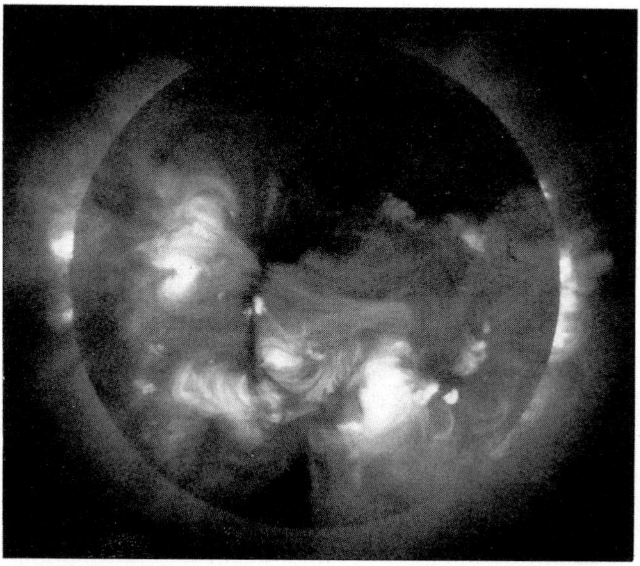

Fig. 7.2. Soft X-ray picture from YOHKOH (Courtesy, Y Uchida)

intermittent heating events are observed in UV from low-lying loops and also in hard X-rays.

Heat is required to balance radiation, conduction and outflow. In a coronal hole this amounts to 600 Wm^{-2} and is required mainly for outflow. In a quiet region it is half this value and is needed to balance conduction and radiation, whereas in an active region it is 5000 Wm^{-2}, again for conduction and radiation. The mechanism is universally thought to be magnetic since the acoustic flux in the transition region is only 10 Wm^{-2} and more strongly magnetic regions are observed to be hotter. Also the Poynting Flux of $S \sim vB^2/\mu$ is typically 10^4 Wm^{-2} for a flow speed of, say, 0.1 km s^{-1} and a field of 100 G and so is more than adequate for coronal requirements. It is highly likely, however, that different mechanisms are acting in different parts of the corona.

Recently, the remarkable NIXT photographs (e.g. Figure 7.1) have shown that active regions consist of many fine-scale loops and that bright points appear to include several interacting loops. X-ray bright points (XBP) were discovered in 1970 from rocket images and were studied with Skylab by Golub et al (1974). Important recent observations of them have been made by Yohkoh . Bright points are uniformly distributed over the solar surface, and their lifetimes vary between 2 and 48 hours, with a mean value of 8 hours. They are situated above pairs of opposite polarity magnetic fragments in the photosphere.

It was natural to assume that the magnetic fragments represent emerging flux and this became the standard explanation for bright points (Heyvaerts et al, 1977; Forbes and Priest, 1984). However, Harvey (1984) showed that two-thirds of the bright points instead lie above so-called "cancelling magnetic features"

(CMF), where pairs of photospheric magnetic fragments are approaching and cancelling (Martin, 1984).

So what is happening in an XBP/CMF event? It cannot be simple submergence of magnetic flux since this would not explain the brightening (which starts well before cancellation). Also, no chromospheric fibrils join the fragments and the fragments are initially widely separated (and therefore unconnected). We need to include the effect of the ambient magnetic field to which the fragments are initially connected, and in doing so have proposed a Converging Flux Model (Priest et al, 1993), which has three phases (Figure 7.3). In the *Pre-interaction Phase* a pair of oppositely directed magnetic fragments in the photosphere are unconnected and approach one another. They are separated by a channel of overlying flux, which is squeezed by the approach until a null point forms in the photosphere. In the *Interaction Phase* the null point moves upwards and coronal reconnection creates an x-ray bright point, whose structure consists of two newly reconnected and heated flux tubes, one a small loop linking the fragments and the other a large loop (as seen in NIXT and Yohkoh images) linking to distant locations. In the *Cancellation Phase* the fragments come into contact and cancel by photospheric reconnection. In the special case when the initial fragments are equal in magnitude the final state consists of two disconnected fields, one above the photosphere and one below.

A simple way of modelling the above processes is to represent the sources by poles of flux $\pm f$ at locations $z = \pm a$, where $z = x + iy$. The field components (B_x, B_y) due to such sources together with a uniform horizontal ambient field (B_o) may be written

$$B_y + iB_x = \frac{if/\pi}{z-a} - \frac{if/\pi}{z+a} + iB_o = iB_o \frac{z^2-b^2}{z^2-a^2} , \tag{7.2}$$

where

$$b = (a^2 - ad)^{\frac{1}{2}} \tag{7.3}$$

is the half-width of the channel and $d = 2f/(\pi B_o)$ we refer to as the "interaction distance".

In the Pre-interaction Phase, the poles are assumed to approach at a speed much slower than the Alfven speed and to make the overlying field evolve through a series of potential states given by (7.2). As the source position (a) decreases, the half-width (b) of the channel decreases while its flux is conserved, until at the interaction distance (a=d) the null point forms at the origin.

When a<d, we have the Interaction Phase with reconnection driven at the null point by the motion of the sources. As a preparation for performing a numerical experiment on such reconnection, a simple model may be set up for continuing evolution through potential states. In equation (2.1) b^2 now becomes negative and so the field vanishes at an X-point on the y-axis at a height $\mid b \mid = (ad - a^2)$, which rises as a decreases to a maximum of $\frac{1}{2}d$ when $a = \frac{1}{2}d$ and then decreases to zero as the sources approach the origin.

If instead there is no reconnection, the topology is preserved and a current sheet forms so that the magnetic energy (W) exceeds the energy (W_o) of the

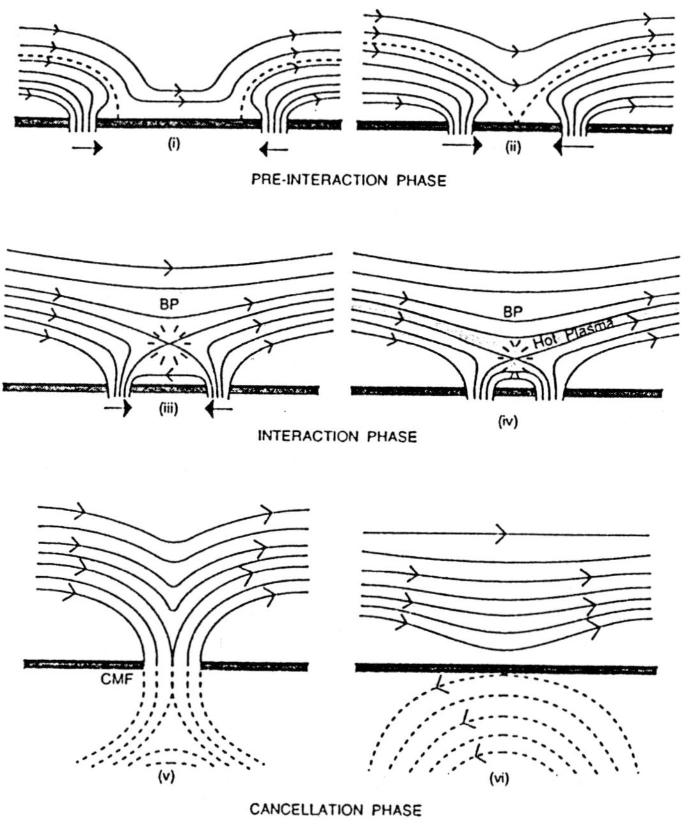

Fig. 7.3. The magnetic field lines for the Converging Flux Model, showing: the Pre-interaction, Interaction and Cancellation Phases

potential state by an amount that can be released by reconnection to give the bright point. The field is given by

$$B_{\mathrm{y}} + iB_{\mathrm{x}} = \frac{iB_o(z^2 + h^2)^{\frac{1}{2}} z}{z^2 - a^2}, \tag{7.4}$$

which tends to a uniform field iB_o at infinity and has a cut (the current sheet) stretching along the y-axis from the origin to $z = ih$. The condition that the flux above the sheet is preserved gives the length of the sheet as $h = (d^2 - a^2)^{\frac{1}{2}}$, which increases from zero to d as a decreases from d to zero. The energy (W) may be calculated and is of order the observed values between 3.10^{20} and 3.10^{21}J (3.10^{27}-3.10^{28} erg) for d = 5-10 Mm.

We have considered particular bright points that were observed on the NIXT flight of July 11, 1991 . In the full disc image there is an active region surrounded by a collection of 5 bright points. The photospheric magnetogram below one of them shows four discrete sources of flux which we have represented by poles, two being of positive polarity and two negative. The resulting field lines in the photospheric plane (Figure 7.4) show that the plane is split into four topologically different regions by four separatrix field lines which intersect at two X-type neutral points. Each region contains only field lines that link two particular poles. When the central source moves to the right, flux is transferred between the regions and the field lines that have reconnected brighten to give shapes that compare well with the bright point. In three dimensions the separatrix field lines become dome-like surfaces and these surfaces intersect in a separator field line that links the two photospheric neutral points. Two field lines before and after reconnection show that, during the process of reconnection in three dimensions, the field lines approach the separator and then reconnect at the two null points before moving away from it (Figure 7.4).

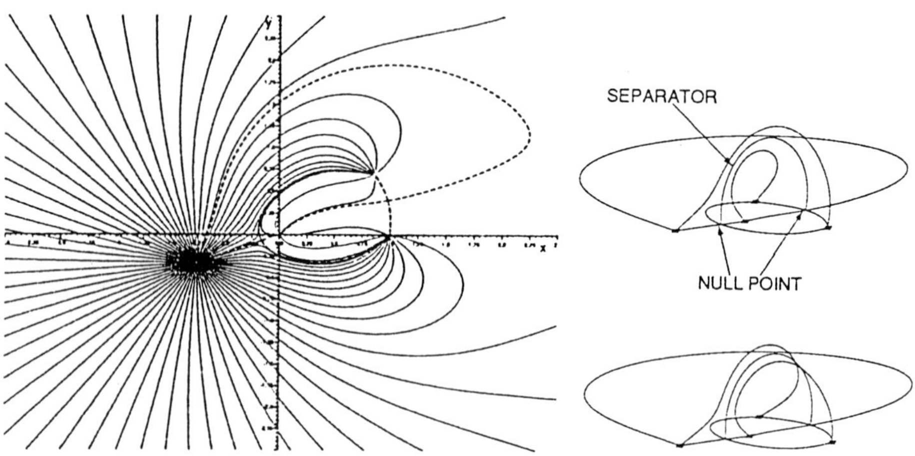

Fig. 7.4. (a) Field lines in the photosphere due to four flux sources below a bright point. (b) Field lines before and after reconnection

The coronal reconnection driven by footpoint motion that we have suggested in the *Converging Flux Model* may also represent an elementary heating event that could heat not just bright points but also coronal loops and, by flux interaction at the edges of supergranule cells, even coronal holes: indeed the flows that are driven by the process could also drive the solar wind.

7.2 Magnetically Open Regions (Coronal Holes) - Heating by Magnetic Waves

Coronal holes are in my view probably heated by Alfven waves (Hollweg, 1983; Roberts, 1984; Goossens, 1991). The main way in which Alfven waves propagating outwards in coronal holes are likely to dissipate is by phase mixing (e.g.

Heyvaerts and Priest, 1983). The simplest way to see this is to consider a unidirectional field $B_0(x)\hat{\mathbf{z}}$ with z directed vertically upwards and the Alfven speed increasing with x . Suppose the footpoints oscillate to and fro in the y- direction and so create a plasma velocity $v(x,z,t)\hat{\mathbf{y}}$ and an extra magnetic field $B_1(x,z,t)\hat{\mathbf{y}}$ in the corona. The ideal MHD equations then reduce to a simple wave equation

$$\frac{\partial^2 v}{\partial t^2} = {v_{\rm A}}^2(x)\frac{\partial^2 v}{\partial z^2} \ , \tag{7.5}$$

with solution of the form $v \sim \exp\left[i(wt - k_z(x))z\right]$, where (7.5) determines the wave number as $k_z(x) = \omega/v_{\rm A}(x)$.

From this we see that there is a continuous spectrum of wavelengths, with waves propagating along different lines with different wavelengths - short wavelengths where the field is weak and long where it is strong. The result is that the phases of the waves on neighbouring field lines become mixed in space. If all of the field lines are shaken in phase at their footpoints, they will become more and more out of phase as they propagate upwards and steep gradients in the x-direction will be generated. Indeed differentiating the expression for v gives $\partial v/\partial x \sim v\ dk_z/dx\ z$, so that the gradient increases linearly with height until eventually it is so strong that dissipation becomes important. Typically, the waves dissipate at a height of a few wavelengths.

Closed magnetic regions such as coronal loops and arcades may be heated by phase mixing of standing waves. For such waves one imposes k and assumes $\omega(x) = k_z v_{\rm A}(x)$, so that we now have a continuous spectrum of wave frequencies rather than wavelengths and the waves phase mix in time rather than space. The usual way of treating such phase mixing is to Fourier analyse in time and have great (and exceedingly complicated) fun with the Fourier inversion in the complex plane. However, Cally (1991) has recently undertaken an equivalent but much simpler treatment by Fourier analysing in space (Figure 7.5). He imposes an initial perturbation $\xi(x,0) = \sin x$ between $x = 0$ and $x = \pi$ and writes the subsequent plasma displacement as

$$\xi(x,t) = \sum_1^\infty a_n(t)\sin nx \ . \tag{7.6}$$

Numerical solution of the resulting equations for $a_n(t)$ show that after several periods the resulting displacement has many oscillations in space as the different harmonics are excited. The energy of the fundamental is converted into that of the first harmonic and then the second, the third and so on as the waves progressively phase mix.

The same process may take place for the other polarization, when the motions are in the direction of the inhomogeneity, but the equations are much more complicated. Thus, if one assumes incompressibility and Fourier analyses in t and z, so that $v(w,z,t) = \tilde{v}(x)\exp i(\omega t - kz)$, then $\tilde{v}(x)$ satisfies

$$\frac{d}{dx}\left[\rho_0(\omega^2 - k^2{v_{\rm A}}^2)\frac{d\tilde{v}}{dx}\right] - k^2\rho_0(\omega^2 - k^2{v_{\rm A}}^2)\tilde{v} = 0 \ , \tag{7.7}$$

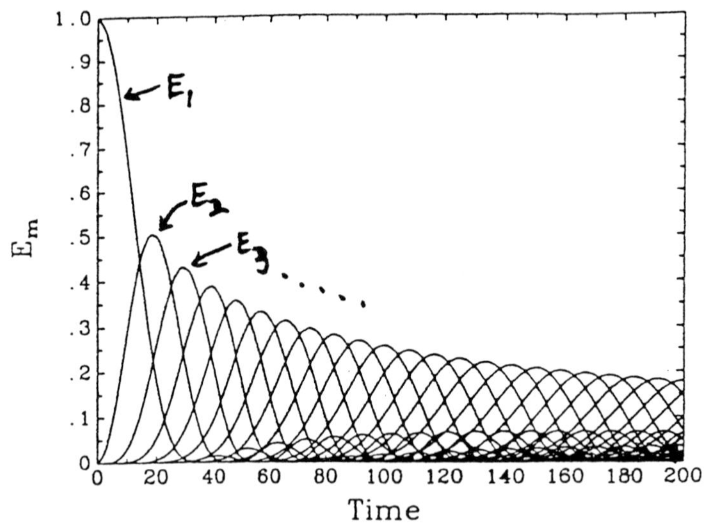

Fig. 7.5. Phase mixing (Cally, 1991)

where $\rho_0 = \rho_0(x), v_A = v_A(x)$ and this contains a singularity where $\omega^2 - k^2 {v_A}^2 = 0$.

This leads to a process of *resonant absorption*, which may heat magnetically closed regions. Acoustic or fast-mode waves may propagate upwards, say in the centre of a supergranulation cell or below a magnetic arcade, and may then dissipate their energy at a critical surface where $v_A(x) = \omega/k$, so that the local Alfven speed equals the phase speed of the wave and the singularity in (7.7) is reached.

7.3 Magnetically Closed Regions (Coronal Loops) - Heating by Current Sheets

In coronal loops and arcades, slow footpoint motions make the coronal magnetic field try to evolve through a series of equilibrium configurations. Often, however, the equilibria are not smooth but contain singularities (current sheets and filaments) where dissipation can occur. There are several ways of forming such sheets, namely driven reconnection and flux interaction, braiding and X-point collapse, so let me say a little about the latter two as follows.

(a) Braiding by Random Footpoint Motions. Parker (1979,1990) has pointed out that, if a series of flux tubes are closely packed together and each twisted in the same direction, then current sheets will form at the boundaries of the tubes (Figure 7.6). If instead the footpoints are braided around one another, a smooth solution exists when the footpoint motions are small, but he argues persuasively that finite-amplitude footpoint displacements lead to nonequilibrium and current sheet formation at the boundaries of the braids.

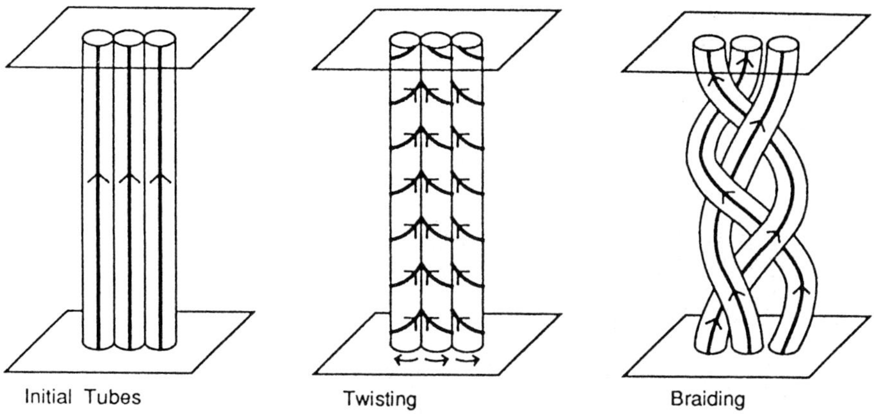

Fig. 7.6. Twisting and braiding of flux tubes

Mikic et al (1988) have conducted an interesting numerical experiment on the braiding of magnetic field lines in three dimensions using an ideal MHD code with zero plasma pressure and 64^3 gridpoints. Random footpoint motions are applied in a series of steps with speeds of $0.03 v_A$ and lead to a series of smooth equilibria with a transfer to small scales and the growth of filamentary currents with the current density growing like exp $(0.01\, t/v_A)$.

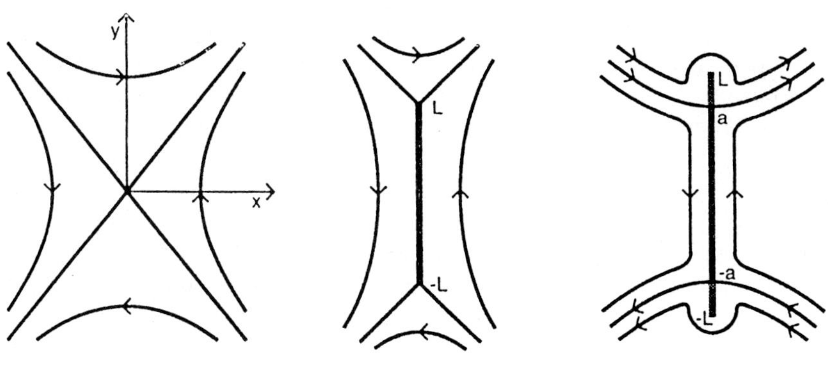

Fig. 7.7. X-point collapse

(b) X-Point Collapse. A simple treatment of the collapse of an X-type neutral point to give a current sheet has been described in Section 6.4. It has been applied to the formation of sheets between interacting dipoles by Priest and Raadu (1975) . Also, a simple model for flux emergence, which may model the minority of bright points which appear to be caused by newly emerging flux, is to start with a dipole in a uniform field such that $B_y + iB_x = iD/z^2 + iB_0$. Then, as more flux emerges, a current sheet is formed, curving from $z = z_1$ to $z = z_1{}^*$ (Tur and Priest, 1976) and having a field $B_y + iB_x = E\left[\left(z^2 - z_1{}^2\right)\left(z^2 - z_1{}^{*2}\right)\right]^{1/2}/z^2$.

All of the above complex-variable solutions model a slow evolution through a series of equilibria. Recently, however, Priest, Titov and Rickard (1994) have discovered some nonlinear, self-similar, compressible solutions for dynamic time-dependent formation of a current sheet from an X-point. When the plasma velocity is much greater than the sound speed and much less than the Alfven speed the dimensionless equation of motion is of the form $\epsilon\rho d\mathbf{v}/dt = \mathbf{j}\times\mathbf{B}$. Expanding $\mathbf{v}$ and $\mathbf{B}$ in powers of $\epsilon(\mathbf{v} = \mathbf{v}_0 + \epsilon\mathbf{v}_1 + \ldots, \mathbf{B} = \mathbf{B}_0 + \epsilon\mathbf{B}_1 + \ldots)$, we then have in two dimensions a series of states with $\mathbf{j}_0 = \mathbf{0}$ around the sheet to lowest order. The plasma velocity $(\mathbf{v}_\perp)$ perpendicular to the magnetic field is given from the motion of the field lines, i.e. from the induction equation

$$\frac{\partial \mathbf{B}_0}{\partial t} = \nabla \times (\mathbf{v}_0 \times \mathbf{B}_0) \ . \tag{7.8}$$

Also the velocity parallel to the magnetic field is determined by the condition that $d\mathbf{v}/dt$ be perpendicular to the zeroth-order magnetic field; this is a consequence of the first-order part of the equation of motion, namely $d\mathbf{v}_0/dt = \mathbf{j}_1 \times \mathbf{B}_0$, which implies that

$$\frac{d\mathbf{v}_0}{dt}.\mathbf{B}_0 = 0 \ . \tag{7.9}$$

What this means physically is that, as the magnetic field lines move, they rotate and the flow along them is given by a balance between the Coriolis and centrifugal forces associated with the rotation of the field lines. The potential magnetic field sequence may not, however, be imposed at will: it has to be determined in a self-consistent way from (7.8) and (7.9).

The simplest solution we have found of this type has magnetic and velocity components

$$B_y + iB_x = \frac{\left(\frac{1}{2}z + \sqrt{\frac{1}{4}z^2 - t}\right)^2}{2\left(\frac{1}{4}z^2 - t\right)}, \qquad v_x + iv_y = \frac{1}{\frac{1}{2}z + \sqrt{\frac{1}{4}z^2 - t}} ,$$

where $z = x + iy$ is the usual complex variable. This is an elegant formalism, which to me is almost as beautiful as some of the wonderful soft X-ray pictures from NIXT and Yohkoh! Individual plasma elements converge on the x-axis along straight lines and they force a current sheet of length $4/\sqrt{t}$ to form (Figure7.8). As the magnetic field collapses, the current sheet grows in length and the magnetic dissipation increases. The ends of the sheet move with speed $1/\sqrt{t}$. They swallow up part of the magnetic flux and cause the remainder of it to pile up around a region of reversed current.

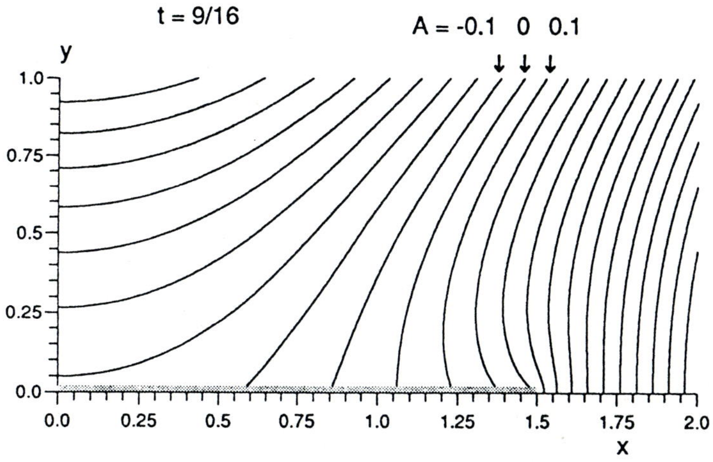

Fig. 7.8. Magnetic field lines at t = 9/16 (Priest, Titov and Rickard, 1993)

7.4 Self-Consistent Model for Heating by MHD Turbulence

Many coronal heating mechanisms, such as braiding and current sheet formation or resistive instabilities or waves, all lead to a state of MHD turbulence, so how can we analyse it? Heyvaerts and Priest (1984) made a start by adapting Taylor's relaxation theory to the coronal environment, in which the field lines thread the boundary rather than being parallel to it. In Taylor's model the global magnetic helicity is conserved, but in our model footpoint motions make the coronal field evolve through a series of linear force-free fields, satisfying $\nabla \times \mathbf{B} = \alpha_0 \mathbf{B}$. The footpoint connections are not preserved, but instead the constant α_0 is determined from the evolution of the *magnetic helicity* $K = \int \mathbf{A}.\mathbf{B}\, dV$, where $\mathbf{B} = \nabla \times \mathbf{A}$, due to the injection by boundary motions according to

$$\frac{dK}{dt} = \int (\mathbf{A}.\mathbf{v})\, \mathbf{B}.\mathbf{dS} \ . \tag{7.10}$$

The resulting heating flux is of the form $F_{\rm H} = B^2 v\, \tau_{\rm d}/(\mu \tau_0)$, where $\tau_{\rm d}$ is the dissipation time and τ_0 the time-scale for footpoint motions. The basic analysis of Heyvaerts and Priest (1984) has been applied to a set of closely packed flux tubes and to an axisymmetric flux tube. Also Vekstein et al (1990) have modelled the simultaneous stressing and relaxing .

Although many mechanisms produce a turbulent state, they are incomplete in the sense that there is a free parameter present, such as $\tau_{\rm d}$ in (7.10) or a correlation time or a relaxation time. In other words, they don't determine the heating flux ($F_{\rm H}$) in terms of photospheric motions alone. Heyvaerts and Priest (1992) have therefore begun a new approach in which they assume photospheric motions inject energy into the corona and maintain it in a turbulent state with a turbulent magnetic diffusivity (η^*) and viscosity (ν^*) . There are two parts to

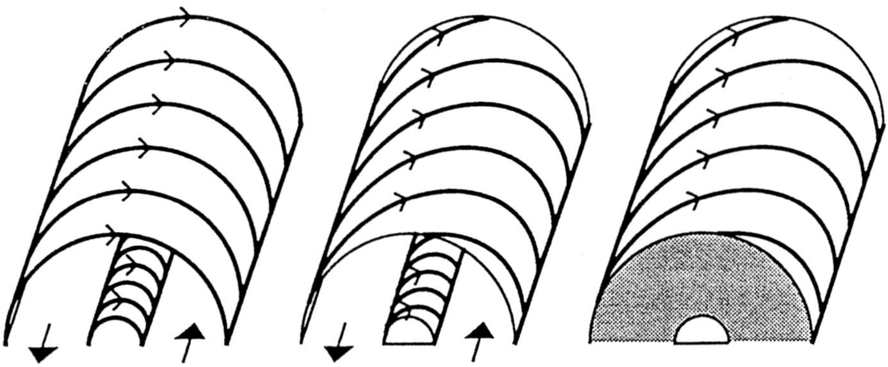

Fig. 7.9. The scenario for Taylor-Heyvaerts-Priest relaxation

their theory. First of all, they calculate the global MHD state driven by boundary motions, which gives F_H in terms of ν^* . Secondly, they invoke cascade theories of MHD turbulence to determine the ν^* and η^* that result from F_H . In other words the circle is completed and F_H is determined independently of ν^* and η^*. They apply their general philosophy to a simple example of one-dimensional random photospheric motions producing a two dimensional coronal magnetic field. They find typically for a quiet-region loop that a density $2 \times 10^{16}\mathrm{m}^{-3}$ and a magnetic field of 30 - 50 G produces a heating F_H of $2.4 - 5.5 \times 10^2 \mathrm{Wm}^{-2}$ and a turbulent velocity of $24 - 33\mathrm{kms}^{-1}$, whereas values of $5 \times 10^{16}\mathrm{m}^{-3}$ and 100 G for an active-region loop give $2 \times 10^3 \mathrm{Wm}^{-2}$ for the heating and 40 km s^{-1} for the turbulent velocities. Given the limitations of the model, these reasonable values are highly encouraging.

7.5 Conclusion

The evidence is very strong for heating by flux convergence and coronal reconnection in x-ray bright points. However, for coronal loops and coronal holes, although there are several viable theories, there is so far no compelling observational evidence for particular ones. The most likely for coronal holes is phase mixing or resonant absorption of propagating Alfven waves, and for coronal loops many small current sheets may be formed in a variety of ways. The resulting wave or current-sheet state may be described by a self-consistent model of MHD turbulence. In future, we eagerly await results from the SOHO satellite which may help to determine just which mechanisms are at work.

8 Prominences

8.1 Observations

A *quiescent prominence* is an immense, dense, cool vertical sheet of plasma in the corona with a density ($n \sim 0.5 - 1.0 \times 10^{17}\mathrm{m}^{-3}$) about 500 times more than the ambient coronal value and a temperature ($5 - 10,000$K) about 500 times less. It changes slowly in shape and is long-lived . It is seen in projection on the disc as a thin dark *filament.* It shows much fine structure and internal motion and often reaches downward in a series of regularly spaced "feet". At some stage it may become unstable, ascend and disappear - an *eruptive prominence.* Generally, it reforms in the same place. They were first observed as "tongues of flame" in the Middle Ages (1239) and then rediscovered at the eclipse of 1842 and first photographed in 1860.

The main questions about these mysterious objects are: how are they formed? how are they supported? what is their magnetic structure? what causes their fine structure and feet? why do they erupt? Reviews of both observations and theories are given in the books by Tandberg-Hanssen (1974), Poland (1986), and Priest (1989), in which the detailed references to prominence papers can be found.

The main observational features of quiescent prominences are as follows:

1. The geometry is that of a thin vertical sheet, lying above a reversal in the line-of-sight magnetic field in the photosphere. The length of the sheet lies between 60 and 600 Mm (typically 200 Mm), while its height ranges between 10 and 100 Mm (typically 50 Mm) and its width is between 4 and 15 Mm (typically 6 Mm).
2. The density is between 10^{16} and 10^{17}m^{-3}, and the temperature is usually between 5000 K and 8000 K. They both tend to decrease with height.
3. The magnetic field above 10 Mm is rather homogeneous, horizontal and inclined at typically 20° to the prominence axis. It has a magnitude of between 3 and 30 Gauss (typically 5G) and usually increases with height by about 50%.
4. In Leroy's (1989) study he found that the high-latitude quiescent prominences have heights above 30 Mm, field strengths between 5 and 10 Gauss and are all of *Inverse Polarity* (for which the magnetic field passes through the prominence in the opposite direction to the underlying field). The low–latitude prominences near active regions are low-lying (below 30 Mm), have field strengths of about 20G and are of *Normal Polarity* (for which the magnetic field has the same polarity as the underlying field). Kim (private correspondence) also finds most of the large quiescent prominences to be of Inverse Polarity, but some of her active-region prominences, which include ones of lower altitude than Leroy, are Normal and some are Inverse.
5. Prominences are long–lived, lasting for 1-300 days, with high-latitude ones enduring typically 140 days. The polarity inversion zones tend to migrate slowly towards the poles and to become stretched in an east-west direction.
6. The observed flows in prominences are much smaller than the free-fall speed $\sqrt{(gh)}$ of about 100 km s^{-1}, and so the plasma is essentially in equilibrium with a rough balance between the magnetic, pressure and gravitational forces as it very slowly dribbles through the magnetic field. Typically one sees a downflow of 0.5 km s^{-1} when observed on the limb and an upflow of 0.5 - 3 km s^{-1} when viewed on the disc. Prominences tend to form where there are converging flows and shearing flows.

7. Plage or active-region prominences are smaller than their quiescent cousins by a factor of 3 or 4. They are lower in height and have larger densities ($\gtrsim 10^{17}\mathrm{m}^{-3}$) and field strengths ($\sim 20-100$G) and strong horizontal flows.
8. On the limb one sees that a prominence consists of vertical threads, typically 5-7 Mm long and 0.3-1 Mm wide, with a filling factor between 0.01 and 0.1.
9. Quiescent prominences appear to reach down to the surface in a series of "feet", spaced by about 30 Mm and located at supergranule boundaries.

When confronted by the above facts, what is the aim of theory? It is certainly not to reproduce on the computer all the features of a beautiful $H\alpha$ picture of a prominence. Rather, the aim is to understand the basic physics and causes of the overall structure and the main properties. Why is it essentially sheet-like? How is it supported? What is the global magnetic structure? In order to make progress, one starts with a simple physical model. For example, a great advance in understanding can be obtained by using a rectangular slab to explain the global properties as outlined in (1) – (7) above. Then later one can attempt to understand the feet (property (9)) and the microstructure in the form of threads (property (8)). But without such slab models one would not have obtained an overall basic understanding within which to consider such complexities. Using the philosophy of mathematical modelling, we shall therefore see how different authors have made theoretical progress by taking different assumptions. Starting with the process of formation (Section 8.2) and the internal structure of the prominence sheet (Section 8.3), we shall move on to consider the external magnetic fields (Section 8.4) and a new Flux Tube Model (Section 8.5).

8.2 Formation of a Prominence

Consider a uniform hot equilibrium ($h_0 = Q\rho_0$), where h_0 is the heating term, $Q\rho_0$ the radiative loss, and Q is a constant. Perturb this equilibrium at a constant pressure, so that the energy equation is

$$c_p \frac{\partial T}{\partial t} = h_0 - Q\rho + \frac{\kappa_\parallel}{\rho}\frac{\partial^2 T}{\partial s^2} \ , \tag{8.1}$$

where s is measured along a field line and $\rho T = \rho_0 T_0$. Then put $T = T_0 + T_1, \rho = \rho_0 + \rho_1$, where $T_1/T_0 \ll 1$ and $\rho_1/\rho_0 \ll 1$, so that (8.1) becomes

$$c_\mathrm{p} \frac{\partial T_1}{\partial t} = \frac{Q\rho_0 T_1}{T_0} + \frac{\kappa_\parallel}{\rho_0}\frac{\partial^2 T_1}{\partial s^2} \ .$$

Finally, assuming T_1 behaves like $\exp(\omega t + 2\pi i s/L)$, we obtain

$$\omega = \frac{Q\rho_0}{c_p T_0} - \frac{\kappa_\parallel 4\pi^2}{c_p \rho_0 L^2} \tag{8.2}$$

for the growth–rate. Thus a *thermal* (or *radiative*) *instability* takes place if

$$L > L_{\mathrm{max}} = 2\pi \left(\frac{\kappa_\parallel T_0}{Q\rho_0^2}\right)^{\frac{1}{2}} \ ,$$

so that the coronal field line is so long that conduction is ineffective and the plasma cools down to prominence temperatures (Parker, 1953).

This simple process may be extended by including: a temperature-dependent radiative loss function (Field, 1965); thermal conduction in the equilibrium (Hood and Priest, 1980) in which case a thermal catastrophe takes place; a magnetic field (Heyvaerts, 1974); the nonlinear development (Hildner, 1974).

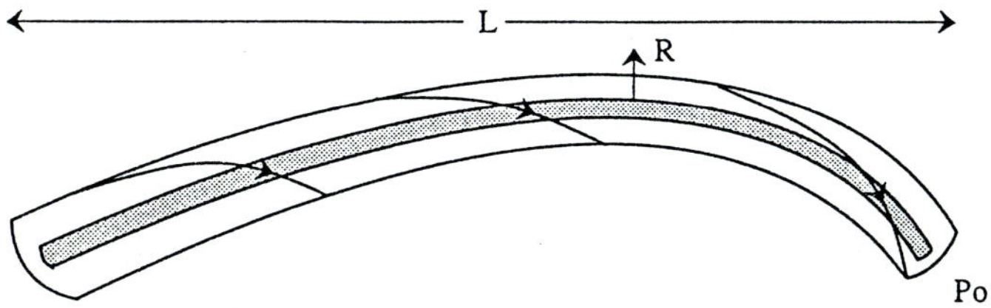

Fig. 8.1. An active-region flux tube with a cool core

The way a prominence may form in a cylindrically symmetric flux tube or coronal loop has been modelled by Hood and Priest (1979). They solve magnetostatic force balance in the radial direction and energy balance along each field line and first of all find a hot equilibrium. Then, as they increase the length, pressure or field line twist, they find that a cool filament forms in the core of the flux tube (Figure 8.1). In a similar way formation in a force-free arcade has been modelled by Priest and Smith (1979): the energy balance equation

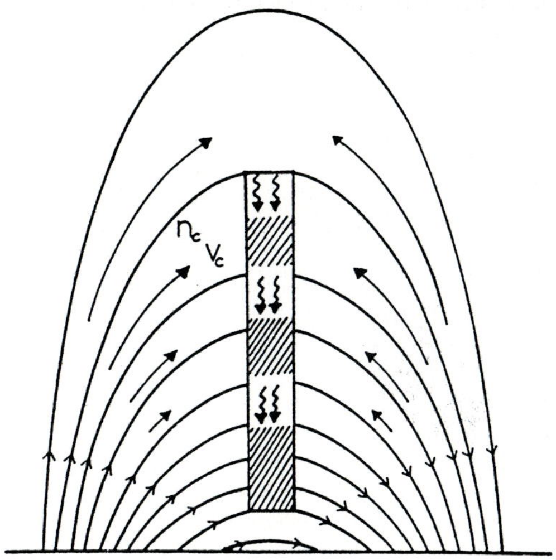

Fig. 8.2. A dynamic model for a prominence

$$\frac{d}{ds}\left(\kappa_{\parallel}\frac{dT}{ds}\right) - \frac{\kappa_{\parallel}}{B}\frac{dB}{ds}\frac{dT}{ds} = \rho^2 Q(T) - h\rho \tag{8.3}$$

is solved along each field line, together with the equation of hydrostatic equilibrium ($dp/dz = -\rho g$). Again a hot equilibrium is obtained and, when the shear of the field reaches a critical value, the summit suddenly cools down to prominence temperatures; as the shear increases, so the range of heights for cool equilibria increases. The same process occurs if the width or pressure of the arcade are too great. This naturally leads to the suggestion of a dynamic process (Figure 8.2) in which plasma is continually draining down through a static field and new material is being sucked up along field lines into the region of nonequlibrium where it cools.

8.3 Internal Structure

Let us first remind ourselves of the magnetohydrostatic equation for equilibrium under a balance between magnetic, pressure and gravitational forces

$$-\nabla(B^2/2\mu) + (\mathbf{B}.\nabla)\mathbf{B}/\mu - \nabla p + \rho g = 0 \ , \tag{8.4}$$

where $\mathbf{j} = \nabla \times \mathbf{B}/\mu$, $\nabla.\mathbf{B} = 0$ and $p = R\rho T$, in the usual notation. Along a magnetic field line the magnetic force vanishes and for a uniform temperature the pressure falls off exponentially ($p = p_0 exp(-y/H)$) where H is the scale height.

Since H is only about 180 km for a temperature of 6000 K this explains the narrow width of a cool prominence sheet. If the prominence plasma is sitting at rest in a curved magnetic flux tube, supported against gravity by the magnetic tension (Figure 8.3), the width of the plasma structure depends on the inclination to the horizontal of the field lines. For a small inclination one needs to travel a large horizontal distance (l) to move a distance H vertically. Thus the half-width (l) may be estimated to be $l = 2(B_x/B_{y0})H$, where B_{y0} is the vertical

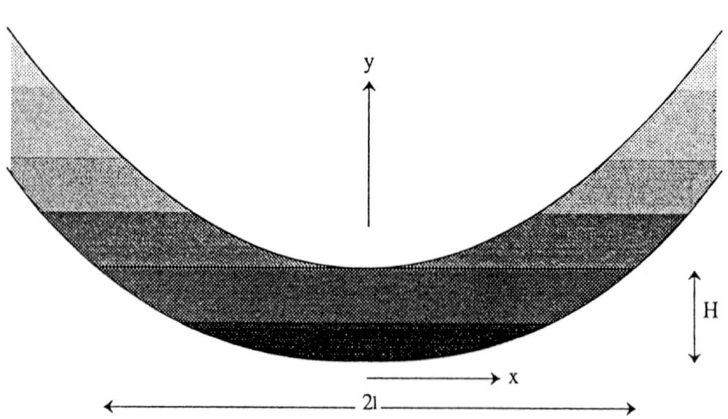

Fig. 8.3. Support of plasma in a curved flux tube

field at a distance (l) from the centre, the factor 2 arising because the field lines are curved rather than straight. This relation may be used to estimate the inclination of the field. With $H=$ 180 km and $l=3$ Mm, say, one finds a value $B_x/B_{y0} \approx 8$, corresponding to an inclination to the horizontal at the edge of the sheet of only a few degrees. In other words the required dip in the magnetic field is extremely small. Outside the prominence the pressure and gravitational forces are dominated by the magnetic field and so (8.4) reduces to the equation for a force-free field. In the models that follow, sometimes the external field is assumed to be potential and sometimes force-free.

Whether the prominence is of Normal or Inverse polarity it may be modelled as a vertical sheet. Kippenhahn and Schluter (1957) set up a simple model for the magnetohydrostatic support of such a sheet by assuming that the temperature (T) and horizontal field components (B_x, B_z) are constant while the vertical field (B_y), pressure (p) and density (ρ) depend on x alone. The horizontal and vertical components of force balance (8.3) then reduce to

$$p+\frac{{B_y}^2}{2\mu}=\frac{{B_{y0}}^2}{2\mu} \tag{8.5}$$

and

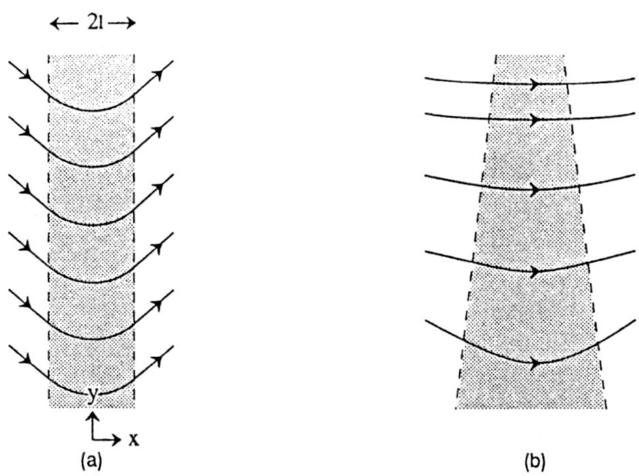

Fig. 8.4. A prominence sheet that is (a) uniform or (b) narrows with height

$$\frac{dB_y}{dx}\frac{B_x}{\mu}-\rho g=0 \ , \tag{8.6}$$

where B_y approaches $\pm B_{y0}$ as x tends to $\pm\infty$ and p approaches zero. The magnetic field therefore plays two roles. According to (8.5) it compresses the plasma sheet laterally and increases the plasma pressure in the sheet by a small amount equal to the external magnetic pressure (${B_{y0}}^2/2\mu$) associated with the vertical field. In addition (8.6) expresses the support of the plasma by the magnetic tension against gravity. The solution of (8.5) and (8.6) is

$$B_y = B_{y0}\, tanh(x/l),\ p = (B_{y0}{}^2/2\mu) sech^2(x/l)\ , \tag{8.7}$$

where the prominence half-width (l) is given by (8.4) as expected.

Several generalisations of this solution have been constructed. Including spatial variations $T(x)$ of the temperature so that $B_y = B_{y0}\, tanh \int_0^x dx/l(x)$ allows a coupling of the magnetohydrostatics and energy balance (Milne et al, 1979). Prominence-like solutions are found when the plasma beta is smaller than a critical value. Another modification due to Ballester and Priest (1987) is to allow slow variations with height $\mathbf{B} = \mathbf{B}_0(x) + \epsilon\mathbf{B}_1(x,y)$, where $\mathbf{B}_0(x)$ is the Kippenhahn-Schluter solution. The result is that the width decreases slowly with height, while the field lines become less curved and the field strength increases, in agreement with observations (Figure 8.3b).

8.4 External Structure

Prominences have been observed for many years but it was only 30 years ago that the first models for magnetic structure were proposed. In particular, the Kippenhahn-Schluter model (1957) has the magnetic topology indicated in Figure 8.5. The prominence is represented here as a sheet with current I at a height h and directed out of the plane, since it produces a change in the direction of the vertical component of the magnetic field from down on the left-hand side to up on the right-hand side. If the photospheric footpoints are line-tied during the formation of the prominence, the preservation of the footpoint position can be modelled by adding an image current ($-I$) a distance h below the photosphere to the original arcade and the prominence sheet. Thus the prominence mass m is supported against gravity both by the line tying (the repulsion $\mu I^2/(4\pi h)$ between I and ($-I$)) and also by the Lorentz force IB acting on I in the original background field B at height h.

An alternative magnetic topology was proposed by Kuperus and Raadu (1974) with the magnetic field passing through the prominence in the opposite direction, as shown in Figure 8.6, where the current I is now directed into the plane. The basic topology is shown on the left-hand side with the outwards spreading of the field lines from the footpoints providing a magnetic tension force upwards. When the lowest field lines are in the form of magnetic loops straddling directly across the polarity inversion line, as on the right-hand side, there is an X-type neutral point below the prominence sheet. Support in the Kuperus-Raadu model is only by line tying ($\mu I^2/(4\pi h)$, as before) since the Lorentz force IB now acts downwards.

Malherbe and Priest (1983) have modelled prominences of both Inverse and Normal Polarity as cuts in the complex plane (Figure 8.7). In addition Anzer (1989) and Demoulin et al (1989) have solved Laplace's equation around a prominence numerically with observed photospheric and prominence fields. For Inverse Polarity Prominences, Amari and Aly (1989) and Demoulin and Priest (1989)

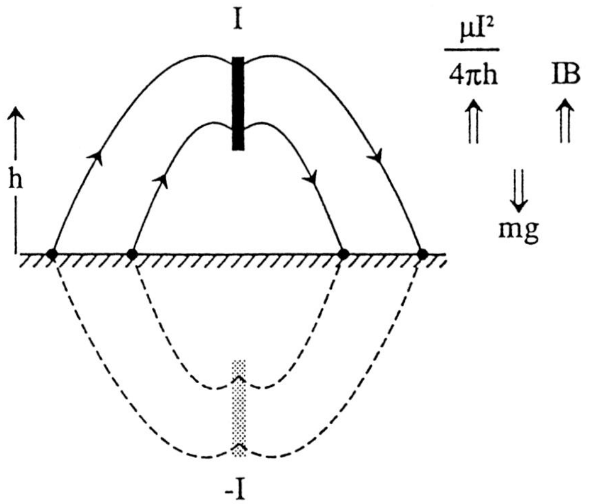

Fig. 8.5. Magnetic topology for prominences of Normal Polarity

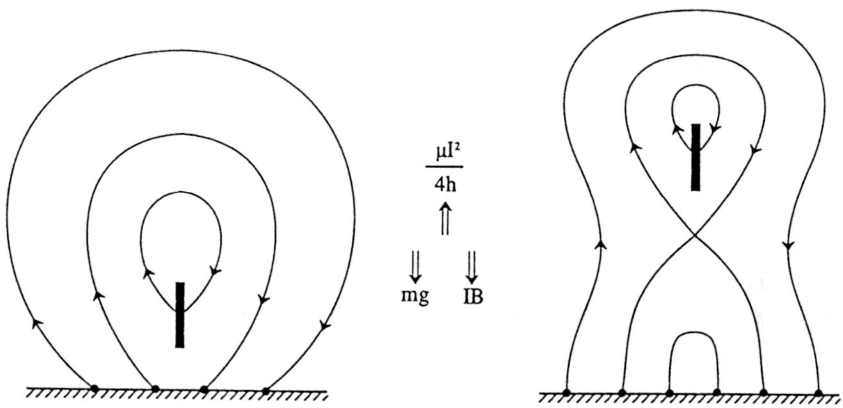

Fig. 8.6. Magnetic topology for prominences of Inverse Polarity

greatly developed an earlier model of Van Tend and Kuperus (1978) and modelled the prominence as a line current in a force-free field.

Recently, Hood and Anzer (1990) have produced a model which includes both the internal and external structure of a prominence (Figure 8.8a). They assume that the field components have the form $(B_x, B_y, B_z) = (X(x), Y(x), Z(x))$ exp $(-y/(2H_c))$ while the pressure and scale height are $p = P(x)/(2\mu)\exp(-y/H_c)$, $H(x) = RT(x)/(\tilde{\mu}\, g)$, where $Z(x) = \alpha X(x)$ and $Y(x) = 2dX/dx\, H_c$.

The configuration is in the form of an arcade of width 2a within which there is a prominence of width $2l$. For simplicity, the temperature is assumed to have a uniform value of T_p in the prominence ($|x| \leq l$) and T_c in the corona ($l \leq x \leq a$), which gives a corresponding step-function for the scale-height $H(x)$. The horizontal and vertical components of the force balance give the following

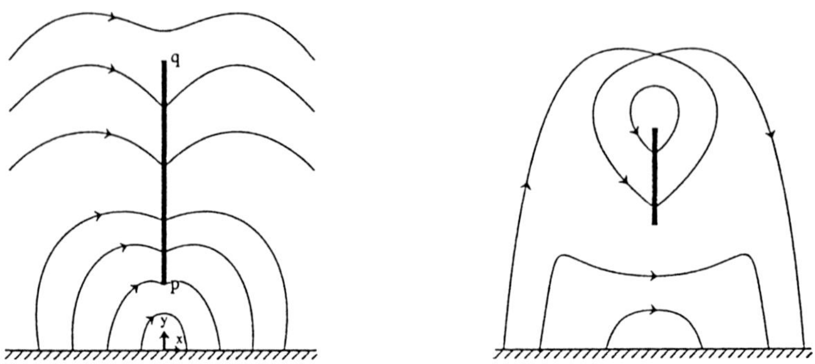

Fig. 8.7. Malherbe-Priest models of Normal Polarity

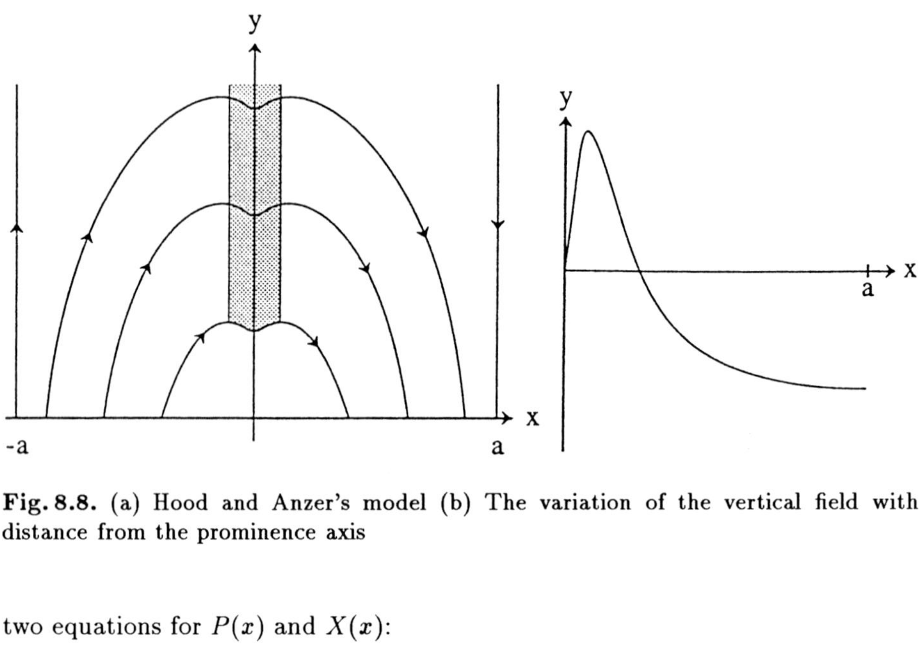

Fig. 8.8. (a) Hood and Anzer's model (b) The variation of the vertical field with distance from the prominence axis

two equations for $P(x)$ and $X(x)$:

$$P + X^2 + Y^2 + Z^2 = P_{\mathrm{T}} \ ,$$

$$2H_{\mathrm{c}}XY' = \left[(1 - H_{\mathrm{c}}/H)Y^2 + (H_{\mathrm{c}}/H)(P_{\mathrm{T}} - X^2 - Y^2) - P_{\mathrm{T}}\right] \ ,$$

where P_{T} is constant and $Y = 2X'H_{\mathrm{c}}$. In the corona $P(x)$ is roughly uniform and the horizontal variation of the vertical field is given approximately by $Y = \cos\left[(1+\alpha^2)^{1/2}(x-a)\right]$ for $l \leq x \leq a$. The full numerical solution is sketched in Figure 8.8b. Hood and Anzer were able to deduce reasonable values for the corona for given prominence conditions.

8.5 A Flux Tube Model

Because of the fact that many prominences are of Inverse Polarity and all the previous Inverse-Polarity models have severe difficulties, we are proposing a Flux Tube Model which can have Inverse Polarity and which also agrees better with many other observational features (Priest, Hood and Anzer, 1989). Previously, Hood and Priest (1979) had suggested that a *plage filament* may be a low-lying twisted flux tube (Figure 8.1), since one can often have motions along such an active-region prominence and it can sometimes end in a sunspot, both of which features contradict the Kippenhahn-Schluter model.

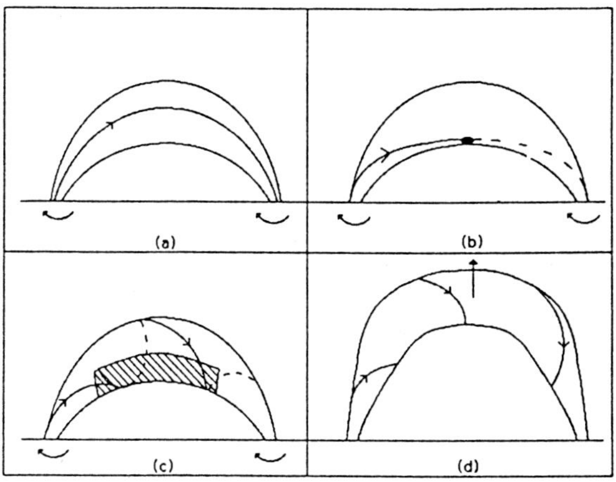

Fig. 8.9. The Twisted Flux Tube model scenario

We are suggesting that for a quiescent or active-region prominence the basic geometry is a large-scale flux tube (figure 8.9). Twist of the tube may be created in several ways – either by general evolutionary footpoint motions (which could produce inverse or normal polarity) or by Coriolis forces (which produce inverse polarity) or by flux cancellation (which produces inverse polarity). Evidence for such twist has been found by many authors and braids seen in plage filaments are one such evidence. Coriolis forces would relentlessly tend to twist up a tube and produce one complete twist in about 35 days. As the twist increases, eventually a dip with upwards curvature is created at the summit (Figure 8.9) and at this point the prominence can begin to form either by condensation (especially for quiescents) or by chromospheric injection (only likely for active-region prominences). The suggestion is therefore that the magnetic mould with the right environment for a prominence to be born (in particular a state of upwards curvature) needs to be created. As the twisting or flux cancellation continues, the prominence grows in length (Figure 8.9) and eventually, when the twist or prominence length is too large (Hood and Priest, 1980; Einaudi and Van Hoven, 1983), the prominence erupts. It undergoes a metamorphosis, like a beautiful butterfly, and reveals its true form as a flux tube for the first time in its life!

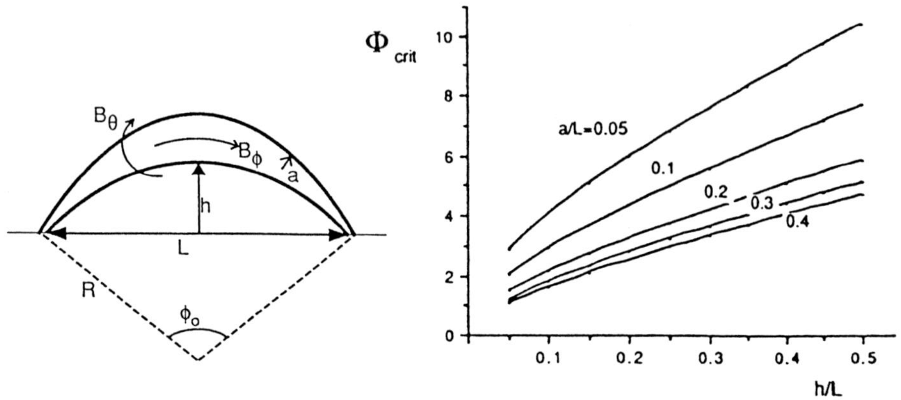

Fig. 8.10. (a) The notation for a flux tube. (b) The critical twist for prominence formation as a function of flux tube height (h) and radius (a)

The critical twist for support for a large aspect-ratio is $\Theta_{\rm crit} = \phi_0 \left(R/a\right)^{\frac{1}{2}}$ in terms of the major (R) and minor (a) radii of curvature (Figure 8.10). The way in which this increases with the summit height (h) and flux tube radius (a) is shown in Figure 8.10b in terms of the footpoint separation L. The way the prominence length increases with twist has also been estimated.

The force-free structure may be modelled by neglecting the large-scale curvature and writing in cylindrical geometry $B_{\rm r} = r^{-1}\partial A/\partial\theta$, $B_\theta = -\partial A/\partial r$, $B_{\rm z} = B_{\rm z}(A)$, where the z-axis is directed along the flux tube (Figure 3.11) and the flux

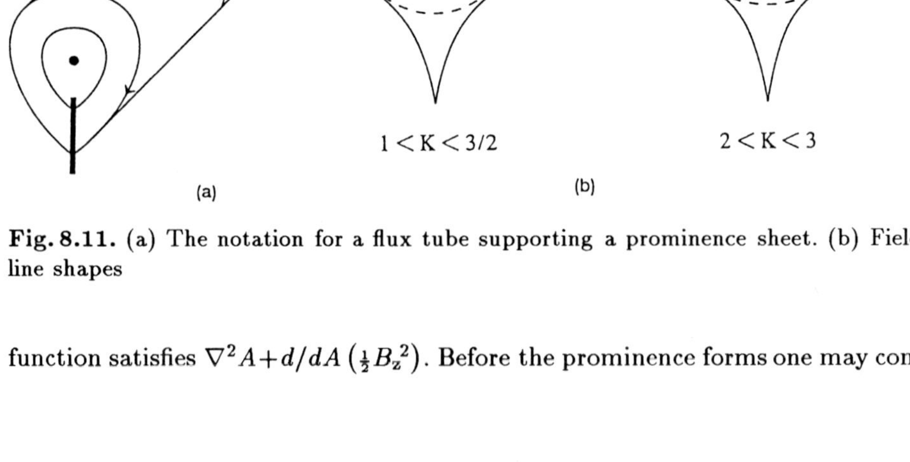

Fig. 8.11. (a) The notation for a flux tube supporting a prominence sheet. (b) Field line shapes

function satisfies $\nabla^2 A + d/dA\left(\frac{1}{2}B_{\rm z}{}^2\right)$. Before the prominence forms one may consider a tube with $B_{0\theta} = B_0 r/(\sqrt{2}l)$, $B_{0z} = B_0\left(1 - r^2/l^2\right)$, for $0 \le r \le a$, where B_0, l and a may be found in terms of the flux (F), twist (Φ) and external pressure (p_e). After the prominence forms it may be treated as a sheet at $\theta = \pi$ and the field may be written as $\mathbf{B}_0(r) + \mathbf{B}_1(r,\theta)$, where $B_{1r} = -\kappa r^{\kappa-1}\sin\kappa\theta$, $B_{1\theta} = -\kappa r^{\kappa-1}\cos K\theta$. The shapes of the field lines when $1 < K < 1.5$ and $2 < K < 3$ are shown in Figure 8.11b on the left and right, respectively, and the variation of the prominence mass with K has been calculated. Unlike the previous inverse-polarity solutions, there are no difficulties at the flux tube axis where there is an 0-type neutral point.

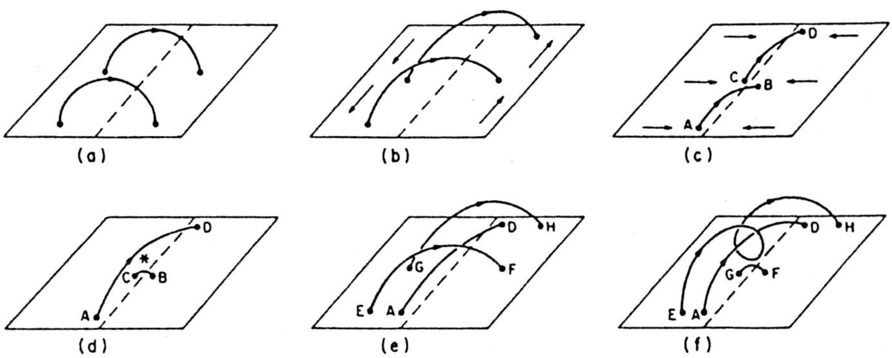

Fig. 8.12. The formation of a twisted flux tube by cancellation in the photosphere (Van Ballegooijen and Martens, 1989)

The suggestion by van Ballegooijen and Martens (1989) that a helical flux tube of the form required in the Twisted Flux Tube model may be created by flux cancellation at the polarity inversion line (Figure 8.12) is particularly appealing since it agrees with the observations of cancelling magnetic features in videomagnetographs by Martin (1986, 1990). As more flux cancels, so the helix tends to rise.

8.6 Conclusion

We have seen the effects of several themes running through the attempts to model prominences. One is the use of potential (or, better, force-free) magnetic fields as external fields to create the magnetic environment of a prominence. Another is regarding the prominence as a line current, or, better, as a current sheet. A third is assuming the temperature is uniform, or, better, a function of position. As the symphony has developed, we have observed how, through ingenuity and mathematical cunning, it has gradually become more realistic, although the earlier simple forms have laid the necessary foundation for later developments. But now we are at a stage where new themes are beginning to be heard, involving fibrils, feet and flux tubes, and I expect them to play a much more prominent part in future. The Flux Tube Model is particularly promising since it contains five new features that account for previously puzzling observations, namely the formation of a magnetic dip in the corona, the flux cancellation in the photosphere, the generation of the correct longitudinal component, the growth of the prominence as its twist increases and consistency with the structure of a prominence during its eruption.

9 Solar Flares

9.1 Introduction

The solar flare is a beautiful, awe-inspiring event which involves many areas of Solar Physics. Today, however, I shall discuss only the magnetohydrodynamics of the flare in order to determine the background environment or mould within which particle acceleration and emissions (from radio through the visible to x-rays and γ-rays) are occurring (Priest, 1981; 1991). After a summary of the observations, I shall describe the theory for the energy conversion process and the magnetic eruption. There is a subtle interaction between the MHD and microscopic plasma physics of the flare. The MHD provides the current sheets, shock waves and fluid turbulence, while the microscopic theory can determine in principle the plasma turbulent transport coefficients and the particle acceleration.

The amount of magnetic energy contained in coronal structures is certainly sufficient for a flare. For example, an arcade of field strength 500 Gauss, radius 20 Mm, length 100 Mm and shear angle 45° contains an excess energy of $6 \times 10^{25} J (6 \times 10^{32}$erg), sufficient for a large flare , whereas a loop of field strength 500G, radius 5 Mm, length 100 Mm and twist 2π contains an excess energy of $7 \times 10^{23} J$, sufficient for a small flare. Also there is enough time to store the energy in excess of potential by slow photospheric motions, injecting the required Poynting flux. For example, footpoints moving at 1km s^{-1} cover a distance of 100 Mm a day.

However, the time-scale for energy release due to magnetic diffusion (ohmic dissipation) over a length-scale l is $\tau_d = l^2/\eta$ when $\eta = (\mu\sigma)^{-1}$ is the magnetic diffusivity and σ is the electrical conductivity; and, for a global length-scale $l \sim 10$ Mm, this time is of order 10^{11} seconds, much too long for a flare. The result is that one needs to create extremely small structures with length-scales of 1 km or less in order to release the energy fast enough. The main questions we therefore need to answer from a magnetohydrodynamic viewpoint are: how does the magnetic structure go unstable and how in detail is the energy released?

A typical large solar flare has three phases. During the *preflare phase* one sees the slow rise of a prominence (for typically half an hour), together with a soft x-ray brightening and the initiation of a coronal mass ejection. At the *rise phase* (for a few minutes) the prominence erupts much more quickly and there is a steep rise in $H\alpha$ and soft x-ray emission; also hard x-ray spikes are present together with impulsive EUV and microwave bursts. Type II and III radio burst may be present and two bright ribbons form in the chromosphere. Throughout the *main phase* for many hours the intensity declines slowly and the ribbons separate, being joined by a rising arcade of loops, with hot x-ray loops located above cool $H\alpha$ loops.

Several new observational features were emphasized during the 1980's. Firstly, the density in the corona is found to increase due to evaporation from the chromosphere. Secondly, the high-temperature part of the flare takes the form of a loop or an arcade. Thirdly, often, but not always, emerging flux is observed close to the prominence before it erupts (Heyvaerts et al, 1977). Fourthly, two main types of flare are observed, namely small simple-loop or compact flares and large two-ribbon flares (Priest, 1981).

The effects on the Earth are considerable. Four years ago in March 1989 there was a historic flare which ejected a great plasmoid that reached the Earth after a couple of days. It compressed the Earth's magnetic field, which drove by induction an electric current through the national electricity grid in Canada. This tripped the safety mechanisms and caused a power cut in the whole of Quebec. Also a great aurora was produced which was seen as far south as Italy and Jamaica. Short-wave radio and tv communications around the globe were disrupted because of the disturbance to the ionsphere, and compass readings were distorted by 10 degrees posing a threat to aeroplane and boat navigation. The SMM satellite was slowed down, its altitude falling by 1 km. Also a Norwegian oil company abandoned surveying because of the effects on the delicate magnetic sensors used by oil explorers to steer the drill heads.

9.2 Energy Release by Magnetic Reconnection

(a) Role of Emerging Flux and Moving Satellite Spots. New magnetic flux emerging vertically through the photosphere or satellite sunspots moving rapidly horizontally are often observed before flares. They may have two roles. The first is the creation of small (simple-loop) flares. As new flux emerges or satellites move they create a current sheet at some height h between the new or satellite flux and the ambient field. A small flare may then begin when h reaches a critical value for the current density to exceed the threshold for strong microturbulence. The critical height has been estimated by solving the energy balance within the sheet (Heyvaerts et al, 1977). The second role is the triggering of large flares when the overlying magnetic configuration contains a lot of stored magnetic energy in excess of potential. The new flux or satellite spots may push up a prominence until the critical height (d) for instability is exceeded. They may also rip away by reconnection some of the field lines lying over the prominence and helping to keep it down.

A numerical experiment on flux emergence has been undertaken by Forbes and Priest (1984), who solve the resistive MHD equations with a flux corrected transport scheme. The numerical box contains initially a uniform horizontal magnetic field and plasma beta of 0.1. Free-floating conditions are adopted on the top and sides, and new oppositely directed flux is forced in through the base.

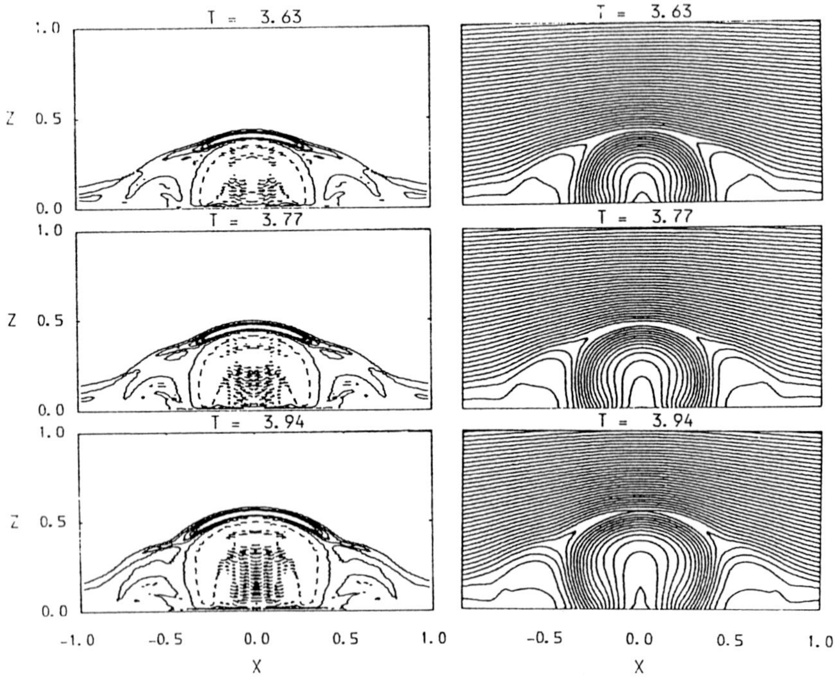

Fig. 9.1. Current density contours and magnetic field lines during emerging flux

The flux emerges and reconnects in a *flux pile-up* regime. After $t = 4$ (in units of the Alfven time) no more flux is forced up through the base, but the new flux continues to rise and pinches off near the base ($t = 4.9$) . It then reconnects and leads ultimately to a potential field as the flow dies down ($t = 14$). Figure 9.1 shows that there is a strong curved current sheet overlying the emerging flux, which is rather long, indicating a regime of flux pile-up. At its end the sheet bifurcates into pairs of slow-mode shocks. The inflow from above and below towards the reconnection point is partially a slow-mode expansion, with the magnetic field strength increasing and the flow diverging as the reconnection point is approached. Figure 9.2a shows the heights of the X- and 0- lines (solid and dashed, respectively) as functions of time, while the electric field (Figure 9.2b) can be seen to be very impulsive around $t = 4$.

(b) Reconnection in Large Flares. The overall picture of a large solar flare (Figure 9.3) is that during the *preflare phase* an active-region prominence and its overlying arcade rises slowly due to some kind of weak eruptive instability or nonequilibrium. At the flare *onset* the field lines that have been stretched out start to break and reconnect, which releases energy impulsively and causes the prominence suddenly to erupt much more rapidly. During the *main phase*, reconnection continues and creates hot X-ray loops and $H\alpha$ loops with $H\alpha$ ribbons at their footpoints. The increase in altitude of the reconnection point

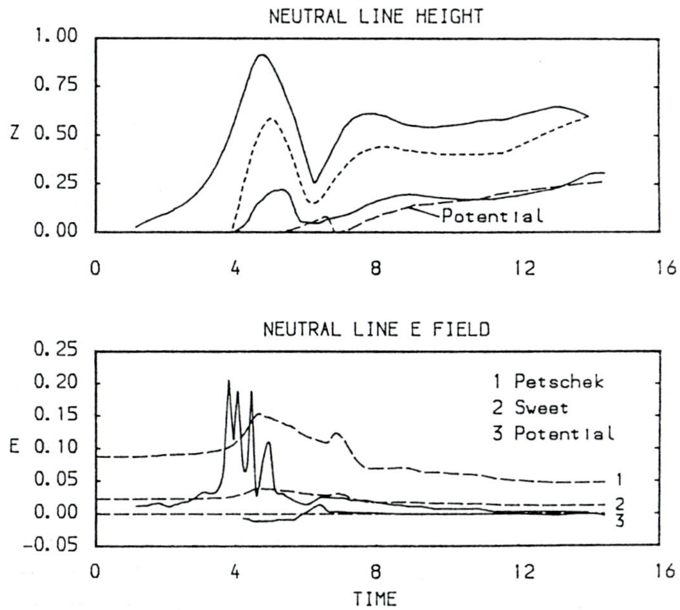

Fig. 9.2. Neutral line height and electric field as functions of time (Forbes and Priest, 1984)

causes the locations of the hot loops to rise and of the chromospheric ribbons to move apart. Thus the role of reconnection in large flares is to release the magnetic energy, both in the impulsive and main phase, and sometimes to trigger this release.

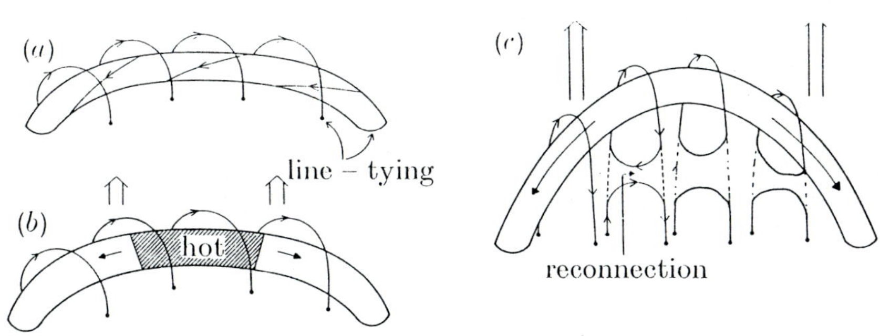

Fig. 9.3. The overall scenario for a large (two-ribbon) flare. (a) Equilibrium, (b) slow rise, (c) fast eruption upwards

The main questions that MHD addresses are the cause of the eruption and the details of the energy conversion process. Numerical experiments have focussed on two problems: first, the details of the closure process, whereby the stretched-out field lines reconnect and close back down; secondly, the global eruption in response to footpoint motions. We summarize some of the work on these in turn below.

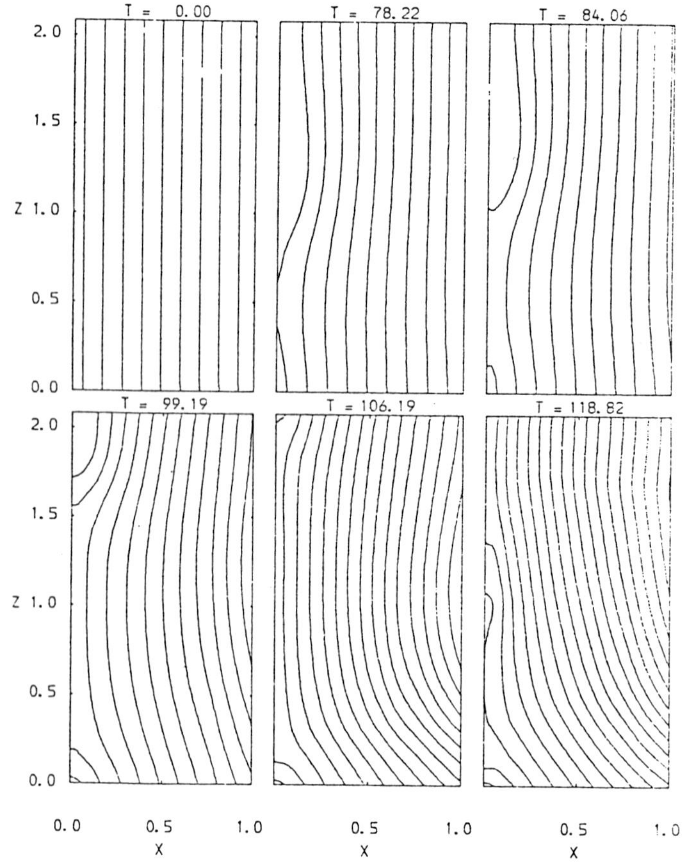

Fig. 9.4. Numerical experiment on the close-down of field lines (Forbes and Priest, 1984) at $t = 0$(a), 78.22 (b), 84.06 (c), 99.19 (d), 106.19 (e), 118.82 (f)

Kopp and Pneuman (1976) first suggested that the reconnection process can create flare loops in a two-ribbon flare, with the loops rising and the ribbons separating as the reconnection point rises. Then Cargill and Priest (1982) showed that the magnetic shocks that propagate from the reconnection site can heat the plasma to the observed temperatures of sometimes 10^7 K. Subsequently, the plasma cools and falls to give the classical $H\alpha$ loops with plasma draining down. Detailed modelling of the positions of the loops has been undertaken by Poletto and Kopp (1988) and a kinematic analysis to deduce the resulting electric fields was presented by Forbes and Priest (1982 a).

Forbes and Priest (1982 b, 1984) set up a numerical experiment with initially vertical field lines in stretched-out equilibrium and anchored at their lower ends to the base of the numerical box due to photospheric line-tying. The experiment showed how the field lines begin to reconnect by the tearing-mode instability and then close down, creating closed loops in a quasi-steady manner (Figure 9.4). The current density contours (Figure 9.5) reveal the location of the diffusion region, and the slow-mode shock waves.

The basic picture of the closure process has recently been refined considerably (Figure 9.5), and recent developments of the numerical experiment have revealed new features of relevance to the observations (Forbes et al, 1989) as follows:

1. The quasi-steady reconnection may be modulated in a time-dependent manner as the reconnection enters an impulsive burst regime (Priest 1986) in which the central diffusion region is so long that it tears (Forbes and Priest 1987), with neutral point pairs being slowly created and then rapidly annihilated. This process is observed in several other numerical experiments (Lee and Fu 1986; Scholer 1989) and may explain the sudden jumps in loop height that are observed, and the impulsive nature of hard X-ray emission.

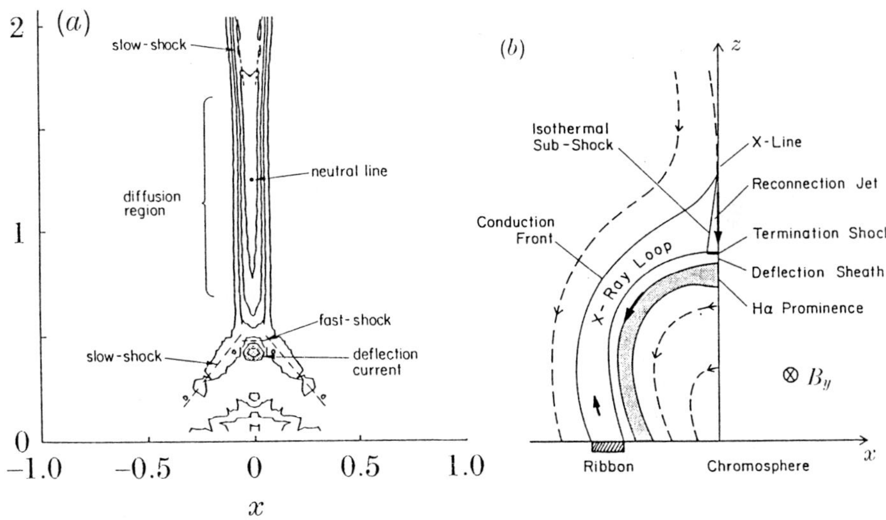

Fig. 9.5. (a) Current density contours in numerical experiment on close-down (Forbes and Priest 1984) at $t = 93.38$. (b) Schematic of creation of flare loops and $H\alpha$ ribbons

2. A fast-mode shock stands in the flow below the reconnection region and slows down the supermagnetosonic stream of plasma as it encounters the obstacle of closed field lines below it (Figure 9.5). At the same time it compresses the plasma. The increase in density drastically reduces the radiative timescale and triggers a thermal condensation which creates cool $H\alpha$ loops below the hot X-ray loops.

3. A reversed deflection current deflects the flow around the stagnation region.

4. The slow-mode shock wave in the presence of thermal conduction splits up into a conduction front (across which the temperature rises), together with an isothermal shock wave (across which the density increases).

5. Evaporation is driven from the chromosphere, both by the conduction front and by high-energy particles which travel from the reconnection site along the separatrix ahead of the conduction front. This greatly enhances the density in the hot X-ray loop, in agreement with observations.

6. When the magnetic field is smaller than 10 G and the field component out of the plane is large enough, a different regime of reconnection is found with submagnetosonic streams of plasma ejected down the reconnection region. The result is that no fast-shock or rapid condensation is created. Such a regime is appropriate to the eruption of quiescent prominences outside active regions.

9.3 Conditions for Flare Occurrence

(a) Observational Pointers. In a large flare the overall scenario (Priest, 1981) is that during the preflare phase a prominence and its highly sheared field rise slowly. Then at flare onset we have the impulsive phase and a rapid eruption of the prominence, which probably occur because of the start of reconnection below the rising prominence. But why does the eruption occur? Why does the flare start?

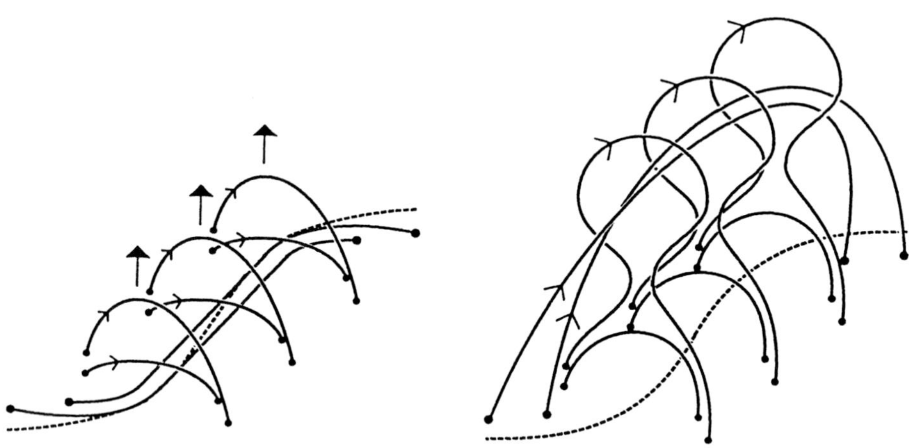

Fig. 9.6. The overall scenario for a two-ribbon flare

Small flares may be driven by photospheric motions with energy released in the corona at the rate of supply. There is for these flares no energy storage and only minor nonpotentiality. For large flares several necessary conditions have been proposed. The obvious one is storage of free energy in a nonpotential field, since the photospheric changes during a flare are small. Also flaring has been associated with strong magnetic gradients.

Two important necessary conditions are "Shear in the Corona" and "Complexity". Shear before flares is often observed in the chromosphere, as shown in $H\alpha$ fibrils and also in the photosphere, as shown in magnetograms near the polarity inversion line. These are suggestive of shear in the overlying corona.

Sara Martin (1980) has long stressed that most major flares have a *filament activation and eruption.* So *what is the role of an active-region or boundary prominence?* Since we know that the magnetic field in a prominence has a shear angle of at least 75°, it is clear that an important role is as an indicator of high shear in the corona. But is a prominence essential to a flare, or is it just a tracer of high shear?

The Emerging Flux Model (Heyvaerts et al, 1977) was developed primarily to explain complexity (in the form of emerging flux) and the presence of widely separated kernels. It suggested that, after new flux emerges, a current sheet becomes turbulent when it reaches a critical height. Also, the type depends on the magnetic environment, so that, if the overlying field contains no excess stored energy, no flare or at most a small one occurs, whereas if a lot of stored energy is present the emerging flux may trigger its release. I am in the process of generalising or modernising this model to give an Interacting Flux Model with either vertically emerging flux or horizontal spot motions and several new aspects Priest (1992). Interacting flux is a regular part of active region evolution with impacting polarities or parasitic polarities or delta spots, and flares often begin at remote footpoints . Indeed the basic structure of a flare is now believed to be an interaction of several bipoles or multiple loops, as seen in $H\alpha$ in X-rays and in radio images.

What is the role of rapid spot motions or rotations, as seen for instance in the relative motion of new and old spots? Sometimes it is to increase shear and sometimes to increase complexity. *Why then is complexity necessary?* Partly, it allows reconnection between separate structures to release excess energy that is previously stored in one or other structure. Partly, when spots move and create a force-free field, new equilibria with different connections and lower energy may become available and the flare then represents a violent transfer to a new equilibrium.

(b) Theoretical Ideas. What can theory tell us about the possible ways in which a flare may start? In principle it seems that one just needs to solve the MHD equations $\mathbf{j} \times \mathbf{B} = 0$ and $\partial \mathbf{B}/\partial t = \nabla \times (\mathbf{v} \times \mathbf{B})$ for an evolution through a series of force-free equilibria due to footpoint motions. There are several possibilities for the evolution of the magnetic energy $W(t)$ when a critical point is reached. First of all, a transcritical or pitchfork bifurcation may occur in which the equilibrium becomes linearly unstable but nonlinearly stable due to the presence of a new nearby stable equilibrium along which the evolution proceeds (Figure 9.7a) with no energy release. Secondly, a subcritical bifurcation may take place due to nonlinear instability or metastability, so that the system jumps down to a new equilibrium with energy release (Figure 9.7b). Thirdly, a state of nonequilibrium may appear with no neighbouring equilibrium, so that a catastrophe occurs with energy release (Figure 9.7c). It was hoped that such behaviour may show up in numerical computations. However, so far, the results have been disappointing.

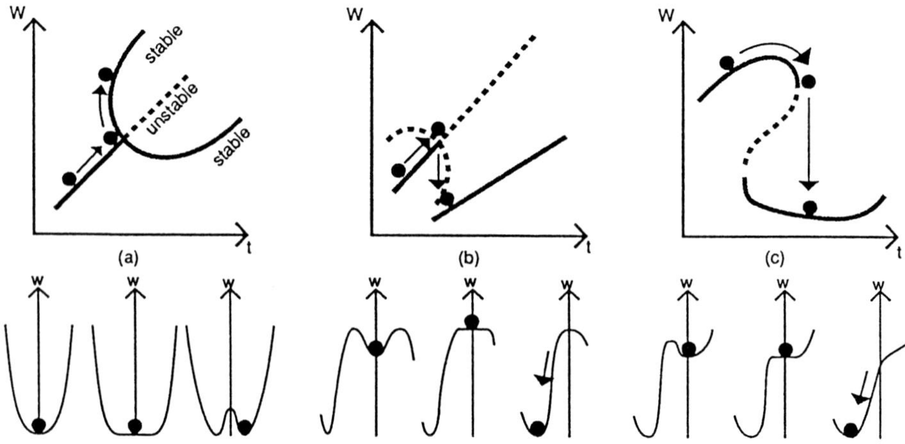

Fig. 9.7. (a) Pitch-fork bifurcation (b) subcritical bifurcation (c) nonequilibrium

Mikic et al (1988) have modelled the global eruption of a coronal arcade numerically. They have a periodic set of arcades and impose a shearing motion of amplitude $0.01v_a$ at the base. With 100 x 256 mesh points their magnetic Reynolds number is 10^4. The arcade evolves through a series of equilibria, and then at some point reconnects and forms a plasmoid which is ejected out of the top of the numerical box (Figure 9.8). However, there are problems with this experiment, since reconnection starts at the edges of the arcade, where it interacts with neighbouring arcades. Also, Biskamp and Welter (1989) find that a single arcade remains stable as it is sheared.

Dahlberg and Antiochos (1991) have conducted a 3D, incompressible, time-dependent, resistive experiment in which they start with a 2D arcade and twist up some of the footpoints at $0.1\ v_A$. The effective Lundquist number is 100, the Reynolds number 5 and the number of gridpoints 32^3. They find that the field expands and the magnetic energy increases by 50%. Then they stop twisting and hold the feet still. The field relaxes and untwists, diffusing in a thin boundary layer, but there is no sign of instability. It is remarkably stable and harder to produce a "flare" than previously realised. Perhaps they need more twist, more complexity, less diffusion or more ingenuity.

So why does eruption occur? One possibility is that it results from the interaction of two separate regions, as in the Emerging Flux Model (Heyvaerts et al 1977) and as Mikic et al (1988) and Biskamp and Welter (1989) model

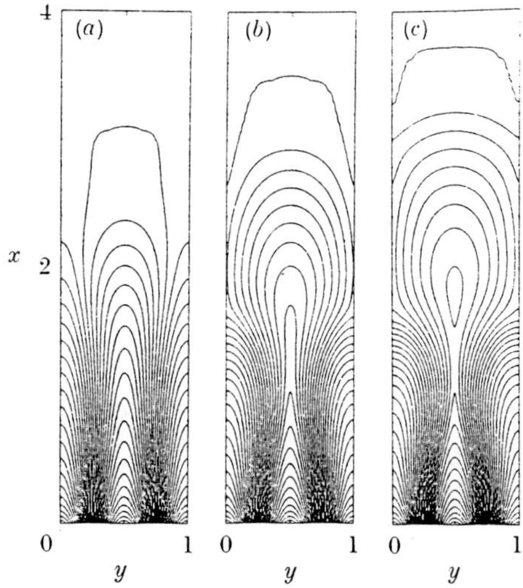

Fig. 9.8. Numerical experiment on global eruption (Mikic et al 1988) at (a) $t = 300\tau_A$, (b) 330 τ_A, (c) 340 τ_A

numerically. This certainly appears to be the case in some, but not all, flares. So is it possible to understand eruption from a single region?

During the 1980s we considered the linear MHD stability of coronal structures and showed that it is photospheric line-tying which provides the stability and enables the field to store magnetic energy in excess of potential as it is sheared by footpoint motions. When a single flux tube is twisted too much it becomes unstable (Hood and Priest 1979; Einaudi and van Hoven 1983). A simple force-free arcade, at least a linear one, is always stable, but, when a magnetic island is present in an arcade, which may support a prominence, it can become unstable if the height of the prominence or its length exceed critical values (Hood and Priest 1980). This is consistent with the observation that parts of polarity-inversion lines where the shear is the greatest are the locations of flares (Hagyard 1990).

(c) Catastrophe Model for Eruption. However, a linear instability result has the limitation of saying nothing about the nonlinear development, and so recently we have been considering the possibility of magnetic non-equilibrium or catastrophe. An analysis by Demoulin and Priest (1989) models a prominence as a sheet of mass and current supported in a linear force-free two-dimensional field with field components $(B_x, B_y, B_z) = (B_x(A), \partial A/\partial z, -\partial A/\partial y)$ where the flux function A is given by $\nabla^2 A + \alpha^2 A = \delta(y)\delta(z-h)$ when the prominence is located at height $z = h$ above the photospheric surface ($z = 0$). The solution of this may be written as the sum of a complementary function and a particular integral (a Green's function) with the normal field component B_z imposed at the photosphere. The condition for prominence equilibrium ($IB_y = mg$) at the prominence of mass m then determines the way the prominence current (I) varies with its height (h). The startling feature is that when the prominence reaches a critical height there is a catastrophe with no neighbouring equilibrium and it erupts. The eruption also occurs if the force-free arcade is sheared too much, in agreement with observations. However, this is only the case when there is a substantial quadrupolar component to the active region: in other words, one needs complexity in the photospheric field, as indeed the observations show.

A similar analysis with a slightly different external field has been carried out by Priest and Forbes (1990). They set up a model for equilibrium and eruption by regarding the prominence as an electric current filament situated at height h in a background active-region modelled by a magnetic line dipole situated below the photosphere. The model is an extension of previous work by Van Tend and Kuperus (1978) and Kuin and Martens (1986). As the filament current or twist increases so the prominence rises slowly through a series of equilibria.

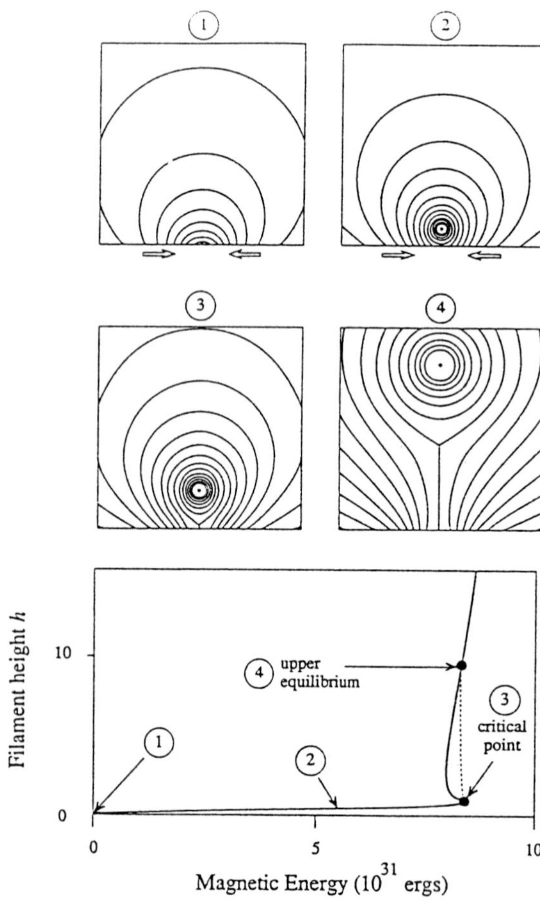

Fig. 9.9. The catastrophe model for prominence eruption of Priest and Forbes

But when a critical twist or current is exceeded there is no neighbouring equilibrium and a magnetic non-equilibrium or catastrophe takes place, with the unbalanced forces causing the prominence to erupt. The velocity (dh/dt) of the prominence as a function of time or height may be obtained by solving its equation of motion $(m\, d^2h/dt^2 = IB^*)$, where m is the prominence mass, I its current and B^* the field at the prominence due to the background field.

When the plasma velocity is sub-Alfvenic the magnetic field (B_x, B_y) around the prominence during the eruption may be represented in terms of the complex variable $Z = x + iy$, where the y axis is vertically upwards, as

$$B_y + iB_z = imh^2 \left[(Z^2 = p^2)(Z^2 + q^2)\right]^{\frac{1}{2}} / pqZ^2(Z^2 + h^2) \ .$$

This possesses a line current at $Z = ih$(the prominence), a line dipole at $Z = 0$ (the background field) and a cut in the complex plane from $Z = ip$ to $Z = iq$ (the prominence current sheet).

When no reconnection is allowed one finds that beyond the non-equilibrium point the magnetic energy declines and the filament speed increases with height. Thus the prominence erupts even if no reconnection is allowed. When reconnection is allowed, however, it is then driven by the eruption. The magnetic work declines more rapidly and the prominence erupts faster, with an energy release eight times faster. The resulting large electric field in the reconnection region may well be important for accelerating fast particles. The creation of the reconnecting current sheet below the erupting prominence is followed numerically (Forbes, 1991), and plots of density and current density contours reveal the presence of a fast-mode shock travelling ahead of the prominence together with slow-mode shocks near the reconnection site.

9.4 Conclusion

As we have seen, there have been many developments in the basic MHD theory for solar flares during the past few years. A combination of numerical experimentation and analytical theory has greatly increased our understanding of the basic process of magnetic reconnection (Chapter 6). Numerical experiments of reconnection in a two-ribbon flare have refined and developed the basic picture of field line closure that creates flare loops and have explained many observational features. Also, the eruption of a prominence at the start of such a flare may well be caused by a process of magnetic non-equilibrium when the length or height of the prominence is too great.

In future, many developments are expected from more sophisticated numerical experiments and from studies of the three-dimensional aspects. Also, one hopes that the microscopic plasma physicists can make use of the MHD studies, so that further progress can be made on the microscopic processes such as particle acceleration, that together we may understand this beautiful phenomenon.

References

Amari, T., and Aly, J. J. *Astron. Astrophys.* **208** (1989) 361.

Anzer, U. *Dynamics and Structure of Quiescent Solar Prominences* p143 (1989) (ed E. R. Priest) Kluwer.
Ballester, J. L. and Priest, E. R. *Solar Phys.* **109** (1987) 335–350
Benz, A. O. and Aschwanden, M. J. *Eruptive Solar Flares* (ed Z. Svestka, B. Jackson, M. Machado), Springer-Verlag, (1992) p 106–115
Biskamp, D. *Phys. Fluids* **29** (1986) 1520–1531
Biskamp, D. and Welter, H. *Solar Phys.* **120** (1989) 49
Cally, P. *J Plasma Phys.* **45** (1991) 453–479
Cargill, P. and Priest, E. R. *Solar Phys.* **76** (1982) 357–375
Chapman, S. *Proc. Roy. Soc.* **A95** (1929) 61
Chapman, S. *Smithsonian Contrib. Astrophys.* **2** (1957) 1
Chupp, E. L. *Ann. Rev. Astron. Astrophys.* **22** (1984) 359
Dahlburg, R. B. and Antiochos, S. *Astrophys. J.* **383** (1991) 420–431
De Jager, C. and Svestka, Z. *Solar Phys.* **100** (1985) 435
Demoulin, P. and Priest, E. R. *Astron. Astrophys.* **214** (1989) 360–368
Demoulin, P., Malherbe, J. M. and Priest, E. R. *Astron. Astrophys.* **211** (1989) 428–440
Einaudi, G. and Van Hoven, G. *Solar Phys.* **88** (1983) 163
Field,G. B. *Astrophys. J.* **142** (1965) 531
Forbes, T. G. *Geophys. Astrophys. Fluid Dyn.* **62** (1991) 15–36
Forbes, T. G. and Priest, E. R. *Solar Phys.* **81** (1982) 303–324
Forbes, T. G. and Priest, E. R. *Planet Space Sci.* **30** (1982a) 1183–1197
Forbes, T. G. and Priest, E. R. *Solar Phys.* **81** (1982b) 303–324
Forbes, T. G. and Priest, E. R. *Solar Phys.* **94** (1984) 315–340
Forbes, T. G., Malherbe, J. M. and Priest, E. R. *Solar Phys.* **120** (1989) 285–308
Golub, L. in *Physics of Solar and Stellar Coronae* (ed J. Linsky and S. Serio), Kluwer (1993) 71–82
Golub, L., Krieger, A. S., Silk, J., Timothy, A. and Vaiana, G. *Astrophys. J,* **189** (1974) L93–L97
Goossens, M. in *Advances in Solar System MHD* (ed E. R. Priest and A. W. Hood), Cambridge (1991) p137–172
Hagyard, M. *Mem. Soc. Ast. It.,* **61** (1990) 337
Harvey, K. L. *Proc. 4th European Meeting on Solar Phys.* ESA SP **220** (1984) 235–236
Heyvaerts, J. *Astron. Astrophys.* **37** (1974) 65
Heyvaerts, J. and Priest, E. R. *Astron. Astrophys* **117** (1983) 220–234
Heyvaerts, J. and Priest, E. R. *Astron. Astrophys.* **137** (1984) 63–78
Heyvaerts, J. and Priest, E. R. *Astrophys. J.* **390** (1992) 297
Heyvaerts, J., Priest, E. R. and Rust, D. M. *Astrophys. J.* **216** (1977) 123–137
Hildner, E. *Solar Phys.* **35** (1974) 123
Hollweg, J. V. in *Solar Wind 5* NASA CP 2280 (1983) 1
Hood, A. W. and Anzer, U. *Solar Phys.* **126** (1990) 117
Hood, A. W. and Priest, E. R. *Astron. Astrophys.* **77** (1979) 233–251
Hood, A. W. and Priest, E. R. *Astron. Astrophys* **87** (1980) 126–131
Kippenhahn, R. and Schluter, A. *Zs. Ap.* **43** (1957) 36–62
Kopp, R. A. and Pneuman, G. W. *Solar Phys.* **50** (1976) 85
Kuin, N. and Martens, P. in *Coronal and Prominence Plasmas* (ed A. Poland), NASA CP-2442 (1986) p241–245
Kuperus, M. and Raadu, M. A. *Astron. Astrophys.* **31** (1974) 189–193
Lee, L. C. and Fu, Z. F. *J Geophys. Res.* **91** (1986) 6807–6815
Leroy, J. L. in *Dynamics and Structure of Quiescent Solar Prominences* (ed. E. R. Priest) (1989) p77–114

Lima, J. and Priest, E. R. *Astron. Astrophys.* **268** (1993) 641–649
Malherbe, J. M. and Priest, E. R. *Astron. Astrophys.* **123** (1983) 80–88
Marten, S. F. in *Small-scale Dynamical Processes in Stellar Atmospheres* (ed S. Keil) Sac. Peak Observatory (1984) p30
Marten, S. F. *Coronal and Prominence Plasmas* (ed. A. Poland) NASA CP 2442, (1986) p73–80
Marten, S. F. *IAU Symp.* **138** (1990) 130–146
Mikic, Z., Barnes, D. C. and Schnack, D. D. *Astrophys. J.* **328** (1988) 830–847
Parker, E. N. *Astrophys. J.* **117** (1953) 431
Parker, E. N. *J. Geophys. Res.* **62** (1957) 509–520
Parker, E. N. *Astrophys. J.* **128** (1958) 664
Parker, E. N. *Cosmical Magnetic Fields* Oxford University Press (1979)
Parker, E. N. *Geophys. Astrophys. Fluid Dyn.* **52** (1990) 183–210
Petschek, H. E. *AAS-NASA Symp. on Phys. of Solar Flares* NASA SP–50, (1964) 425–439
Pneuman, G. W. and Kopp, R. A. *Solar Phys.* **18** (1971) 258
Poland, A. I. *Coronal and Prominence Plasmas* NASA CP 2442 (1986)
Poletto, G. and Kopp, R. A. *Solar Phys.* **116** (1988) 163–178
Priest, E. R. (ed) *Solar Flare Magnetohydrodynamics* Gordon and Breach New York (1981)
Priest, E. R. *Solar Magnetohydrodynamics* Reidel Dordrecht (1982)
Priest, E. R. *Rep. Prog. Phys.* **48** (1985) 955–1090
Priest, E. R. *Mit. Astron. Ges.* **65** (1986) 41–51
Priest, E. R. *Dynamics and Structure of Quiescent Solar Prominences* Kluwer Holland (1989)
Priest, E.R. in *Physics of Magnetic Flux Ropes* (ed C Russell, E Priest and L Lee), AGU Monograph, (1990) p1–22
Priest, E. R. *Phil. Trans. Roy. Soc. Lond. A.* **336** (1991) 363–380
Priest, E. R. and Forbes, T. G. *J. Geophys. Res.* **91** (1986) 5579–5588
Priest, E. R. and Forbes, T. G. *Solar Phys.* **126** (1990) 319–350
Priest, E. R. and Forbes, T. G. *J. Geophys. Res.* **97** (1992a) 1521–1531
Priest, E. R. and Forbes, T. G. *J. Geophys. Res.* **97** (1992b) 16757–16772
Priest, E. R., Hood, A. W. and Anzer, U. *Astrophys. J.* **344** (1989) 1010–1025
Priest, E. R. and Hood, A. W. *Advances in Solar System MHD* Cambridge Uni. Press (1991)
Priest, E. R. and Lee, L. C. *J. Plasma Phys.* **44** (1990) 337–360
Priest, E. R., Parnell, C. E. and Martin, S. F. *Astrophys. J.* in press (1993)
Priest, E. R. and Raadu, M. A. *Solar Phys.* **43** (1975) 177–188
Priest, E. R. and Smith, E. A. *Solar Phys.* **64** (1979) 267
Priest, E. R., Titov, V. S., and Rickard, G. K. *Phil. Trans. Roy. Soc. Lond.* (1994)
Priest, E. R., Titov, V. S., Vekstein, G. E. and Rickard, G. K. *J. Geophys. Res* (1994)
Roberts, B. in *Hydromagnetics of Sun* ESA SP–220 137
Russell, C. T., Priest, E. R. and Lee, L. C. *Magnetic Flux Ropes* A,. Geophys. Union (1990)
Scholer, M. *J. Geophys. Res.* **94** (1989) 8805–8812
Sonnerup, B. U. O. and Priest, E. R. *J. Plasma Phys.* **14** (1975) 417–431
Sweet, P. A. *IAU Symp.* **6** (1958) 123-134
Tandberg-Hanssen, E. *Solar Prominences* D Reidel, Dordrecht, Holland (1974)
Tsuganos, K. in *Advances in Solar System MHD* (ed. E. R. Priest and A. W. Hood) Cambridge, (1991) p257

Ulmschneider, P., Priest, E. R. and Rosner, R. *Mechanisms of Chromospheric and Coronal Heating* Springer-Verlag, Berlin (1991)
Weber, E. J. and Davis, L. *Astrophys. J.* **148** (1967) 217

Kinetic Plasma Physics

Don B.Melrose

Department of Theoretical Physics
The University of Sydney
Sydney, NSW 2006, Australia

In the kinetic theory of plasmas, particles are described by distribution functions whose evolution is determined by kinetic equations. The spectrum of fluctuations in the plasma includes distributions of weakly damped waves. The kinetic equation for the particles may be averaged over these fluctuations to find averaged kinetic equations, sometimes called the quasilinear equations. These kinetic equations describe the slow evolution of the waves, which may be damped or may grow, and of the particles, which diffuse in momentum space. In this lecture course, a general derivation of the quasilinear equations is given in the first two lectures, and these equations are used to treat resonant scattering and related processes is the next three lectures. The last four lectures are concerned with radiation processes in various astrophysical sources. Radiation processes may also be treated using the theory outlined in the first two lectures.

1 Distributions of Particles and Waves

In order to derive a general form for the quasilinear equations, one needs a general theory for the properties of waves in plasmas, and for the interaction between particles and waves. The theory summarized here is discussed in more detail in Melrose (1980a, 1986) and Melrose and McPhedran (1991), cf. also Kaplan and Tsytovich (1973).

1.1 Linear Response of a Plasma

Consider a distribution of particles of species α (α = e for electrons, α = i for ions) with charge q_α and mass m_α. The distribution may be described in terms of the density of representative points for particles in the 6-dimensional $\boldsymbol{x}$–$\boldsymbol{p}$ phase space, where $\boldsymbol{x}$ is the position vector and $\boldsymbol{p}$ is the momentum of the particle. A kinetic theory description of a plasma is based on a set of kinetic equations that describes how the distribution functions evolves due wave-particle and particle-particle interactions.

The Vlasov Equations

The basic equations of the kinetic theory of plasmas are called the *Vlasov equations.* These consist of the following:
1) A set of equations which is formally identical to a set of Boltzmann equations for each species of particles:

$$\left\{\frac{\partial}{\partial t}+\boldsymbol{v}\cdot\frac{\partial}{\partial \boldsymbol{x}}+q_\alpha[\boldsymbol{E}(t,\boldsymbol{x})+\boldsymbol{v}\times\boldsymbol{B}(t,\boldsymbol{x})]\cdot\frac{\partial}{\partial \boldsymbol{p}}\right\}f_\alpha(\boldsymbol{p},t,\boldsymbol{x}) = \left.\frac{df_\alpha(\boldsymbol{p},t,\boldsymbol{x})}{dt}\right|_{\text{coll}}, \tag{1.1}$$

where $f_\alpha(\boldsymbol{p},t,\boldsymbol{x})$ is the single particle distribution function for species α. The right hand term describes the effect of collisions in the plasma, and this term is neglected in a 'collisionless' plasma.
2) A pair of equations giving the charge and current density in terms of the single particle distribution functions:

$$\rho(t,\boldsymbol{x}) = \sum_\alpha q_\alpha \int d^3\boldsymbol{x}\, f_\alpha(\boldsymbol{p},t,\boldsymbol{x}), \tag{1.2}$$

$$\boldsymbol{J}(t,\boldsymbol{x}) = \sum_\alpha q_\alpha \int d^3\boldsymbol{x}\, \boldsymbol{v}\, f_\alpha(\boldsymbol{p},t,\boldsymbol{x}), \tag{1.3}$$

where the sum is over all species of particle.
3) Maxwell's equations: In SI units these are

$$\operatorname{curl}\boldsymbol{E} = -\partial\boldsymbol{B}/\partial t, \tag{1.4}$$

$$\operatorname{curl}\boldsymbol{B} = \mu_0\boldsymbol{J} + (1/c^2)\,\partial\boldsymbol{E}/\partial t, \tag{1.5}$$

$$\operatorname{div}\boldsymbol{E} = \rho/\varepsilon_0, \tag{1.6}$$

$$\operatorname{div}\boldsymbol{B} = 0. \tag{1.7}$$

The source terms ρ and $\boldsymbol{J}$ are identified with those in (1.2) and (1.3).

If there is a magnetostatic field, $\boldsymbol{B}_0$, then the plasma is said to be *magnetized.* In a magnetized plasma, the unperturbed distribution function can depend only on the components of momentum parallel, $p_\parallel$, and perpendicular, $p_\perp$, to $\boldsymbol{B}_0$, respectively. This is because the particles spiral about the magnetic field lines so that the gyrophase is a linear function of time; however, the unperturbed part, $f^{(0)}$, of the distribution function must be time independent, and so cannot depend on the gyrophase. For the same reason, no static, parallel component of electric field is allowed in a time-independent system. A static, perpendicular component, $\boldsymbol{E}_0$, of the electric field implies that all particles drift with velocity $\boldsymbol{E}_0\times\boldsymbol{B}_0/B_0^2$, and this may be removed by making a Lorentz transformation to a frame with zero static electric field.

In the Vlasov equations $\boldsymbol{E}$ and $\boldsymbol{B}$ are the *self-consistent* fields, which are functionals of the distribution functions. Despite its apparent simplicity, the set of equations of the form (1.1) is a set of nonlinear, coupled, integro-differential equations for the distribution functions f_α.

Fourier Transforms

It is convenient to imagine a separation into two time scales. On a shorter time scale there are fluctuations in the plasma, which include waves in various wave modes. These fluctuations are treated by Fourier transforming. The longer time scale is that on which the distribution functions evolve after averaging over the fluctuating spectrum. The quasilinear equations describe the evolution on this longer time scale.

For a function, $F(t, \boldsymbol{x})$, that involves the fluctuations, the Fourier transform is defined by

$$\tilde{F}(\omega, \boldsymbol{k}) = \int dt\, d^3\boldsymbol{x}\, e^{\mathrm{i}(\omega t - \boldsymbol{k}\cdot\boldsymbol{x})}\, F(t, \boldsymbol{x}). \tag{1.8}$$

The inverse transform is

$$F(t, \boldsymbol{x}) = \int \frac{d\omega d^3\boldsymbol{k}}{(2\pi)^4}\, e^{-\mathrm{i}(\omega t - \boldsymbol{k}\cdot\boldsymbol{x})}\, \tilde{F}(\omega, \boldsymbol{k}). \tag{1.9}$$

In the following the tilde on Fourier transformed quantities is omitted.

The Fourier transformed form of Maxwell's equations (1.4)–(1.7) is

$$\boldsymbol{k} \times \boldsymbol{E}(\omega, \boldsymbol{k}) = \omega \boldsymbol{B}(\omega, \boldsymbol{k}), \tag{1.10}$$

$$\mathrm{i}\boldsymbol{k} \times \boldsymbol{B}(\omega, \boldsymbol{k}) = \mu_0 \boldsymbol{J}(\omega, \boldsymbol{k}) - \mathrm{i}\omega \boldsymbol{E}(\omega, \boldsymbol{k})/c^2, \tag{1.11}$$

$$\boldsymbol{k} \cdot \boldsymbol{E}(\omega, \boldsymbol{k}) = -\mathrm{i}\rho(\omega, \boldsymbol{k})/\varepsilon_0. \tag{1.12}$$

$$\boldsymbol{k} \cdot \boldsymbol{B}(\omega, \boldsymbol{k}) = 0 \tag{1.13}$$

Equation (1.13) is redundant, being implied by (1.10).

Linearized Vlasov Equations

In *weak turbulence* theory one solves the Vlasov equations by first expanding the distribution functions in powers of the fluctuating fields:

$$f_\alpha(\boldsymbol{p}, t, \boldsymbol{x}) = f_\alpha^0(\boldsymbol{p}) + f_\alpha^{(1)}(\boldsymbol{p}, t, \boldsymbol{x}) + \cdots, \tag{1.14}$$

where $f_\alpha^0(\boldsymbol{p})$ is the unperturbed distribution function, $f_\alpha^{(1)}(\boldsymbol{p}, t, \boldsymbol{x})$ is linear in $\boldsymbol{E}$ or $\boldsymbol{B}$, and so on. After linearizing and Fourier transforming, for unmagnetized particles (1.1) becomes

$$-\mathrm{i}(\omega - \boldsymbol{k}\cdot\boldsymbol{v}) f_\alpha^{(1)}(\omega, \boldsymbol{k}) + q_\alpha[\boldsymbol{E}(\omega, \boldsymbol{k}) + \boldsymbol{v} \times \boldsymbol{B}(\omega, \boldsymbol{k})] \cdot \frac{\partial f_\alpha^{(0)}(\boldsymbol{p})}{\partial \boldsymbol{p}} = 0. \tag{1.15}$$

It is then trivial to solve for $f_\alpha^{(1)}$. On using (1.10) to reexpress $\boldsymbol{B}(\omega, \boldsymbol{k})$ in terms of $\boldsymbol{E}(\omega, \boldsymbol{k})$, this gives $f_\alpha^{(1)}(\omega, \boldsymbol{k})$ as a linear function of $\boldsymbol{E}(\omega, \boldsymbol{k})$.

The Linear Response

The linear response of a plasma may be introduced by Fourier transforming the linear current density obtained from (1.3), with the distribution function identified as the linear term $f_\alpha^{(1)}$. The resulting linear current may be written in the form

$$J^{(1)}(\omega, \boldsymbol{k}) = \sigma_{ij}(\omega, \boldsymbol{k}) E_j(\omega, \boldsymbol{k}), \tag{1.16}$$

where $\sigma_{ij}(\omega, \boldsymbol{k})$ is the conductivity tensor. The charge density is included implicitly in the left hand side of (1.16), through the Fourier transform of the charge continuity condition $\omega\rho(\omega, \boldsymbol{k}) = \boldsymbol{k} \cdot \boldsymbol{J}(\omega, \boldsymbol{k})$. Also the right hand side includes the magnetic field as well as the electric field through (1.10).

It is often more convenient to describe the response in term of the equivalent dielectric tensor,

$$K_{ij}(\omega, \boldsymbol{k}) = \delta_{ij} - \frac{i}{\omega\varepsilon_0}\sigma_{ij}(\omega, \boldsymbol{k}), \tag{1.17}$$

where δ_{ij} is the unit tensor. The equivalent dielectric tensor K_{ij} (or the conductivity tensor σ_{ij}) completely describes the linear response of the medium. Thus, if one refers to, for example, a 'magnetoionic medium', a 'cold plasma' or a 'magnetized thermal plasma', one means that $K_{ij}(\omega, \boldsymbol{k})$ has the specific form implied by, respectively, the magnetoionic model, the cold plasma model or the magnetized thermal plasma model used to calculate it.

1.2 The Formal Theory of Waves

Formally a 'wave' corresponds to a plane-wave solution of a wave equation. Here the wave equation is that derived from Maxwell's equations by Fourier transforming in space and time, and including the linear response using (1.16).

The Wave Equation

The set (1.10)–(1.13) of Fourier transformed Maxwell's equations needs to be reduced to a single wave equation. This is achieved by noting that (1.13) is redundant, interpreting (1.10) as defining a subsidiary vector $\boldsymbol{B}(\omega, \boldsymbol{k})$ in terms of $\boldsymbol{E}(\omega, \boldsymbol{k})$, and interpreting (1.12) as defining a subsidiary quantity $\rho(\omega, \boldsymbol{k})$ in terms of $\boldsymbol{J}(\omega, \boldsymbol{k})$. The current density $\boldsymbol{J}$ is separated into a part that describes the linear response, as given by (1.16), and a remaining 'extraneous' current that acts as a source term.

If the extraneous current is ignored, the remaining equation reduces to the homogeneous wave equation

$$\Lambda_{ij}(\omega, \boldsymbol{k}) E_j(\omega, \boldsymbol{k}) = 0, \tag{1.18}$$

with

$$\Lambda_{ij}(\omega, \boldsymbol{k}) = \frac{|\boldsymbol{k}|^2 c^2}{\omega^2}(\kappa_i\kappa_j - \delta_{ij}) + K_{ij}(\omega, \boldsymbol{k}). \tag{1.19}$$

where $\boldsymbol{\kappa} = \boldsymbol{k}/|\boldsymbol{k}|$ is the unit vector along $\boldsymbol{k}$.

The Dispersion Equation

The homogeneous wave equation (1.18) may be regarded as three simultaneous equations for the three cartesian components of $\boldsymbol{E}(\omega, \boldsymbol{k})$. From this viewpoint, $\Lambda_{ij}(\omega, \boldsymbol{k})$ in (1.19) is the matrix of coefficients. The condition for a solution to exist is that the determinant of this matrix vanish. This determinant may written as

$$\Lambda(\omega, \boldsymbol{k}) = \det\left[\Lambda_{ij}(\omega, \boldsymbol{k})\right], \tag{1.20}$$

where "det" denotes the determinant. In this way, the condition for a solution of the homogeneous wave equation to exist gives the *dispersion equation*

$$\Lambda(\omega, \boldsymbol{k}) = 0. \tag{1.21}$$

Dispersion Relations

One solves the dispersion equation in various ways by making different choices of the independent and dependent variables. One choice of dependent variable is ω, in which case one solves (1.21) for ω as a function of $\boldsymbol{k}$. The dispersion relation for an any specific wave mode, labeled mode M, is then written in the form

$$\omega = \omega_M(\boldsymbol{k}), \tag{1.22}$$

where one has $\Lambda(\omega_M(\boldsymbol{k}), \boldsymbol{k}) = 0$. It is often convenient to write $\boldsymbol{k}$ as $k\boldsymbol{\kappa}$, where $k = |\boldsymbol{k}|$ is the wave number, and $\boldsymbol{\kappa}$, which is the unit vector along $\boldsymbol{k}$, is referred to as the *direction of wave propagation*, the *normal to the wave front* or the *wave normal direction*.

Another choice of independent variable that is often made in electromagnetic theory is the *refractive index* $n = kc/\omega$. One usually solves for the square of the refractive index as a function of ω and of appropriate angles. In this case the dispersion relation for waves in the mode M is written

$$\frac{k^2c^2}{\omega^2} = n_M^2(\omega, \boldsymbol{\kappa}). \tag{1.23}$$

The Polarization Vector

When the dispersion relation for the wave mode M is satisfied, there exists a solution of the homogeneous wave equation. One then solves the wave equation (1.18) for the electric field $\boldsymbol{E}$. The *polarization vector*, $\boldsymbol{e}_M$, for the mode M is defined by normalizing this solution to unity. In general a polarization vector has both longitudinal and transverse parts. A strictly *longitudinal* polarization vector corresponds to $\boldsymbol{e} = \boldsymbol{\kappa}$. Waves whose polarization vector is longitudinal are called *longitudinal waves*.

In general the polarization vector for an arbitrary wave mode in a magnetized plasma can be chosen of the form

$$e_M = \frac{L_M \boldsymbol{\kappa} + T_M \boldsymbol{t} + i\boldsymbol{a}}{(L_M^2 + T_M^2 + 1)^{1/2}}. \tag{1.24}$$

In a coordinate system with $\boldsymbol{B} = B(0,0,1)$ along the z-axis and $\boldsymbol{k}$ in the x-z plane at angle θ to $\boldsymbol{B}$, one has

$$\boldsymbol{\kappa} = (\sin\theta, 0, \cos\theta), \quad \boldsymbol{t} = (\cos\theta, 0, -\sin\theta), \quad \boldsymbol{a} = (0,1,0). \tag{1.25}$$

The parameter L_M describes the longitudinal part of the polarization, and T_M describes the transverse part of the polarization. In general, the transverse polarization is elliptical, and T_M is the axial ratio of the ellipse, with $T_M = 1$ ($T_M = -1$) for right (left) circularly polarization, $T_M = 0$ for linear polarization along $\boldsymbol{a}$, and $T_M = \infty$ for linear polarization along $\boldsymbol{t}$.

Ratio of Electric to Total Energy

One other characteristic property of an arbitrary wave mode appears in the theory developed in lecture 2. This is the ratio, R_M, of the electric to total energy in the waves. The total energy consists of electric and magnetic energies and the kinetic energy associated with the forced particle motion in the waves (sometimes referred to as "sloshing about"). For waves with polarization vector in the form (1.24), one has

$$R_M = \frac{L_M^2 + T_M^2 + 1}{2(T_M^2 + 1)\, \partial(\omega n_M)/\partial\omega}. \tag{1.26}$$

1.3 Particular Wave Modes

Several different wave modes appear in applications discussed in these lectures. The following are brief descriptions of relevant waves (e.g., Stix 1962; Akhieser et al. 1967; Lifshitz and Pitaevskii 1981).

Waves in Unmagnetized Plasmas

In an unmagnetized plasma containing electrons (charge $-e$, mass m_e) and one species of ion (charge $Z_i e$, mass m_i) the plasma parameters include the electron number density, n_e, the ion number density, $n_i = n_e/Z_i$, where charge neutrality is assumed, and the temperatures, T_e of the electrons, and T_i of the ions, which need not be equal. These parameters determine the plasma frequency, $\omega_p = (e^2 n_e/\varepsilon_0 m_e)^{1/2}$, the ion plasma frequency, $\omega_{pi} = (Z_i^2 e^2 n_i/\varepsilon_0 m_i)^{1/2}$, and the thermal speeds $V_e = (T_e/m_e)^{1/2}$ and $V_i = (T_i/m_i)^{1/2}$, where Boltzmann's constant is set equal to unity. The Debye length of electrons is $\lambda_{De} = V_e/\omega_p$, and the ion sound speed is $v_s = \omega_{pi}\lambda_{De}$.

Three different wave modes are identified in a treatment of waves in an isotropic thermal plasma: a transverse mode and two longitudinal modes.

Transverse waves : Transverse electromagnetic waves (label T) have their electric vector orthogonal to $\boldsymbol{k}$, and they reduce to electromagnetic waves in vacuo at high frequencies, $\omega \gg \omega_{\mathrm{p}}$. Their dispersion relation is

$$\omega_{\mathrm{T}}(\boldsymbol{k}) = (\omega_{\mathrm{p}}^2 + k^2c^2)^{1/2}, \quad n = (1 - \omega_{\mathrm{p}}^2/\omega^2)^{1/2}, \quad \boldsymbol{e}_{\mathrm{T}} \cdot \boldsymbol{\kappa} = 0. \tag{1.27}$$

The refractive index (1.27) is less than unity and it has a zero, referred to as a *cutoff*, at $\omega - \omega_{\mathrm{p}}$. At frequencies with $n^2 < 0$, corresponding to $\omega \leq \omega_{\mathrm{p}}$ here, the waves are evanescent; all wave energy incident on a region of evanescence is reflected.

Langmuir waves : Langmuir waves (label L) are electron plasma oscillations. Their dispersion relation is

$$\omega_{\mathrm{L}}(\boldsymbol{k}) \approx \omega_{\mathrm{p}} + 3k^2V_{\mathrm{e}}^2/2\omega_{\mathrm{p}}, \quad \boldsymbol{e}_{\mathrm{L}} = \boldsymbol{\kappa}. \tag{1.28}$$

These waves exist only for phase speeds at least several times V_{e}, being heavily Landau damped at lower phase speeds. Landau damping is the absorption process corresponding to Cerenkov emission, as discussed in lecture 2.

Ion sound waves : Ion sound waves (label S) exist only below ω_{pi}, where their dispersion relation is

$$\omega_{\mathrm{S}}(\boldsymbol{k}) \approx kv_{\mathrm{s}}/(1 + k^2\lambda_{De}^2)^{1/2}, \quad \boldsymbol{e}_{\mathrm{S}} = \boldsymbol{\kappa}. \tag{1.29}$$

Except for $T_{\mathrm{e}} \gg T_{\mathrm{i}}$ or $Z_{\mathrm{i}} \gg 1$, these waves are strongly Landau damped by thermal ions, and effectively do not exist.

The Magnetoionic Waves

The magnetoionic theory (the name arises from a now outdated use of the word ion) is the theory of waves in a cold, magnetized electron gas. This corresponds to a model of a plasma in which only the motion of electrons in a uniform magnetic field is taken into account, and the thermal motions of these electrons are neglected. The ions are usually assumed to form a uniform positively charged background, but in fact they play no role whatsoever. There are only two independent parameters in the theory, ω_{p}, and the electron cyclotron frequency, $\Omega_{\mathrm{e}} = eB/m_{\mathrm{e}}$.

The dispersion equation in this case may be written as a quadratic equation for n^2. The two solutions are referred to as the ordinary and extraordinary modes, with each of these having two branches separated by a stop band, which is a region of evanescence, as illustrated in Figure 1.1. The two higher frequency branches are referred to as the o-mode and x-mode. These may be regarded as magnetically split versions of the transverse waves in an isotropic plasma. They have cutoffs at ω_{p} and at

$$\omega_{\mathrm{x}} = \tfrac{1}{2}\Omega_{\mathrm{e}} + \tfrac{1}{2}(4\omega_{\mathrm{p}}^2 + \Omega_{\mathrm{e}}^2)^{1/2}, \tag{1.30}$$

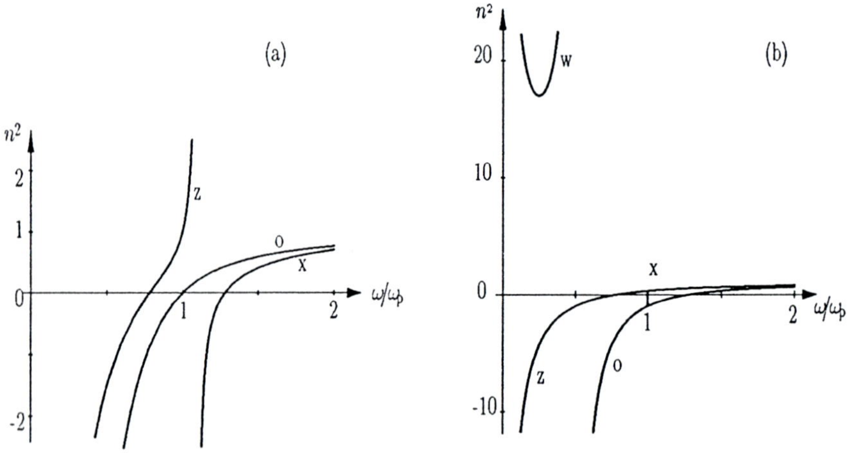

Fig. 1.1. Dispersion curves for the magnetoionic waves: (a) the squares of the refractive indices for frequencies near the plasma frequency and an oblique angle θ, showing the o-mode, the x-mode and the z-mode; (b) the squares of the refractive indices on a different scale (for $\theta = 0$) showing the whistler mode in the upper left.

respectively. The two lower frequency branches are referred to as the z-mode and the whistler mode. Both these wave modes have *resonances*, corresponding to $n^2 \to \infty$, at ω_p and Ω_e, respectively, for $\omega_p > \Omega_e$. (Note that "resonance" is used in a different sense in lecture 2.) Near their resonance, waves in the z-mode may be regarded as magnetized forms of Langmuir waves; thermal effects then need to be included in a kinetic theory treatment. The role of whistler waves is discussed in detail in lecture 5.

The Hydromagnetic waves

The hydromagnetic waves in a collisionless plasma are closely analogous to the MHD (magnetohydrodynamic) waves in a magnetized, compressible fluid, as discussed in the lectures by ER Priest. There are two characteristic speeds in compressible MHD: the (isentropic) sound speed, c_s, and the Alfvén speed, $v_A = B/(\mu_0 \eta)^{1/2}$, where η is the mass density. In MHD theory, there are three modes: the Alfvén mode and the fast and slow magnetoacoustic waves. Alfvén waves are torsional waves in the magnetic field, and the other two modes are modified magnetic sound and gas sound waves. For hydromagnetic waves in a collisionless plasma, the sound speed, c_s, in MHD theory is replaced by the ion sound speed, v_s, as introduced above. Here only the Alfvén and magnetoacoustic modes for $v_A \gg v_s$ are discussed.

The Alfvén mode : At low frequencies, $\omega \ll \Omega_i$, Alfvén waves (label A) have dispersion relation

$$\omega_{\mathrm{A}}(\boldsymbol{k}) = kv_{\mathrm{A}}|\cos\theta|, \quad v_{\mathrm{A}} = \frac{(\Omega_{\mathrm{e}}\Omega_{\mathrm{i}})^{1/2}}{\omega_{\mathrm{p}}}c, \quad \boldsymbol{e}_{\mathrm{A}} = (1,0,0), \tag{1.31}$$

where θ is the angle between $\boldsymbol{k}$ and $\boldsymbol{B}$. The Alfvén mode has a resonance at $\omega = \Omega_{\mathrm{i}}$, and does not exist above this frequency. They become increasingly elliptically polarized as $\omega/\Omega_{\mathrm{i}}$ increases, approaching circular polarizations for $\omega \to \Omega_{\mathrm{i}}$, with the same handedness (left) as spiraling ions. The polarization is also approximately circular for $\theta \ll \omega/\Omega_{\mathrm{i}} \ll 1$.

The magnetoacoustic mode : The magnetoacoustic mode, for $\omega \ll \Omega_{\mathrm{i}}$, has dispersion relation

$$\omega_{\mathrm{m}}(\boldsymbol{k}) = kv_{\mathrm{A}}, \quad \boldsymbol{e}_{\mathrm{m}} = (0,i,0). \tag{1.32}$$

The dispersion relation becomes modified for $\omega \approx \Omega_{\mathrm{i}}$, where the magnetoacoustic mode for $\omega \ll \Omega_{\mathrm{i}}$ becomes the whistler mode for $\Omega_{\mathrm{i}} < \omega < \Omega_{\mathrm{e}}$. The polarization is approximately circular, and opposite to that of the Alfvén mode, for $\omega \approx \Omega_{\mathrm{i}}$, and for $\theta \ll \omega/\Omega_{\mathrm{i}} \ll 1$.

1.4 Distributions of Waves

In describing the level of excitation of waves in a specific wave mode, it is convenient to adopt a quantum mechanical notation. This involves regarding the waves as a collection of wave quanta. Each quantum has energy $\hbar\omega$ and momentum $\hbar\boldsymbol{k}$.

The level of excitation of the waves may then be described in terms of the occupation number, denoted $N_M(\boldsymbol{k})$, for wave quanta in the mode M with dispersion relation $\omega = \omega_M(\boldsymbol{k})$. Classically, $\hbar N_M(\boldsymbol{k})$ is the wave action, and $\hbar\omega_M(\boldsymbol{k})N_M(\boldsymbol{k})$ is the wave energy density, for waves in the range $d^3\boldsymbol{k}/(2\pi)^3$. Thus the wave energy density in the mode M, W_M, is related to the occupation number by

$$W_M = \int \frac{d^3\boldsymbol{k}}{(2\pi)^3}\, T_M(\boldsymbol{k}), \tag{1.33}$$

where

$$T_M(\boldsymbol{k}) = \hbar\omega_M(\boldsymbol{k})N_M(\boldsymbol{k}) \tag{1.34}$$

is the *effective temperature* in the waves (in the Rayleigh-Jean's limit). For convenience, Boltzmann's constant is set equal to unity, so that temperatures are measured in energy units. Useful numbers relating energy units to kelvins are $1\,\mathrm{eV} = 1.16 \times 10^4\,\mathrm{K}$, $m_{\mathrm{e}}c^2 = 511\,\mathrm{keV} = 5.9 \times 10^9\,\mathrm{K}$, and $1\,\mathrm{J} = 7.24 \times 10^{22}\,\mathrm{K}$.

Low frequency waves in plasmas are often highly excited compared with the thermal level $T_M(\boldsymbol{k}) = T_{\mathrm{e}}$, where T_{e} is the electron temperature. For photons at $\omega \gg \omega_{\mathrm{p}}$, one refers to thermal emission for $T(\boldsymbol{k}) = T_{\mathrm{e}}$, and nonthermal emission for $T(\boldsymbol{k}) \gg T_{\mathrm{e}}$. Quantum effects need to be taken into account for photons with energy $\hbar\omega \gtrsim T_{\mathrm{e}}$, when one needs to use the Planck distribution

$$N_M(\boldsymbol{k}) = \frac{1}{\exp[\hbar\omega_M(\boldsymbol{k})/T_e] - 1}. \tag{1.35}$$

In hard X-ray sources, where both the photon energies and the electron temperatures are $\gtrsim 10\,\mathrm{keV}$, (1.35) implies that the thermal spectrum the (classical) Rayleigh-Jeans spectrum, $N_M(\boldsymbol{k}) \approx T_e/\hbar\omega_M(\boldsymbol{k})$ for $\hbar\omega_M(\boldsymbol{k}) \ll T_e$, passes over into the Wien spectrum, $N_M(\boldsymbol{k}) \approx \exp[-\hbar\omega_M(\boldsymbol{k})/T_e]$ for $\hbar\omega_M(\boldsymbol{k}) \gg T_e$. Even for a very modest level of excitation of low-frequency waves, in the sense that the wave energy density is very low compared with the thermal energy density in the particles, the effective temperature can be very large in the sense $T_M(\boldsymbol{k}) \gg T_e$.

2 Wave-Particle Resonance

The interaction between a distribution of particles and a distribution of waves may be separated into two parts: the "sloshing about" of particles in the waves, and resonant wave-particle interactions. Wave-particle resonance allows the emission of radiation, the growth or damping of the waves, and the scattering and acceleration of particles. These processes may be described using the set of kinetic equations called the quasilinear equations. General forms of the quasilinear equations may be derived in various ways (e.g., Hall and Sturrock 1967; Hasselmann and Wibberenz 1968); here they are derived using the Einstein coefficients (e.g., Tsytovich 1966; Kaplan and Tsytovich 1973; Melrose 1968, 1980a, 1986).

2.1 Resonance in Unmagnetized Plasma

The resonance condition for a particle with velocity $\boldsymbol{v}$ in an unmagnetized plasma follows by considering the phase $\omega t - \boldsymbol{k}\cdot\boldsymbol{x}$, cf. (1.9), at the position of the particle, which we take to be $\boldsymbol{x} = \boldsymbol{x}_0 + \boldsymbol{v}t$, where $\boldsymbol{x}_0$ is the position of the particle at $t = 0$. The resonance condition corresponds to a time-independent phase:

$$\omega - \boldsymbol{k}\cdot\boldsymbol{v} = 0, \tag{2.1}$$

which is called the *Cerenkov condition*. This condition corresponds to a resonance in the sense that when (2.1) is satisfied the particle sees the electric field of the wave as a static electric field in its (the particle's) rest frame. To show this, consider a wave that has frequency ω and wave vector $\boldsymbol{k}$ in the laboratory frame. By applying a Lorentz transformation from the laboratory frame to the rest frame of the particle one obtains the corresponding quantities ω', $\boldsymbol{k}'$ that describe the wave in the rest frame. One finds

$$\omega' = \gamma(\omega - \boldsymbol{k}\cdot\boldsymbol{v}), \tag{2.2}$$

where γ is the Lorentz factor of the particle. Resonance, in the sense that the fields appear static in the rest frame of the particle, corresponds to $\omega' = 0$, implying (2.1) in the laboratory frame.

The Cerenkov Cone

The Cerenkov condition can be satisfied only for waves with phase speed less than the speed of light, which corresponds to a refractive index, n, greater than unity. This is seen by writing the Cerenkov condition (2.1) in the form

$$1 - n\boldsymbol{\kappa} \cdot \boldsymbol{\beta} = 1 - n\beta\cos\chi = 0, \tag{2.3}$$

with $\boldsymbol{\beta} = \boldsymbol{v}/c$, and where χ is the angle between $\boldsymbol{k}$ and $\boldsymbol{v}$. It follows from $\beta < 1$ and $\cos\chi \le 1$ that (2.3) can be satisfied only for $n \ge 1/\beta > 1$.

Cerenkov emission occurs only when the resonance condition (2.1) is satisfied, and hence is possible only for particles with $\beta \ge 1/n$. Emission at the threshold $\beta = 1/n$ is at $\chi = 0$ and the angle χ increases with the speed of particles to a maximum value $\chi_{\rm max} = \arccos(1/n)$ for highly relativistic particles. It follows that Cerenkov emission is confined to the surface of a cone, as illustrated in Figure 2.1a, with the cone half-angle χ determined by the parameter $n\beta$, which is the ratio of the speed of the particle to the phase speed of the wave. The Cerenkov emission pattern may also be viewed as analogous to the bow wave of a ship or a supersonic aircraft, as illustrated in Figure 2.1b. The half-angle of the bow wave is $\theta' = \pi/2 - \chi = \arcsin(1/\beta n)$.

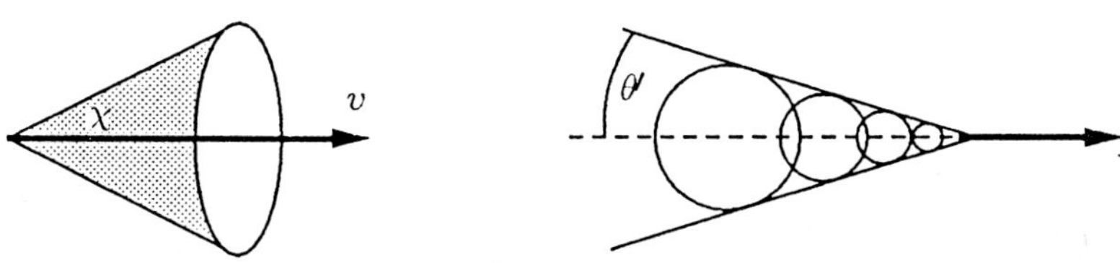

(a) (b)

Fig. 2.1. (a) The Cerenkov cone half angle χ is between the direction of motion of the particle and the wave normal direction. (b) Cerenkov emission forms a bow shock with half-angle $\theta' = \pi/2 - \chi$. Each circle represents the wavefront, at a given time, due to emission by the particle at an earlier time when it was located at the center of the circle.

Semiclassical Interpretation

A different interpretation of the Cerenkov condition emerges from a quantum mechanical viewpoint. In this case the Cerenkov condition is interpreted as a consequence of conservation of energy and momentum on a microscopic scale.

The semiclassical interpretation of emission is as a process in which a particle emits a quantum of energy. This wave quantum has energy $\hbar\omega$ and momentum $\hbar\boldsymbol{k}$. Let the initial energy and momentum of the particle be ε, $\boldsymbol{p}$ and let the corresponding quantities after emission of the wave quantum be ε', $\boldsymbol{p}'$. Conservation of energy and momentum then requires

$$\varepsilon' = \varepsilon - \hbar\omega, \quad \boldsymbol{p}' = \boldsymbol{p} - \hbar\boldsymbol{k}. \tag{2.4}$$

The relativistically correct formula for the energy is

$$\varepsilon = \varepsilon(\boldsymbol{p}) = \left(m^2c^4 + \boldsymbol{p}^2c^2\right)^{1/2}, \quad \varepsilon' = \varepsilon(\boldsymbol{p}') = \left(m^2c^4 + \boldsymbol{p}'^2c^2\right)^{1/2}. \tag{2.5}$$

On inserting $\boldsymbol{p}' = \boldsymbol{p} - \hbar\boldsymbol{k}$ into the expression for ε' in (2.5) to obtain the classical limit one expands in powers of $\hbar$. This may be achieved by introducing a differential operator and making a Taylor series expansion:

$$\varepsilon - \hbar\omega = [1 - \hat{D} + \tfrac{1}{2}\hat{D}^2 + \cdots]\varepsilon(\boldsymbol{p}), \quad \hat{D} = \hbar\boldsymbol{k}\cdot\frac{\partial}{\partial\boldsymbol{p}}. \tag{2.6}$$

This gives

$$\omega = \boldsymbol{k}\cdot\boldsymbol{v} - \frac{\hbar}{2\varepsilon}\left[k^2c^2 - (\boldsymbol{k}\cdot\boldsymbol{v})^2\right] + \cdots. \tag{2.7}$$

The lowest order terms reproduce the Cerenkov condition (2.1). Thus the Cerenkov condition is interpreted as the requirement that energy conservation be compatible with momentum conservation on a microscopic level. It is only when both energy and momentum are conserved that the particle can emit a wave quantum.

The next order term in $\hbar$ in (2.6) is interpreted in terms of the quantum recoil of the emitting particle due to the emission of the wave quantum. This term allows one to take into account the back reaction of the emission (or absorption) on the energy and momentum of the particle. An unsatisfactory feature of classical electromagnetism is that it does not conserve energy and momentum. This weakness is patched up for emission in vacuo by inventing a radiation reaction force. However, the radiation reaction depends on the properties of the medium in which the emission occurs, and conservation of energy and momentum needs to be taken into account in other ways in a medium. Moreover, the concept of radiation reaction applies to spontaneous emission, and it cannot be used in its simplest form when considering absorption or maser emission. Conceptually, the appeal to quantum ideas, as in (2.4)–(2.5), is the simplest and most direct way of imposing energy and momentum conservation.

Bump-in-tail Instability

Particles close to resonance with a wave tend to be dragged into resonance, like a surfer catching a wave. Hence if the particle is slightly slower than the wave it tends to be accelerated, gaining energy from the wave. On the other hand, a particle that is slightly faster than the wave tends to be slowed down and so to give energy to the wave. It follows that if the number of particles is a decreasing function of velocity at the phase velocity, $\boldsymbol{v}_\phi = \omega\boldsymbol{k}/k^2$ of the wave, then the resonant particles extract energy from the wave. This leads to a form of collisionless damping of the wave, called *Landau damping*. Landau damping is the dominant damping mechanism for waves (such as Langmuir waves and ion sound waves) in an unmagnetized, thermal plasma. Landau damping may be interpreted as the absorption process corresponding to Cerenkov emission. That is, Landau damping of waves may be attributed to absorption by thermal particles due to the inverse of Cerenkov emission.

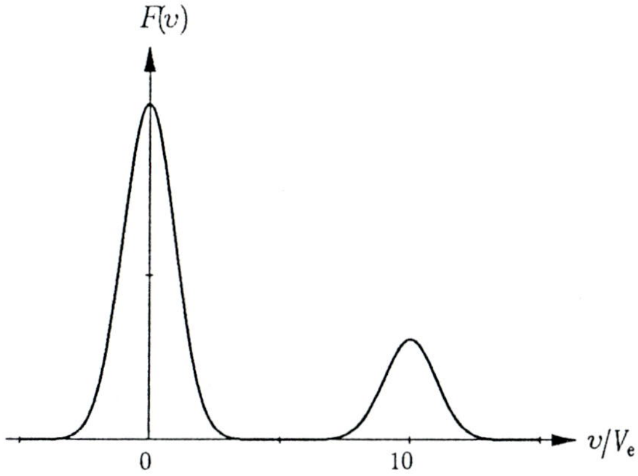

Fig. 2.2. The large peak corresponds to a thermal distribution of particles, which has a negative gradient in v so that the resonance interaction leads to Landau damping of waves. The small peak corresponds to a bump-in-tail distribution which has a range with $dF(v)/dv > 0$. In the bump-in-tail instability, waves with $v_\phi = v$ grow for $dF(v)/dv > 0$.

On the other hand, if the number of particles increases with increasing velocity at the resonant velocity, then the energy flow is from the particles to the waves and the waves grow. This is a particular example of a microinstability, usually called the bump-on-tail instability. Due to this type of instability, a beam of electrons with speed much greater than the thermal speed of electrons in the plasma causes Langmuir waves (with $\omega \approx \omega_p$ and $k \approx \omega_p/v$) to grow. Such a distribution function is illustrated in Figure 2.2. The one-dimensional velocity distribution $F(v)$ is obtained from $f(\boldsymbol{p})$ and by integrating over the momentum components perpendicular to the streaming direction making the nonrelativistic approximation $\boldsymbol{p} = m\boldsymbol{v}$.

Langmuir waves generated through such a streaming instability play an essential role in the theory of solar radio bursts, as discussed in lecture 8 below.

2.2 Resonance in Magnetized Plasma

The resonance condition in the presence of a magnetic field is the Doppler condition

$$\omega - s\Omega - k_\parallel v_\parallel = 0, \tag{2.8}$$

where s is an integer, $\Omega = |q|B/m\gamma$ is the relativistic gyrofrequency, and $k_\parallel$, $v_\parallel$ are the components of $\boldsymbol{k}$, $\boldsymbol{v}$ parallel to $\boldsymbol{B}$. Resonances at $s > 0$ are said to be via the *normal Doppler effect*, and those at $s < 0$ are said to be via the *anomalous Doppler effect*. Resonances at $s \leq 0$ are possible only for waves with refractive index greater than unity.

From a classical viewpoint, (2.8) corresponds to the wave frequency being an integral multiple of the gyrofrequency in the inertial frame in which the gyrocenter of the particle is at rest. In this inertial frame, the particle has a circular orbit, and each time it completes an orbit it encounters the wave with the same phase. As a result, the effect of the acceleration in successive orbits is cumulative, and does not average to zero over many orbits.

Semiclassical Interpretation

In a quantum mechanical treatment the energy eigenvalues of a relativistic particle (ignoring its spin) are

$$\varepsilon = \varepsilon_n(p_\parallel) = (m^2c^4 + p_\parallel^2c^2 + 2n|q|B\hbar c^2)^{1/2}. \tag{2.9}$$

The quantum numbers are the continuous parameter $p_\parallel$ and the discrete parameter $n = 0, 1, 2, \ldots$. The classical limit corresponds to $\hbar \to 0$, $n \to \infty$, with $\hbar n \to p_\perp^2/2|q|B$.

When a particle emits a wave quantum in a magnetic field, momentum perpendicular to the magnetic field is not conserved in general. Let $p'_\parallel$ and n' be the quantum numbers after emission of the wave quantum. Conservation of parallel momentum and of energy require

$$p'_\parallel = p_\parallel - \hbar k_\parallel, \qquad \varepsilon' = \varepsilon_{n'}(p'_\parallel) = \varepsilon_n(p_\parallel) - \hbar\omega. \tag{2.10}$$

Now write $n' = n - s$. On taking the classical limit $\hbar \to 0$, the fractional changes in $p_\parallel$, ε, n become infinitesimal. Then an expansion in the small quantities gives

$$\varepsilon - \hbar\omega = \left(1 - \hat{D}_s + \tfrac{1}{2}\hat{D}_s^2 + \cdots\right)\varepsilon(p_\parallel, p_\perp), \qquad \hat{D}_s = \hbar\left(\frac{s\Omega}{v_\perp}\frac{\partial}{\partial p_\perp} + k_\parallel\frac{\partial}{\partial p_\parallel}\right), \tag{2.11}$$

with change of notation $\varepsilon_n(p_\parallel) \to \varepsilon(p_\parallel, p_\perp) = (m^2c^4 + p_\parallel^2c^2 + p_\perp^2c^2)^{1/2}$ in the classical limit. To lowest order in $\hbar$, (2.11) reduces to (2.8). The term of order $\hbar^2$, that is, the term $\frac{1}{2}\hat{D}_s^2$, leads to a correction to the resonance condition associated with the quantum recoil of the particle.

Resonance Ellipse

The resonance condition (2.8) is amenable to a graphical interpretation. It is convenient to plot the resonance curve in $v_\perp$–$v_\parallel$ space for given ω and $k_\parallel$. The resonance condition for each harmonic defines a *resonance ellipse*. The resonance ellipse corresponds to all the values of $v_\perp$ and $v_\parallel$ for which resonance with a wave at given ω, $k_\parallel$ and s is possible. That is, a given wave resonates with all particles that lie on the resonance ellipse that it defines. Similarly, a given particle resonates with all waves that define resonance ellipses which pass through the representative point of the particle in $v_\perp$–$v_\parallel$ space.

The resonance ellipse is centered on the $v_\parallel$ axis at $v_\parallel = v_c$, with semi-major axis v_R perpendicular to the $v_\parallel$ axis, and with eccentricity e. These parameters are given by

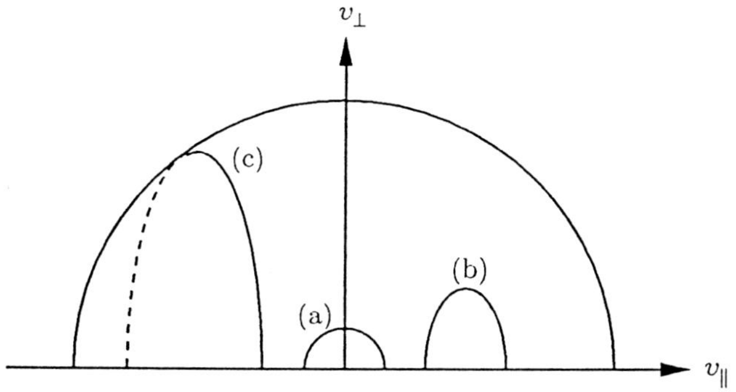

Fig. 2.3. Examples of resonance ellipses: (a) a semicircle centered on the origin, (b) an ellipse inside $v = c$, (c) an ellipse touching $v = c$.

$$\frac{v_c}{c} = \frac{\omega k_\parallel c}{s^2\Omega_0^2 + k_\parallel^2 c^2}, \quad \left(\frac{v_R}{c}\right)^2 = \frac{s^2\Omega_0^2 + k_\parallel^2 c^2 - \omega^2}{s^2\Omega_0^2 + k_\parallel^2 c^2}, \quad e^2 = \frac{k_\parallel^2 c^2}{s^2\Omega_0^2 + k_\parallel^2 c^2}. \tag{2.12}$$

Some examples of resonance ellipses are illustrated in Figure 2.3: (a) corresponds to a resonance with a perpendicularly propagating wave ($k_\parallel = 0$), (b) corresponds to a resonance for a nearly perpendicularly propagating wave ($k_\parallel \ll \omega/c$), and (c) corresponds to a resonance with large $k_\parallel$, in which case the resonance ellipse may touch the unit circle with the dashed part of the ellipse nonphysical.

2.3 The Probability of Spontaneous Emission

When either the resonance conditions (2.1) or (2.8) is satisfied, the particles emit or absorb the waves. A convenient general way of describing the emission and absorption is in terms of the probability of spontaneous emission of a wave quantum in a specific wave mode. Detailed derivations of this probability are described elsewhere (Melrose 1980a, 1986; Melrose and McPhedran 1991), and are not repeated here.

The probability is per unit time and per unit volume of $\boldsymbol{k}$-space (element $d^3\boldsymbol{k}/(2\pi)^3$) for spontaneous emission by the particle of a wave quantum in the wave mode M, with energy $\hbar\omega_M(\boldsymbol{k})$ and momentum $\hbar\boldsymbol{k}$. For an unmagnetized particle of charge q, mass m and momentum $\boldsymbol{p}$, a detailed calculation gives

$$w_M(\boldsymbol{p}, \boldsymbol{k}) = \frac{2\pi q^2 R_M(\boldsymbol{k})}{\varepsilon_0 \hbar \omega_M(\boldsymbol{k})} \left|\boldsymbol{e}_M(\boldsymbol{k}) \cdot \boldsymbol{v}\right|^2 \delta\big(\omega_M(\boldsymbol{k}) - \boldsymbol{k} \cdot \boldsymbol{v}\big). \tag{2.13}$$

This probability contains the resonance condition (2.1) through the δ-function.

The probability of emission at the sth harmonic for a particle with momentum components $p_\parallel$ and $p_\perp$ in a uniform magnetic field $\boldsymbol{B}$ is

$$w_M(\boldsymbol{p},\boldsymbol{k},s)=\frac{2\pi q^2R_M(\boldsymbol{k})}{\varepsilon_0\hbar\omega_M(\boldsymbol{k})}\left|\boldsymbol{e}_M^*(\boldsymbol{k})\cdot\boldsymbol{V}(\boldsymbol{k},\boldsymbol{p};s)\right|^2\delta\big(\omega_M(\boldsymbol{k})-s\Omega-k_\parallel v_\parallel\big),$$
$$\boldsymbol{V}(\boldsymbol{k},\boldsymbol{p};s)=\Big(v_\perp\frac{s}{k_\perp R}J_s(k_\perp R),-i\eta v_\perp J_s'(k_\perp R),v_\parallel J_s(k_\perp R)\Big),$$
$$\Omega=\frac{\Omega_0}{\gamma},\quad \Omega_0=\frac{|q|B}{m},\quad \eta=\frac{q}{|q|},\quad R=\frac{v_\perp}{\Omega}=\frac{p_\perp}{|q|B}. \tag{2.14}$$

This probability contains the Doppler condition (2.8) through the δ-function.

The Einstein Coefficients

The emission and absorption may be treated in terms of the probability of spontaneous emission by appealing to the Einstein coefficients. The Einstein coefficients relate spontaneous emission, stimulated emission and true absorption on a microscopic level.

Consider a quantum description of the states of the particle, as used in (2.9). Let the state of the particle before emission of the wave quantum have a set of quantum numbers $\{q\}$ (with $\{q\}=\{p_\parallel,n\}$ for a magnetized particle), which changes to $\{q'\}$ after emission of the wave quantum, as in (2.10). Suppose that the quantum mechanical occupation numbers of these states are n_q and $n_{q'}$, respectively. Let $w_{qq'}(\boldsymbol{k})$ be the probability of spontaneous emission $\{q\}\to\{q'\}$ for a single particle. Then, according to the Einstein coefficients, the probabilities of spontaneous emission $\{q\}\to\{q'\}$, stimulated emission $\{q\}\to\{q'\}$ and of true absorption $\{q'\}\to\{q\}$ for the distribution of particles are

$$w_{qq'}(\boldsymbol{k})n_q,\quad w_{qq'}(\boldsymbol{k})n_qN_M(\boldsymbol{k}),\quad w_{qq'}(\boldsymbol{k})n_q'N_M(\boldsymbol{k}),$$

respectively, where $N_M(\boldsymbol{k})$ is the occupation number of the wave quanta.

Kinetic Equations

The rate of change of the occupation number for the wave quanta may be found by adding the contributions from spontaneous and stimulated emission, and subtracting the contribution from true absorption. This gives

$$\frac{dN_M(\boldsymbol{k})}{dt}=\sum_{q,q'}w_{qq'}(\boldsymbol{k})\left\{n_q[1+N_M(\boldsymbol{k})]-n_{q'}N_M(\boldsymbol{k})\right\}. \tag{2.15}$$

The rate of change of the occupation number n_q for the particles may be found by adding the rates of increase due to absorption $\{q'\}\to\{q\}$ from lower energy states and emission $\{q''\}\to\{q\}$ from higher energy states, and subtracting the rates of decrease due to the inverse processes. This gives

$$\frac{dn_q}{dt}=\int\frac{d^3\boldsymbol{k}}{(2\pi)^3}\Big(\sum_{q''}w_{q''q}(\boldsymbol{k})\left\{n_{q''}[1+N_M(\boldsymbol{k})]-n_qN_M(\boldsymbol{k})\right\}$$
$$-\sum_{q'}w_{qq'}(\boldsymbol{k})\left\{n_q[1+N_M(\boldsymbol{k})]-n_{q'}N_M(\boldsymbol{k})\right\}\Big). \tag{2.16}$$

2.4 The Quasilinear Equations

The quasilinear equations are obtained by rewriting (2.15) and (2.16) in a more appropriate notation. This involves replacing the occupation number n_q for the particles by the classical distribution function, and replacing the sum over quantum states by an integral over momentum space.

The Absorption Coefficient

The kinetic equation (2.15) for the waves may be written in the form

$$\frac{dN_M(\boldsymbol{k})}{dt} = \alpha_M(\boldsymbol{k}) - \gamma_M(\boldsymbol{k})N_M(\boldsymbol{k}), \tag{2.17}$$

where the $\alpha_M(\boldsymbol{k})$ is an emission coefficient that describes the effect of spontaneous emission, and the other term defines the absorption coefficient, $\gamma_M(\boldsymbol{k})$. An explicit value for the absorption coefficient follows by rewriting the relevant terms in (2.15) and using the differential operator, $\hat{D}$, given by (2.6) in the unmagnetized case, to take the classical limit. This gives

$$\gamma_M(\boldsymbol{k}) = -\int d^3\boldsymbol{p}\, w_M(\boldsymbol{p},\boldsymbol{k})\hat{D}f(\boldsymbol{p}), \quad \hat{D} = \hbar\boldsymbol{k}\cdot\frac{\partial}{\partial\boldsymbol{p}}. \tag{2.18}$$

For most emission processes, it is possible for the absorption coefficient to be positive or negative. Negative absorption corresponds to maser action. The bump-in-tail instability discussed above is an example of maser action.

In the magnetized case, (2.18) is replaced by

$$\gamma_M(\boldsymbol{k}) = -\sum_{s=-\infty}^{\infty}\int d^3\boldsymbol{p}\, w_M(\boldsymbol{p},\boldsymbol{k},s)\hat{D}_s f(\boldsymbol{p}),$$
$$\hat{D}_s = \hbar\left(\frac{s\Omega}{v_\perp}\frac{\partial}{\partial p_\perp} + k_\parallel\frac{\partial}{\partial p_\parallel}\right). \tag{2.19}$$

The differential operator may be rewritten in terms of spherical polar coordinate, p, α in momentum space, with $p_\parallel = p\cos\alpha$, $p_\perp = p\sin\alpha$:

$$\hat{D}_s = \frac{\hbar\omega}{v}\left(\frac{\partial}{\partial p} + \frac{\cos\alpha - n_M\beta\cos\theta}{p\sin\alpha}\frac{\partial}{\partial\alpha}\right). \tag{2.20}$$

A simple way of deriving the coefficients in (2.20) is to note that one has $\hat{D}_s = (\Delta p)\partial/\partial p + (\Delta\alpha)\partial/\partial\alpha$, with $\Delta p = \hat{D}_s p$, $\Delta\alpha = \hat{D}_s\alpha$.

Diffusion in Momentum Space for Unmagnetized Particles

The quasilinear equation for unmagnetized particles follows from (2.16) with $q \to \boldsymbol{p}$, $q' \to \boldsymbol{p} - \hbar\boldsymbol{k}$, $q'' \to \boldsymbol{p} + \hbar\boldsymbol{k}$, and with the probability given by (2.13). Hence (2.16) reduces to

$$\frac{df(\boldsymbol{p})}{d\boldsymbol{p}} = \frac{\partial}{\partial p_i}\left[A_i(\boldsymbol{p})f(\boldsymbol{p}) + D_{ij}(\boldsymbol{p})\frac{\partial f(\boldsymbol{p})}{\partial p_i}\right], \tag{2.21}$$

with

$$\begin{pmatrix} A_i(\boldsymbol{p}) \\ D_{ij}(\boldsymbol{p}) \end{pmatrix} = \int \frac{d^3\boldsymbol{k}}{(2\pi)^3}\, w_M(\boldsymbol{p},\boldsymbol{k}) \begin{pmatrix} \hbar k_i \\ \hbar^2 k_i k_j N_M(\boldsymbol{k}) \end{pmatrix}. \tag{2.22}$$

The term involving A_i describes the back reaction of the particles to the spontaneous emission. This term necessarily implies a net flow of energy from the particles to the waves, causing the particles to move systematically to smaller p. The other term describes the net effect of the induced processes of stimulated emission and true absorption. This term corresponds to a diffusion in momentum space. In general, the effect of such diffusion on a group of particles initially localized around some $\boldsymbol{p}$ is twofold: the mean $\boldsymbol{p}$ of these particles changes systematically, and the spread in momentum of these particles increases systematically. If absorption dominates over stimulated emission, then the mean p increases, and if these particles are involved in maser action then their mean p decreases.

Diffusion in Momentum Space for Magnetized Particles

The quasilinear diffusion equation for magnetized particles follows from (2.16) in a similar way, with the probability now given by (2.14). One finds that the particle distribution function evolves according to

$$\begin{aligned} \frac{df(\boldsymbol{p})}{dt} &= \frac{1}{p_\perp}\frac{\partial}{\partial p_\perp}\left[p_\perp A_\perp(\boldsymbol{p})f(\boldsymbol{p})\right] + \frac{\partial}{\partial p_\parallel}\left[A_\parallel(\boldsymbol{p})f(\boldsymbol{p})\right] \\ &+ \frac{1}{p_\perp}\frac{\partial}{\partial p_\perp}\left\{p_\perp\left[D_{\perp\perp}(\boldsymbol{p})\frac{\partial}{\partial p_\perp} + D_{\perp\parallel}(\boldsymbol{p})\frac{\partial}{\partial p_\parallel}\right]f(\boldsymbol{p})\right\} \\ &+ \frac{\partial}{\partial p_\parallel}\left\{\left[D_{\parallel\perp}(\boldsymbol{p})\frac{\partial}{\partial p_\perp} + D_{\parallel\parallel}(\boldsymbol{p})\frac{\partial}{\partial p_\parallel}\right]f(\boldsymbol{p})\right\}. \end{aligned} \tag{2.23}$$

The form (2.23) is the standard form for evolution of a distribution function in cylindrical polar coordinates in momentum space. The corresponding result for spherical polar coordinates in momentum space is

$$\begin{aligned} \frac{df(\boldsymbol{p})}{dt} &= \frac{1}{p^2}\frac{\partial}{\partial p}\left[p^2 A_p(\boldsymbol{p})f(\boldsymbol{p})\right] + \frac{1}{\sin\alpha}\frac{\partial}{\partial\alpha}\left[\sin\alpha\, A_\alpha(\boldsymbol{p})f(\boldsymbol{p})\right] \\ &+ \frac{1}{\sin\alpha}\frac{\partial}{\partial\alpha}\left\{\sin\alpha\left[D_{\alpha\alpha}(\boldsymbol{p})\frac{\partial}{\partial\alpha} + D_{\alpha p}(\boldsymbol{p})\frac{\partial}{\partial p}\right]f(\boldsymbol{p})\right\} \\ &+ \frac{1}{p^2}\frac{\partial}{\partial p}\left\{p^2\left[D_{p\alpha}(\boldsymbol{p})\frac{\partial}{\partial\alpha} + D_{pp}(\boldsymbol{p})\frac{\partial}{\partial p}\right]f(\boldsymbol{p})\right\}. \end{aligned} \tag{2.24}$$

Explicit expressions for the coefficients, A_Q, that describe the changes due to spontaneous emission, and the diffusion coefficients, $D_{QQ'}$, that describe the effect of the induced processes, are given by

$$A_Q(\boldsymbol{p}) = \sum_{s=-\infty}^{\infty} \int \frac{d^3\boldsymbol{k}}{(2\pi)^3}\, w_M(\boldsymbol{p},\boldsymbol{k},s)\,\Delta Q, \qquad \Delta Q = \hat{D}_s Q,$$

$$D_{QQ'}(\boldsymbol{p}) = \sum_{s=-\infty}^{\infty} \int \frac{d^3\boldsymbol{k}}{(2\pi)^3}\, w_M(\boldsymbol{p},\boldsymbol{k},s)\,\Delta Q\,\Delta Q'\,N_M(\boldsymbol{k}), \tag{2.25}$$

with Q, Q' identified as $p_\perp$, $p_\parallel$, p, or α. One finds

$$\Delta p_\perp = \frac{s\Omega}{v_\perp}, \quad \Delta p_\parallel = \hbar k_\parallel, \quad \Delta\alpha = \frac{\hbar(\omega\cos\alpha - k_\parallel v)}{pv\sin\alpha}, \quad \Delta p = \frac{\hbar\omega}{v}. \tag{2.26}$$

In the lectures by JG Kirk, the variable $\mu = \cos\alpha$ is used rather than α. In this case (2.24) is replaced by

$$\begin{aligned}\frac{df(\boldsymbol{p})}{dt} &= \frac{1}{p^2}\frac{\partial}{\partial p}\left[p^2 A_p(\boldsymbol{p}) f(\boldsymbol{p})\right] + \frac{\partial}{\partial\mu}\left[A_\mu(\boldsymbol{p}) f(\boldsymbol{p})\right] \\ &+ \frac{\partial}{\partial\mu}\left\{\left[D_{\mu\mu}(\boldsymbol{p})\frac{\partial}{\partial\mu} + D_{\mu p}(\boldsymbol{p})\frac{\partial}{\partial p}\right] f(\boldsymbol{p})\right\} \\ &+ \frac{1}{p^2}\frac{\partial}{\partial p}\left\{p^2\left[D_{p\mu}(\boldsymbol{p})\frac{\partial}{\partial\mu} + D_{pp}(\boldsymbol{p})\frac{\partial}{\partial p}\right] f(\boldsymbol{p})\right\},\end{aligned} \tag{2.27}$$

with, in (2.25),

$$\Delta\mu = -\frac{\hbar(\omega\cos\alpha - k_\parallel v)}{pv}. \tag{2.28}$$

3 Resonant Scattering

Resonant scattering causes galactic cosmic rays to diffuse through the interstellar plasma and solar cosmic rays to diffuse through the solar wind. It also limits the trapping of particles in the Earth's radiation belts, and plays an essential role in both diffusive shock acceleration and stochastic or Fermi acceleration. A feature of resonant scattering is that the particles themselves may generate the waves that cause their scattering.

3.1 Pitch-Angle Scattering

During resonant interactions between waves and particles, energy and momentum are transferred between the waves and the particles. In a magnetized plasma, provided that the waves are of sufficiently low frequency, significant momentum transfer can occur with negligible energy transfer. This limit corresponds to pitch-angle scattering.

Low-frequency Limit

In the limit of arbitrarily low frequencies,the wave frequency is much less than the cyclotron frequency, and the resonance condition (2.8) reduces to

$$s\Omega \approx -k_{\|}v_{\|}, \quad |k_{\|}v_{\|}| \gg \omega. \tag{3.1}$$

The conditions (3.1) requires $|v_{\|}| \gg \omega/|k_{\|}|$, or $v \gg v_{\phi}$, where $v_{\phi} = \omega/k$ is the phase speed of the wave.

There are threshold speeds that must be exceeded before resonant interactions between ions or electrons and low-frequency waves are possible. For ions to resonate with Alfvén or magnetoacoustic waves, with dispersion relation $\omega \approx kv_{\rm A}$, the inequality $|k_{\|}v_{\|}| \gg \omega$ requires $v \gg v_{\rm A}$, where $v_{\phi} = v_{\rm A}$ is the phase speed. The low-frequency limit corresponds to $\omega \ll \Omega_{\rm i}$. Then, for nonrelativistic ions ($\Omega \approx \Omega_{\rm i}$) at $s = \pm 1$, the approximate (3.1) reduces to $|v_{\|}| \approx \Omega_{\rm i}/|k_{\|}|$, and hence requires $|v_{\|}| \approx \omega/|k_{\|}|$. Ignoring trigonometric factors, because hydromagnetic waves have $\omega/k \sim v_{\rm A}$, this requires $v \gg v_{\rm A}$. Hence, the resonant interaction is possible only for ions with speed greater than the Alfvén speed.

For electrons, it is convenient to write the condition (3.1) for $s = \pm 1$ in the form $eB = |k_{\|}p_{\|}|$. Then the low-frequency limit, $\omega \ll \Omega_{\rm i}$, implies that the condition for electrons to resonate with these waves is $|p_{\|}| > m_{\rm i}v_{\rm A}$. In practice this usually requires that the electrons be relativistic to resonate with hydromagnetic waves. Nonrelativistic electrons can resonate with whistlers, for which the threshold condition is $v \gg 43v_{\rm A}$, as discussed in lecture 5.

Changes During Emission and Absorption

It is straightforward to see that the low-frequency approximation corresponds to no change in p. Consider the change in $p^2 = p_{\|}^2 + p_{\perp}^2$. On emission of a wave quantum, p^2 changes by, cf. (2.26),

$$\Delta p^2 = 2p_{\|}\Delta p_{\|} + 2p_{\perp}\Delta p_{\perp} = 2\hbar p_{\|}k_{\|} + 2\hbar p_{\perp}\left(\frac{s\Omega}{v_{\perp}}\right) = 2\hbar m\gamma(k_{\|}v_{\|} + s\Omega), \tag{3.2}$$

which is negligible when (3.1) is satisfied. The change in α due to emission follows from (2.26):

$$\Delta\alpha \approx -\frac{k_{\|}v_{\|}}{p_{\perp}} \approx \frac{s\Omega}{p_{\perp}}. \tag{3.3}$$

The sign of the change is such that one has $\alpha \rightarrow \alpha - \Delta\alpha$ on emission and $\alpha \rightarrow \alpha + \Delta\alpha$ on absorption of a wave quantum. Pitch-angle scattering corresponds to this limit, in which α changes at approximately constant p.

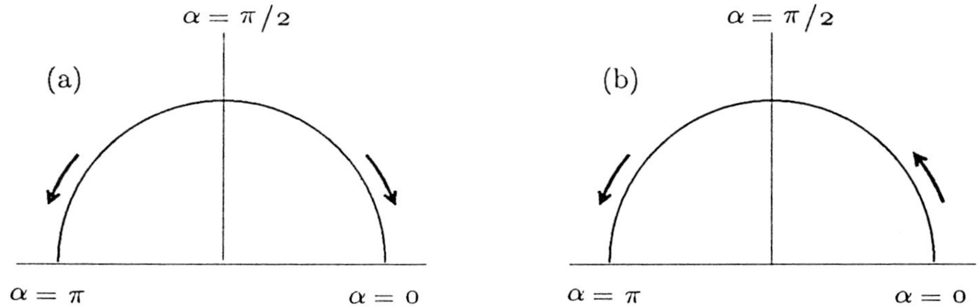

Fig. 3.1. Direction of motion around a path of constant momentum (a semicircle) in momentum space due to emission of a wave quantum. The arrows indicate the path of a particle (a) due to emission at $s > 0$, and (b) due to emission with $k_\parallel > 0$.

Diffusion at Constant p

Changes at constant p correspond to motion around a sphere in momentum space, which becomes a semicircle in $p_\perp$-$p_\parallel$ space, as illustrated in Figure 3.1. Emission can occur at $s > 0$ or $s < 0$, which cause $p_\perp$ to decrease or increase, respectively, and at $k_\parallel > 0$ or $k_\parallel < 0$, which cause $p_\parallel$ to decrease or increase, respectively. The condition (3.1) implies that for $v_\parallel > 0$, corresponding to the right hand quadrants in Figure 3.1a and 3.1b, only $s > 0$ and $k_\parallel < 0$ or $s < 0$ and $k_\parallel > 0$ are allowed. For $v_\parallel < 0$, corresponding to the left hand quadrants, only $s > 0$ and $k_\parallel > 0$ or $s < 0$ and $k_\parallel < 0$ are allowed. These properties are used below to infer the conditions under which an anisotropic distribution of particles tends to cause waves to grow.

Absorption leads to changes whose sign is opposite to those for emission. Diffusion occurs as a result of emission and absorption causing the particles to move back and forth around the semicircle $p =$ constant in Figure 3.1. There appears to be a problem in principle in crossing $\alpha = \pi/2$, because (3.1) cannot be satisfied for sufficiently small $v_\parallel$. However, in practice there is no evidence that there is a serious impediment to ions scattering across $\alpha = \pi/2$. The mechanism that allows scattering across $\alpha = \pi/2$ is not entirely clear. A plausible mechanism is that very low frequency turbulence causes compressions and rarefactions of B that act as magnetic mirrors, reflecting particles with $\alpha \approx \pi/2$.

The Diffusion Coefficients

When the changes in p are neglected, only one of the diffusive terms in (2.24) or (2.27) remains, and this corresponds to a diffusion in pitch angle α. This is called pitch-angle scattering, and may be described by, cf. (2.24),

$$\frac{df}{dt} = \frac{1}{\sin\alpha}\frac{\partial}{\partial\alpha}\left(\sin\alpha\, D_{\alpha\alpha}\frac{\partial f}{\partial\alpha}\right), \tag{3.4}$$

with $D_{\alpha\alpha}$ called the pitch-angle diffusion coefficient. Scattering tends to make the particles isotropic. The condition for a distribution of particles, $f(p, \alpha)$, to be *isotropic* is $\partial f(p, \alpha)/\partial\alpha = 0$. For $\partial f(p, \alpha)/\partial\alpha \neq 0$ the distribution is *anisotropic*.

The actual ratio of the rates of diffusion in pitch angle and in momentum may be estimated from (2.26). One has

$$D_{\alpha\alpha} : D_{pp}/p^2 = (\Delta\alpha)^2 : (\Delta p)^2/p^2. \tag{3.5}$$

Then using (2.25) one finds

$$D_{\alpha\alpha} : D_{pp}/p^2 = (\omega\cos\alpha - k_{\parallel}v)^2 : (\omega\sin\alpha)^2 \approx (k_{\parallel}v)^2 : (\omega\sin\alpha)^2, \tag{3.6}$$

where the low-frequency approximation is made. The ratio (3.6) is necessarily large because one already requires $k_{\parallel}v \gg \omega$, implying $(k_{\parallel}v\omega)^2 \gg 1$.

The terms involving $D_{\alpha p}$, $D_{p\alpha}$ and D_{pp} in (2.24) need to be included for some purposes even in the low-frequency limit. The term involving $D_{\alpha p}$ leads to the neutral streaming speed $\sim v_{\mathrm{A}}$ for galactic cosmic rays, as discussed in lecture 4. The term involving D_{pp} describes the effect of energy exchange between the particles and the waves. In principle, hydromagnetic waves may be damped by relativistic particles, leading to acceleration of the particles. Such damping can be due to two different processes. Resonances at high harmonics, $|s| \gg 1$ lead to so-called gyroresonant absorption. Astrophysical applications of gyroresonant acceleration have been considered (e.g., Lacombe 1977; Barbosa 1979). Resonance at $s = 0$ leads to a form of Fermi acceleration, or magnetic pumping, which applies to magnetoacoustic waves, but is not possible for Alfvén waves (which are non-compressive). This mechanism is discussed briefly in lecture 5.

Magnetic Fluctuations

Physically, large momentum transfer with small energy transfer occurs when the Lorentz force $q\boldsymbol{v} \times \boldsymbol{B}_{\mathrm{w}}$ due to the magnetic field in the wave is much larger than the electric force $q\boldsymbol{E}_{\mathrm{w}}$ due to the wave. The Lorentz force does no work, and only causes the direction of the momentum to change. With curl $\boldsymbol{E}_{\mathrm{w}} = -\partial\boldsymbol{B}_{\mathrm{w}}/\partial t$ implying $\boldsymbol{k} \times \boldsymbol{E}_{\mathrm{w}} = \omega\boldsymbol{B}_{\mathrm{w}}$ for a plane wave, this condition requires $kv \gg \omega$, which is essentially the same as the condition for the ω-term to be omitted from the resonance condition (3.1).

Thus pure pitch-angle scattering, as described by (3.4), corresponds to an approximation in which the spectrum of waves is replaced by a spectrum of magnetic fluctuations (e.g., Jokipii 1971).

3.2 Scattering of Ions by Hydromagnetic Waves

Resonant scattering of ions by Alfvén waves and by magnetoacoustic waves is possible provided that the threshold condition, $v > v_A$, is satisfied.

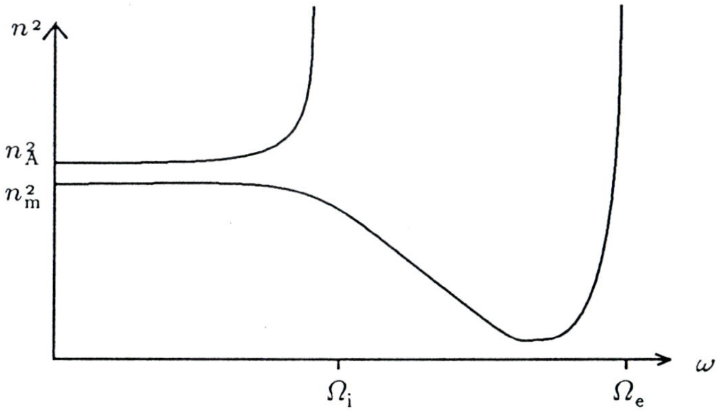

Fig. 3.2. Schematic plots showing the dependence of the square of the refractive index on frequency for the Alfvén mode (upper curve) and the magnetoacoustic mode (lower curve). The Alfvén mode only exists for $\omega < \Omega_i$, and the magnetoacoustic mode at $\omega \lesssim \Omega_i$ becomes the whistler mode for $\Omega_i \lesssim \omega < \Omega_e$.

Properties of the Hydromagnetic Waves

Relevant wave properties are for Alfvén waves (A) and magnetoacoustic waves (m) for $\omega \ll \Omega_i$ are, cf. (1.13) and (1.32),

$$n_A^2 = \frac{c^2}{v_A^2 \cos^2\theta}, \quad \boldsymbol{e}_A = (1, 0, 0), \quad R_A = \frac{v_A^2}{2c^2}; \tag{3.7}$$

$$n_m^2 = \frac{c^2}{v_A^2}, \quad \boldsymbol{e}_m = (0, i, 0), \quad R_m = \frac{v_A^2}{2c^2}. \tag{3.8}$$

The dispersion relations in (3.7(and (3.8) correspond to the approximately horizontal portions of the curves in Figure 3.2.

In fact Alfvén waves only exist for $\omega < \Omega_i$, and they have a resonance at $\omega = \Omega_i$, as indicated in Figure 3.2. Approximate wave properties for $\theta \approx 0$ are

$$n_A^2 = \frac{c^2}{v_A^2}\frac{\Omega_i}{\Omega_i - \omega}, \quad \boldsymbol{e}_A = \frac{1}{\sqrt{2}}(1, -\mathrm{i}, 0); \tag{3.9}$$

$$n_m^2 = \frac{c^2}{v_A^2}\frac{\Omega_i}{\Omega_i + \omega}, \quad \boldsymbol{e}_m = \frac{1}{\sqrt{2}}(1, \mathrm{i}, 0). \tag{3.10}$$

For $\omega \lesssim \Omega_i$ the modified Alfvén waves are referred to as ion cyclotron waves. Such ion cyclotron waves are circularly polarized in the sense that ions gyrate. The two modes are approximately circularly polarized at nearly all angles of emission for $\omega \sim \Omega_i$, and they are also approximately circularly polarized for a small range of angles near $\sin\theta = 0$. Specifically, the linear polarizations in (3.7)–(3.8) are replaced by the circular polarizations in (3.9)–(3.10) for $\theta \lesssim (\omega/\Omega_i)^{1/2}$.

In most astrophysical plasma the dominant ions are protons, and then Ω_i is to be interpreted as the proton cyclotron frequency Ω_p. As ω approaches Ω_i from below the dispersion relation becomes strongly modified due to the strong interaction between the waves and the ions near the ion cyclotron frequency. The dispersion relation for the magnetoacoustic mode is affected relatively little by the ion cyclotron resonance, and continues to frequencies $\omega > \Omega_i$ where it becomes the whistler mode, as illustrated schematically in Figure 3.2.

Approximations to the Probability of Emission

In treating the emission and absorption, a relevant simplifying assumption is the small gyroradius approximation, such that the argument of the Bessel functions is $\ll 1$ in the probability (2.14). (This assumption is not necessarily well justified, but adopting it should not lead to serious error.) Making this approximation, only $s = 0$ and $s = \pm 1$ give nonzero probabilities, cf. (2.14), and $s = 0$ is not relevant here. For $s = \pm 1$, (2.14) reduces to

$$\boldsymbol{V}(\boldsymbol{k}, \boldsymbol{p}; \pm 1) \approx \pm\tfrac{1}{2} v_\perp (1, \mp i\eta, 0), \tag{3.11}$$

$$w_M(\boldsymbol{p}, \boldsymbol{k},) = \frac{2\pi q^2 R_M(\boldsymbol{k})}{\varepsilon_0 \hbar \omega_M(\boldsymbol{k})} \left| e^*_{Mx}(\boldsymbol{k}) \mp i\eta e^*_{My}(\boldsymbol{k}) \right|^2 \delta[\omega_M(\boldsymbol{k}) - s\Omega - k_\parallel v_\parallel]. \tag{3.12}$$

With these approximations, the simplest useful approximation for the probability for Alfvén waves is

$$w_A(\boldsymbol{p}, \boldsymbol{k}, s) = \frac{\pi q^2}{4\varepsilon_0} \frac{v_A^2}{c^2} \frac{v \sin^2\alpha}{\hbar\omega |\cos\alpha \cos\theta|} \delta(k + s\Omega_0/\gamma v \cos\alpha \cos\theta). \tag{3.13}$$

The approximation (3.13) applies when the polarization vectors are given by (3.7)–(3.8). At small angles $\theta \lesssim (\omega/\Omega_i)^{1/2}$, where the polarization vectors are circular, cf. (3.9)–(3.10), an extra factor $\frac{1}{2}(1 \pm \eta s)^2$ is to be included in the numerator in (3.13), with the upper sign for the Alfvén mode and the lower sign for the magnetoacoustic mode. It follows that in this limit where the waves are circularly polarized, resonant interaction between ions ($\eta = 1$) and (i) Alfvén waves and (ii) magnetoacoustic waves is effective only at (i) $s = 1$ and (ii) $s = -1$.

The Absorption Coefficient for Hydromagnetic Waves

With the approximations (3.13) to the probability of emission, the absorption coefficient (2.19), applied to Alfvén waves, reduces to

$$\begin{aligned}\gamma_A(\omega, \theta) = \frac{2\pi q^2 v_A^2}{\varepsilon_0 k c} \int_{-1}^{+1} d\cos\alpha \, \frac{\sin^2\alpha}{|\cos\alpha|} \Bigg[\frac{p^2 v_A}{c} \\ \times \left(\frac{\cos\theta}{|\cos\theta|} \frac{v}{v_A \sin\alpha} \frac{\partial}{\partial\alpha} - p\frac{\partial}{\partial p} \right) f(p, \alpha) \Bigg]_{p=p_R},\end{aligned} \tag{3.14}$$

with $p_{\rm R} = m\Omega_0/k|\cos\alpha\cos\theta|$. The sum over $s = \pm 1$ is performed as explained in section 3.3. The absorption coefficient for magnetoacoustic waves, $\gamma_{\rm m}(\omega,\theta)$, differs from (3.14) only in the omission of the factor $|\cos\theta|$.

The Pitch-Angle Diffusion Coefficient

The pitch-angle diffusion coefficient, $D_{\alpha\alpha}$, in (3.4) takes a simple form when evaluated for Alfvén waves with the approximations made in (3.7). Including also the coefficient $D_{\alpha p}$, (2.25) implies

$$\begin{bmatrix} D_{\alpha\alpha} \\ D_{\alpha p} \end{bmatrix} = \sum_{s=-\pm 1} \int \frac{d^3\boldsymbol{k}}{(2\pi)^3}\, w_{\rm A}(\boldsymbol{p},\boldsymbol{k},s)\, N_M(\boldsymbol{k}) \begin{bmatrix} (\hbar k\cos\theta/p\sin\alpha)^2 \\ (\hbar k\cos\theta/p\sin\alpha)(\hbar\omega/pv) \end{bmatrix}, \tag{3.15}$$

where the approximation (3.1) is used. Writing $W_{\rm A}(k_\parallel)dk_\parallel$ for the energy density in the Alfvén waves in the range $k_\parallel$ to $k_\parallel + dk_\parallel$, one has

$$\int dk_\parallel\, W_{\rm A}(k_\parallel) = \int \frac{d^3\boldsymbol{k}}{(2\pi)^3}\, \hbar\omega\, N_{\rm A}(\boldsymbol{k}). \tag{3.16}$$

Then one finds

$$D_{\alpha\alpha} = \frac{\pi q^2}{4\varepsilon_0\gamma\varepsilon p^2|\cos\alpha|} W_{\rm A}(k_{\rm R}), \quad D_{\alpha p} = -\frac{\cos\theta}{|\cos\theta|}\frac{v_{\rm A}}{v}\, p\sin\alpha\, D_{\alpha\alpha}, \tag{3.17}$$

where

$$k_{\rm R} = \Omega_{\rm i}/\gamma p|\cos\alpha| \tag{3.18}$$

is the resonant wavenumber.

3.3 Generation of the Resonant Waves

Resonant scattering of particles occurs whenever resonant waves are present, that is, when there waves with $k_\parallel = k_{\rm R}$, as given by (3.18). These waves may be generated in any way. However, in many applications it is thought that the resonant waves are generated by the particles themselves. This requires that the absorption coefficient for the waves be negative.

Conditions for Wave Growth

Any anisotropic distribution of particles with $v \gg v_{\rm A}$ tends to cause growth of the hydromagnetic waves with which they resonate. However, the anisotropy must be sufficiently large for growth to occur. This is due to the final term in (3.14) being proportional to $\partial f/\partial p$, which derivative is negative for all relevant distributions of relativistic particles. (An exception is for relativistic pairs in a pulsar magnetosphere, cf. lecture 9, but resonant scattering in the form under discussion here is not relevant in the superstrong pulsar magnetic fields.) Hence, the term involving $\partial f/\partial p$ in (3.14) always leads to a net damping. The other term in (3.14), involving $\partial f/\partial\alpha$ tends to cause negative absorption for some waves and positive absorption for other waves. The waves subject to negative absorption grow provided that the term involving $\partial f/\partial\alpha$ exceeds that involving $\partial f/\partial p$ in magnitude.

The waves that tend to grow depend on the nature of the anisotropy. To understand this, note that (i) emission at $k_\parallel > 0$ decreases $p_\parallel$, (ii) emission at $s > 0$ decreases $p_\perp$, and (iii) emission at $s < 0$ increases $p_\perp$. Wave growth is driven by processes that reduce the relevant anisotropy. Hence, a forward streaming motion is reduced by emission of waves with $k_\parallel > 0$, an excess of perpendicular momentum over parallel momentum is reduced by emission at $s > 0$, and a deficiency of perpendicular momentum over parallel momentum is reduced in magnitude by emission at $s < 0$.

Streaming Anisotropy

Consider a streaming motion along the magnetic field lines. Such a motion can be reduced due to the waves carrying off forward momentum. Hence, in this case one expects the growing waves to have $\cos\theta > 0$ ($k_\parallel > 0$). According to Figure 3.1, this requires that particles in the forward hemisphere ($\cos\alpha > 0$) emit at $s = -1$, and those in the backward hemisphere emit at $s = +1$. The sum over $s = \pm 1$ in (3.14) then corresponds to the ranges $-1 < \cos\alpha < 0$ and $0 < \cos\alpha < 1$, respectively. This sum is performed implicitly when carrying out the integral over all $\cos\alpha$ in (3.14).

A forward streaming distribution has an excess of particles with $\cos\alpha > 0$ and a deficiency of particles with $\cos\alpha < 0$. This implies that $\partial f/\partial\alpha$ is negative for both $\cos\alpha > 0$ and $\cos\alpha < 0$. The sign of the term involving $\partial f/\partial\alpha$ in (3.14) is negative for $\cos\theta > 0$, confirming that waves emitted in the forward direction ($\cos\theta > 0$) tend to grow due to a streaming instability. Note that the excess particle momentum is carried off by the waves with little transfer of energy to the waves. The ratio of the energy density to the momentum density in the waves is $\sim \omega/k \sim v_\mathrm{A}$, whereas the corresponding ratio for particles is $\sim v \gg v_\mathrm{A}$.

The back reaction of the waves on the particles is described by pitch-angle scattering. Emission of waves in the forward direction clearly reduces the forward streaming speed. Note that the emission here is a maser instability, and not spontaneous emission, and that the back reaction is diffusive and not systematic. Spontaneous emission, which is entirely negligible in the present context, would cause a systematic loss of forward momentum by the particles. In contrast, maser emission involves many individual stimulated emission and true absorption events, with a small excess of stimulated emission over true absorption. As a result there are many changes $\Delta\alpha$ with opposite sign, causing the particles to diffuse in α. The small excess of stimulated emission over true absorption leads to the net transfer of forward momentum from the particles to the waves.

A forward streaming anisotropy may also be reduced by absorption of waves in the backward direction. Such waves cannot be generated by the streaming particles themselves, but they may be present due to some other generation mechanism, such as turbulent motions in the plasma. The pitch-angle diffusion coefficient (3.17) applies either to emission of forward waves or to absorption of backward waves. The appropriate value of $D_{\alpha\alpha}$ is a sum over these two contributions. Hence, $W_{\rm A}(k_{\rm R})$ in (3.17) should be interpreted as $W_{\rm A}(k_{\rm R})+W_{\rm A}(-k_{\rm R})$, with the former contribution due to emission of forward waves and the latter contribution due to absorption of backward waves.

Compressional and Rarefactional Anisotropies

Another type of anisotropy is that generated by a compression in the magnetic field. As B increases, $p_\perp^2/B$ remains constant, implying that $p_\perp^2$ increases at constant $p_\parallel$. As a result the distribution function develops an anisotropy with an excess of $p_\perp$ over $p_\parallel$ (a positive anisotropy). Similarly, a rarefaction in the field leads to a deficiency of $p_\perp$ over $p_\parallel$ (a negative anisotropy). A rarefactional-type anisotropy also results from synchrotron losses, which reduce $p_\perp^2$ at constant $p_\parallel$, cf. lecture 7.

As argued above, a compressional or positive anisotropy is reduced by emission at $s>0$, and a rarefactional or negative anisotropy is reduced by emission at $s<0$. If one modifies the argument given above for a streaming anisotropy to determine the contributions to growth and damping from particles in the forward and backward hemisphere, one finds that they cancel. That is, say for waves with $\cos\theta>0$, one finds that the contribution of the term involving $\partial f/\partial\alpha$ in (3.14) for $\cos\alpha>0$ is equal and opposite to that for $\cos\alpha<0$. This cancellation applies only in the approximation in which the waves are assumed linearly polarized. There is always a small elliptical component, and as discussed following (3.13), the waves become circularly polarized at sufficiently small angles. In the circularly polarized limit, the probability of emission by ions is given by (3.13) for Alfvén waves only at $s=1$ and for magnetoacoustic only at $s=-1$. For $\theta=0$, the probability of emission by ions is zero for Alfvén waves at $s=-1$ and for magnetoacoustic at $s=1$.

It follows that compressional and rarefactional anisotropies tend to generate circularly polarized waves. Ions with a compressional anisotropy, which is reduced by emission at $s=1$, cause circularly (or elliptically) polarized Alfvén waves to grow. Similarly, ions with a rarefaction anisotropy, which is reduced by emission at $s=-1$, cause circularly (or elliptically) polarized magnetoacoustic waves to grow. The hydromagnetic waves become nearly circularly polarized in two limiting cases: for $\omega\approx\Omega_{\rm i}$ for any θ, and for $\theta\lesssim(\omega/\Omega_{\rm i}$ for any $\omega\lesssim\Omega_{\rm i}$. A compressional anisotropy can cause growth only of waves in one or other of these limiting cases.

The effect of all resonant waves on the particles is not sensitive to the direction of wave propagation or to the polarization of the waves. Thus, linearly polarized waves, if present, are as effective as circularly polarized waves in causing pitch-angle diffusion, and hence in reducing the anisotropy. The restriction to circularly (or elliptically) polarized waves applies only when the anisotropic particles are required to generate the waves that scatter them.

3.4 Spatial Diffusion

Resonant scattering causes diffusion in pitch angle. The relation between diffusion in pitch-angle and spatial diffusion along the magnetic field lines includes some subtleties. First, consider how the equation for diffusion in pitch angle may be used to derive an equation for spatial diffusion along the field lines.

Diffusion Equations

The left hand side of the equation (3.4), which describes diffusion in pitch angle, may be written in the form

$$\left(\frac{\partial}{\partial t}+v\cos\alpha\frac{\partial}{\partial z}\right)f=\frac{1}{\sin\alpha}\frac{\partial}{\partial\alpha}\left(\sin\alpha\, D_{\alpha\alpha}\frac{\partial f}{\partial\alpha}\right), \tag{3.19}$$

where any force (other than due to the waves) on the particles is neglected, and it is assumed that the only relevant spatial variation is along the z axis.

We seek to construct a spatial diffusion equation that describes diffusion in one dimension along the field lines. Denoting the distance along the field lines as z, this equation is of the form

$$\frac{df}{dt}=\frac{\partial}{\partial z}\left(\kappa\frac{\partial f}{\partial z}\right), \tag{3.20}$$

where κ is the one-dimensional spatial diffusion coefficient. Provided that the anisotropy is small, a simple way of relating κ in (3.20) to $D_{\alpha\alpha}$ in (3.19) involves expanding in Legendre polynomials. A different procedure that continues to apply when the anisotropy is large is given in the lectures by JG Kirk.

Expansion in Legendre Polynomials

To relate (3.20) to (3.19), one expands the pitch-angle distribution in Legendre polynomials:

$$f(p,\alpha)=\sum_{n=0}^{\infty}f_n(p)P_n(\cos\alpha),\quad P_0(x)=1,\quad P_1(x)=x,\quad \ldots. \tag{3.21}$$

Here only the two leading terms in the expansion are retained. The evolution of $f_n(p)$ follows from (3.19) by inserting the expansion (3.21), multiplying by $P_n(\cos\alpha)$, integrating over $\cos\alpha$, and using the orthogonality relations for the Legendre polynomials. One finds

$$\frac{\partial f_0}{\partial t}+\frac{v}{3}\frac{\partial f_1}{\partial z}=0,\quad \frac{\partial f_1}{\partial t}+\frac{v}{3}\frac{\partial f_0}{\partial z}=-\tfrac{1}{2}f_1\int_{-1}^{+1}d\cos\alpha\,\sin^2\alpha\, D_{\alpha\alpha}, \tag{3.22}$$

and so on. Ignoring the term $\partial f_1/\partial t$, the second equation determines f_1.

The Spatial Diffusion Coefficient

On substituting for f_1 in the first of equations (3.22), one obtains an equation of the form (3.20). Writing $f_0 \to f$, the spatial diffusion coefficient is identified as

$$\kappa = \frac{v^2}{9}\left[\tfrac{1}{2}\int_{-1}^{+1} d\cos\alpha\,\sin^2\alpha\, D_{\alpha\alpha}\right]^{-1}, \qquad f_1 = -\frac{3\kappa}{v}\frac{\partial f_0}{\partial z}. \tag{3.23}$$

Thus diffusion in pitch angle implies spatial diffusion along the field lines.

Relation Between Pitch-Angle and Spatial Diffusion

One subtle feature of resonant scattering causing spatial diffusion is that the pitch-angle and spatial diffusion coefficients are essentially inverses of each other, in the sense that κ varies inversely with $D_{\alpha\alpha}$. This may be understood qualitatively by comparing a case where pitch-angle scattering is weak with one where it is strong. For strong pitch-angle scattering the particles repeatedly turn around on a short timescale, due to their diffusing rapidly from $0 < \alpha < \pi/2$ to $\pi/2 < \alpha < \pi$. In the presence of such scattering, particles can be confined to a relatively small spatial region for a relatively long time. Hence a large spatial gradient may be maintained for a relatively long time. Now consider this case from the viewpoint of spatial diffusion. A large spatial gradient can persist against spatial diffusion only if the spatial diffusion is weak. Strong spatial diffusion smears out gradients rapidly, and large gradients imply weak diffusion. Hence strong pitch-angle diffusion implies weak spatial diffusion, and the converse is also valid.

Another way of understanding this is in terms of the scattering mean free path, λ. The mean free path is related to the distance in which the particles turn around, in the sense defined above. When pitch-angle scattering is strong, this distance is short, and when pitch-angle scattering is weak this distance is long. The spatial diffusion coefficient is proportional to the the mean free path ($\kappa = \lambda v/3$), again implying weak spatial diffusion when the pitch-angle scattering is strong.

Growth of Waves due to a Spatial Gradient

As already remarked, one of the important features of resonant scattering is that the anisotropic particles themselves may generate the waves required to scatter them. That the resonant waves grow due to the effect of a spatial gradient may be seen as follows. On substituting the expansion (3.21) into (2.19), retaining only the two leading terms in (3.21) and only $s = \pm 1$ in (2.19), and inserting the expression (3.23) for f_1, one finds

$$\gamma_M(\boldsymbol{k}) = -\sum_{s=\pm 1}\int d^3\boldsymbol{p}\; w_M(\boldsymbol{p},\boldsymbol{k},s)\,\hat{D}_s f(\boldsymbol{p}),$$

$$\hat{D}_s f(\boldsymbol{p}) = \frac{\hbar\omega}{v}\left(\frac{\cos\alpha - n_M\beta\cos\theta}{p\sin\alpha}\frac{3\kappa}{v}\frac{\partial f_0}{\partial z} + \frac{\partial f_0}{\partial p}\right). \tag{3.24}$$

In practice $\partial f_0/\partial p$ is negative and leads to a positive contribution to absorption; that is, the final term in (3.24) opposes growth of the waves that cause the resonant scattering. The relevant waves have large refractive indices, and for $n_M\beta \gg 1$ the term involving the spatial derivative contributes to wave growth for $\cos\theta(\partial f_0/\partial z) < 0$. Thus the waves that tend to grow propagate in the direction opposite to the gradient ($\cos\theta < 0$ for $\partial f_0/\partial z > 0$).

This example may be used to illustrate a general property of resonant scattering when the waves are generated by the anisotropic particles themselves: the momentum transfer from the particles to the waves is always such as to reduce the anisotropy of the particles. In the present case, the waves generated are propagating in the direction of the spatial gradient. Now, according to (3.23), a positive spatial gradient implies a negative f_1, and this implies that $f \approx f_0 + f_1 \cos\alpha$ has more particles in the backward hemisphere ($\cos\alpha < 0$) than in the forward hemisphere ($\cos\alpha > 0$). The gradient is reduced by any process that reduces this excess, that is, any process that transfers particles from the backward to the forward hemisphere. Emission of waves in the forward direction reduces the backward momentum of the particles, and emission of waves in the backward direction reduces the forward momentum of the particles. Hence, the inferred backward emission reduces the anisotropy, and hence the spatial gradient. More generally, in any case where a maser operates, the emission tends to reduce the feature that is causing the wave growth. This reduction may be treated quantitatively using the quasilinear equation, and hence is called *quasilinear relaxation.*

4 Scattering of Cosmic Rays

The theory of resonant scattering was developed in the early 1960s, initially in connection with the scattering of particles in the Earth's radiation belts (lecture 5), and with cosmic rays in the solar wind (e.g., Jokipii 1971; Lee 1983). One of the first major successes of plasma astrophysics, as a then new branch of astrophysics, came from the application of these ideas to the propagation of cosmic rays through the galaxy (e.g., Wentzel 1974; Cesarsky 1980).

4.1 Galactic Cosmic Rays

The observed properties of galactic cosmic rays include their energy spectrum, their isotopic contribution, and their anisotropy (e.g., reviews by Hillas 1984; Wdowczyk and Wolfendale 1989; Gaisser 1989; Longair 1992; and the lectures by GJ Kirk). From these observations one can infer the lifetime of cosmic rays, that is the typical time that a cosmic ray spends in the galaxy before it escapes. Both the inferred lifetime and the observed anisotropy imply that cosmic rays must diffuse slowly through the galactic disk.

The Spectrum of Cosmic Rays

Cosmic rays are high energy particles, mostly protons, that continuously bombard the Earth. Their energy, ε, is usually measured in electron volts, $1\,\mathrm{eV} = 1.602 \times 10^{-19}\,\mathrm{J}$. The energy spectrum, $N(\varepsilon)$, is defined such that $N(\varepsilon)d\varepsilon$ is the number density of cosmic rays in the energy range between ε and $\varepsilon + d\varepsilon$.

The energy spectrum observed at the Earth has a peak at about $10^9\,\mathrm{eV} = 1\,\mathrm{GeV}$. Below this peak the spectrum of cosmic rays varies with the 11-year solar cycle, and on shorter time scales it varies in association with solar activity. It is accepted that the spectrum at $\ll 1\,\mathrm{GeV}$ is strongly modified from what it is in the interstellar medium (ISM) as the particles propagate to the Earth against the solar wind. The solar wind forms a cavity in the ISM called the heliosphere, and the cosmic rays in the ISM must propagate from the edge of the heliosphere to the Earth to be observed. The spectrum at $\gtrsim 1\,\mathrm{GeV}$ is not greatly affected by the solar wind, and falls off as a power law

$$N(\varepsilon)d\varepsilon = K\varepsilon^{-a}d\varepsilon, \tag{4.1}$$

where K is a constant and where a is the spectral index. One finds $a = 2.6$ for $\varepsilon \lesssim 10^{15}\,\mathrm{eV}$, and $a \approx 3$ at $\varepsilon \gtrsim 10^{15}\,\mathrm{eV}$.

Axford (1992) referred to galactic cosmic rays below and above the knee as GCR1 and GCR2, respectively, where GCR denotes galactic cosmic ray. The very highest energy component, above about $3 \times 10^{18}\,\mathrm{eV}$, has a spectrum similar to the GCR1 component. Axford referred to this as the EGCR (extragalactic cosmic ray) component, for reasons discussed below.

The Composition of Cosmic Rays

The isotopic composition of the cosmic rays corresponds to ionic abundances similar to the Sun, other stars and the ISM. A notable difference is that the abundances of the light elements (Li, Be, B) are much higher in cosmic rays.

The fact that the isotopic abundances of cosmic rays is so close to normal cosmic abundances is strongly indicative of a source of the cosmic rays in the ISM, or perhaps in stellar winds. Other conceivable sources, such as the ejecta from supernovae and particles accelerated in pulsar magnetospheres, would have very different abundances.

The presence of the light elements (Li, Be, B) are attributed to spallation, that is, they are interpreted as the daughter nuclei produced as the result of nuclear interactions between the cosmic rays and particles in the ISM. The relative abundance of the light elements (when combined with the known cross sections for spallation interactions) then provides information on the column density of matter that a typical cosmic ray has passed through. Using the inferred column density of $\approx 6\,\mathrm{g\,cm^{-2}}$ (e.g., Wdowczyk and Wolfendale 1989), one may estimate the path length given the mean density along the path of the cosmic ray. Assuming a mean number density $n_e \approx 0.1\,\mathrm{cm^{-3}}$, a typical cosmic ray has propagated through the ISM over a distance $\approx 4 \times 10^{25}\,\mathrm{cm} \approx 4 \times 10^7\,\mathrm{ly}$. Hence, the life time is a few times $10^7\,\mathrm{yr}$.

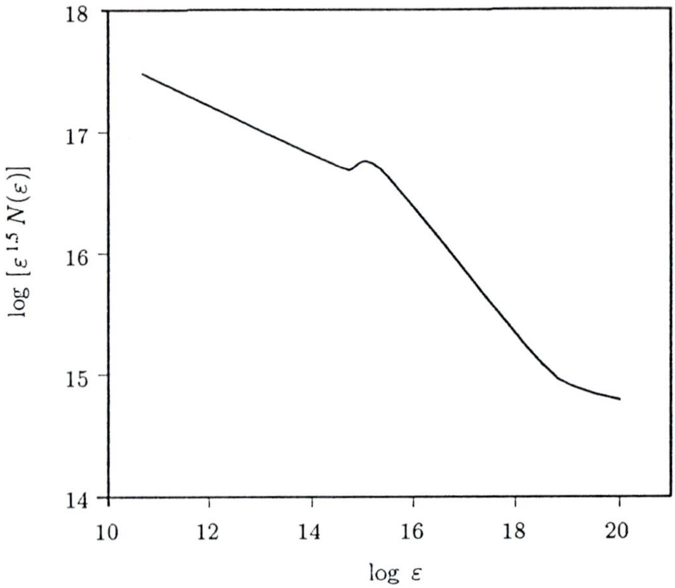

Fig. 4.1. The galactic cosmic rays spectrum is shown. Note that the spectrum is multiplied by a factor $\varepsilon^{1.5}$ to show the knee and the change in slope near 10^{15} more clearly. The units of $\varepsilon^{1.5} N(\varepsilon)$ are $(\mathrm{eV})^{1.5}\mathrm{m}^{-2}\mathrm{s}^{-1}\mathrm{sr}^{-1}$.

Another estimate of the lifetime of cosmic rays comes from the observed abundance of the unstable spallation product ^{10}Be, which implies a lifetime $\approx 2 \times 10^7$ yr (Wdowczyk and Wolfendale 1989).

This estimated lifetime is much longer than the typical time that one would estimate for a cosmic ray to propagate at the speed of light from a source region in the galactic plane to the edge of the galactic disk. Specifically, propagation at close to the speed of light for several hundred parsecs (1 pc = 3.02 ly) along a galactic magnetic field approximately in the plane of the disk, cf. Figure 4.1, would allow escape from the disk in $\approx 10^3$ yr. This suggests that cosmic rays are trapped in the galaxy for about 10^7 yr, and are in balance between a steady input and a slow escape. This confinement time is so long that the only plausible explanation for it is that the cosmic rays diffuse very slowly, due to an efficient form of scattering.

Properties of the ISM

The estimate of the mean density of along the path of a cosmic ray is dependent on the assumed model for the ISM. It is now thought that the ISM consists of a hot ($> 10^6$ K), low density $n_e \approx 3 \times 10^{-3}\,\mathrm{cm}^{-3}$ interconnected region, plus cooler, denser clouds of various types (e.g., Field 1986; Spitzer 1990). The mean density along the path of a cosmic ray depends on the fraction of its path through

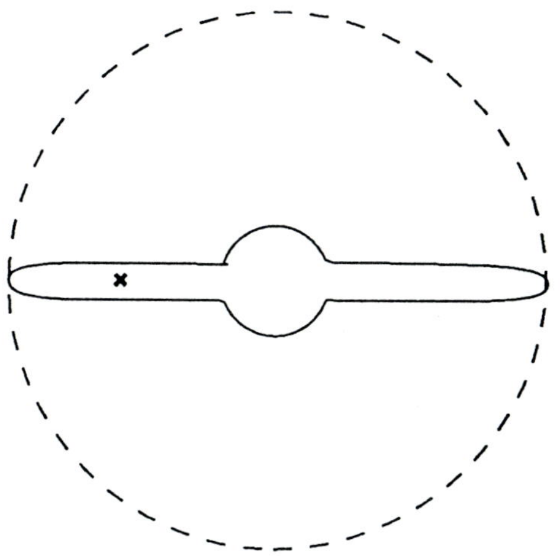

Fig. 4.2. The gross structure of the galaxy is indicated schematically. The galactic disk is of radius $\approx 15\,\mathrm{kpc}$ and of half-thickness $\approx 200\,\mathrm{pc}$. The Sun, indicated by the cross, is $\approx 10\,\mathrm{kpc}$ from the galactic center. The galactic halo is roughly spherical, indicated by the dashed line.

each of these components. The estimate $n_e \approx 0.1\,\mathrm{cm}^{-3}$ used above is somewhat arbitrary.

The Alfvén speed in the ISM, and the gyroradii of cosmic rays play a role in the theory discussed below. Both depend on B, $v_A \propto n_e/B^{1/2}$ and $R_g \propto B$. The strength of the magnetic field in the ISM is probably about $5\,\mu\mathrm{G} = 5 \times 10^{-10}\,\mathrm{T}$.

Anisotropy of Cosmic Rays

The observed anisotropy of cosmic rays provides direct evidence that the cosmic rays are streaming slowly. The anisotropy, which is less than one part in 10^3, may be expanded in Legendre polynomials, as in (3.21). The first order term, expressed as a fraction of the zeroth order term, gives the ratio v_{CR}/v, of the streaming speed to the particle speed. For example, assuming an observed ratio of 0.06% at $\sim 10^{14}\,\mathrm{eV}$ (Longair 1992) implies a streaming speed $v_{CR}/c = 6\times10^{-4}$, cf. (4.3) below. At such a bulk speed, the flow of the cosmic rays would cover $\sim 10^4\,\mathrm{ly} \sim 3\,\mathrm{kpc}$ in the mean lifetime of a typical cosmic ray. Individual cosmic rays may propagate much larger distances, but nevertheless this suggests that individual cosmic rays remain localized to relatively small regions of the galaxy before ultimately escaping.

Origin of Cosmic Rays

It is now widely accepted that the acceleration of cosmic rays at energies below the knee at $\sim 10^{15}\,\mathrm{eV}$ in the spectrum, cf. Figure 4.1, is due to diffusive shock acceleration due to shock waves associated with supernova explosions. (The acceleration of cosmic rays is discussed in detail in the lectures by JG Kirk.)

Diffusive shock acceleration requires efficient scattering of the particles, and resonant scattering is an essential ingredient in the theory. This mechanism is thought to accelerate ions from the ambient thermal plasma, which accounts for the ionic composition of cosmic rays being close to normal cosmic abundances. There is direct evidence supporting the suggestion for acceleration of thermal ions in this way at shocks in the solar wind (e.g., Reames 1992). However, the details of how thermal ions become suprathermal, and exceed the threshold $v \gtrsim v_{\mathrm{A}}$ for resonant scattering, referred to here as the pre-acceleration problem, is one of the uncertain points in the theory. With this and other provisos, the acceleration of the GCR1 component seems to be explained by diffusive shock acceleration at supernova-generated shocks in the ISM.

The origin of cosmic rays at energies above the knee (GCR2) is less well understood. The mechanism favored for the GCR1 component becomes slower with increasing ε, and appears to be ineffective for the GCR2 component, as discussed in the lectures by JG Kirk.

The origin of the electron component in cosmic rays is more problematical. The ratio of electrons to ions is about three percent at a given energy. The presence of the electrons is important from an observational viewpoint because of the synchrotron radiation that they emit. By mapping the distribution of radio emission, one can map the distribution of cosmic rays throughout the galaxy. Why the ratio of electrons to ions has the observed value is unclear. It may be associated with the relative efficiency with which thermal electrons and thermal ions are accelerated at shocks. However, the pre-acceleration problem for electrons is more severe than for ions, as discussed at the end of lecture 5.

Highest Energy Cosmic Rays

At the highest energies the gyroradii of cosmic rays are larger than the thickness of the galactic disk. The gyroradius of a relativistic particle is given by

$$R_{\mathrm{g}} = 1.7 \times 10^{-3}\,\frac{\gamma\beta\sin\alpha}{B}\,\mathrm{m} = 1.0 \times 10^{-15}\left(\frac{\varepsilon\sin\alpha}{1\,\mathrm{eV}}\right)\left(\frac{B}{1\,\mu\mathrm{G}}\right)^{-1}\,\mathrm{pc}, \tag{4.2}$$

where the first expression is in SI units with $v = \beta c$, and where the second expression applies to highly relativistic particle ($\beta \approx 1$) in units most often used in astrophysics. The strength of the interstellar magnetic field is probably about $5\,\mu\mathrm{G} = 5 \times 10^{-10}\,\mathrm{T}$. Hence cosmic rays with energies $\gtrsim 10^{18}\,\mathrm{eV}$ have gyroradii larger than the thickness of the galactic disk, and so could not be confined by the galactic magnetic field. This suggests that the highest energy cosmic rays are of extragalactic origin, and hence the designation EGCR.

It is thought that the EGCR component is accelerated at shock in extragalactic jets (e.g., Wdowczyk and Wolfendale 1989). However, for the very highest energies, $\gtrsim 10^{20}\,\mathrm{eV}$, limits on the maximum energy in diffusive shock acceleration (Jokipii 1982,1987; Achterberg 1990) appear to be violated if the particles are protons rather than Fe-ions (Hillas 1984; Wdowczyk and Wolfendale 1989). How this inconsistency is to be resolved is unclear.

4.2 Self-Confinement of Cosmic Rays

Galactic cosmic rays are said to be self-confined if they generate the Alfvén waves that scatter them (e.g., Kulsrud and Pearce 1969; Wentzel 1974; Skilling 1975a,b,c; Melrose 1980b, Cesarsky 1980).

Simple Model for a Streaming Distribution

A streaming motion may be described in terms of the $n = 1$ term in the expansion (3.21) of the distribution function in Legendre polynomials. Let $v_{\rm CR}$ be the streaming velocity for the cosmic rays; in general $v_{\rm CR}$ is a function of p. Then, on retaining only the terms $n \le 1$ in (3.21), one has

$$f(p,\alpha) = f_0(p) + f_1(p)\cos\alpha, \quad f_0(p) = K_{\rm CR}\left(\frac{p}{p_0}\right)^{-b}, \quad f_1(p) = \frac{v_{\rm CR}}{v} f_0(p). \tag{4.3}$$

The spectrum $f_0(p)$ in (4.3) is the same as the spectrum (4.1), but written in a different notation. The relation between the energy spectrum and the distribution function is

$$N(\varepsilon)d\varepsilon = 4\pi f_0(p)\,p^2 dp. \tag{4.4}$$

Consequently, for relativistic particles, the power law indices a in (4.1) and b in (4.3) are related by $b = a + 2$, implying $b = 4.6$ for galactic cosmic rays. The normalization constant $K_{\rm CR}$ is related to the number density of cosmic rays above a given energy, say above some normalization value p_0:

$$n_{\rm CR}(p > p_0) = 4\pi \int_{p_0} dp p^2\, f_0(p) = \frac{4\pi K_{\rm CR}\, p_0^2}{b-3}. \tag{4.5}$$

If one chooses $p_0 = m_p c = 0.94 \times 10^9\,\mathrm{eV}/c$, observations give

$$n_{\rm CR}(p > p_0) = 2 \times 10^{-10}\,\mathrm{cm}^{-3}, \qquad K_{\rm CR} = 2.5 \times 10^{-11} p_0^2\,\mathrm{cm}^{-3}. \tag{4.6}$$

Generation of Alfvén Waves by Streaming Cosmic Rays

The absorption coefficient (3.14) for Alfvén waves may be evaluated directly for the distribution (4.3). The absorption coefficient for magnetoacoustic waves differs from (3.14) only in the omission of the factor $|\cos\theta|$, and as the important waves have $\theta \approx 0$ nearly along the streaming direction, this difference is unimportant and is ignored. The absorption coefficient reduces to

$$\begin{aligned}\gamma_{\mathrm{A,m}}(\omega,\theta) &= -\frac{3(b-3)}{2b(b-2)}\,\frac{\pi q^2 v_{\mathrm{A}}}{\varepsilon_0 k_0 p_0 c^2}\,n_{\mathrm{CR}}(p>p_0)\left(\frac{k|\cos\theta|}{k_0}\right)^{b-3}\\ &\quad\times\left(\frac{\cos\theta}{|\cos\theta|}\,v_{\mathrm{CR}}-\frac{b}{3}\,v_{\mathrm{A}}\right)\\ &= -2.7\times10^{-7}\,\mathrm{s}^{-1}\left(\frac{n_e}{1\,\mathrm{cm}^{-3}}\right)^{-1/2}\left(\frac{p}{p_0}\right)^{-1.6}\\ &\quad\times\left(\frac{\cos\theta}{|\cos\theta|}\,\frac{v_{\mathrm{CR}}}{c}-1.5\,\frac{v_{\mathrm{A}}}{c}\right). \end{aligned} \tag{4.7}$$

In the latter form numerical values are inserted for galactic cosmic rays, and $|\cos\theta|$ is approximated by unity.

The Neutral Streaming Speed

It follows from (4.7) that the waves grow in the forward streaming direction ($\cos\theta > 0$ for $v_{\mathrm{CR}} > 0$) provided that the streaming speed exceeds a *neutral streaming speed*, $1.5v_{\mathrm{A}}$. Qualitatively, the neutral streaming speed may be attributed to the fact that the scattering tends to isotropize the particles in a frame in which the scattering centers are at rest. This frame corresponds to one moving with $\sim v_{\mathrm{A}}$ when the scattering is due to forward-propagating hydromagnetic waves.

The streaming speed inferred from observation may be smaller than the value $\approx 1.5v_{\mathrm{A}}$ predicted by the theory. For example, this would be the case if the cosmic rays arriving at the Earth are diffusing away from a large number of different sources. The spatial distribution of the sources can then lead to an approximate isotropy in the cosmic rays. A relevant qualitative point is that resonant scattering need not be effective everywhere through the galactic disk. Scattering only needs to be effective somewhere along a prospective escape path for a cosmic ray. In figurative language, if the galaxy is viewed as a leaky box for cosmic rays, scattering is only required to plug the holes in the leaky box in order to confine the cosmic rays effectively.

Effectiveness of Self-Confinement

The growth rate (4.7) is fast enough to account for growth of the waves that resonate with lower energy cosmic rays, specifically the GCR1 component. For example, for $n_e = 10^{-2}\,\mathrm{cm}^{-3}$ and $v_{\mathrm{CR}} - 1.5v_{\mathrm{A}} = 30\,\mathrm{km\,s}^{-1}$, one finds a growth time $1/\gamma_{\mathrm{A,m}} \approx 200(p/p_0)^{1.6}\,\mathrm{yr}$. Thus the waves needed to scatter the cosmic rays with $p \approx 10^9\,\mathrm{eV}/c$ grow in a few hundred years, which is less than the propagation time of free-streaming cosmic rays to the edge of the galactic disk.

There is an approximate one-to-one relation between the resonant wave number k and the momentum of the resonant particles, $kp = |q|B$, so that the higher the energy or the momentum of a cosmic ray, the smaller the frequency or the wave number of the hydromagnetic wave with which it resonates. According to

(4.7), the growth time increases rapidly with p. This implies that there is insufficient time for the resonant waves to grow for particles with sufficiently high energy. For example, for $p \gg 10^{15}\,\mathrm{eV}/c$ the growth time exceeds the lifetime of the cosmic rays. This suggests that self-confinement can be effective only for the GCR1 component of cosmic rays.

In an early discussion of self-confinement (Kulsrud and Pearce 1969) the then-existing model for the ISM implied that it is partially ionized. Hydromagnetic waves are strongly damped (e.g., due to charge exchange interactions) in a partially ionized plasma. At the time this imposed a severe constraint on the theory. It is now assumed that the scattering of cosmic rays occurs only in the hot region of the ISM. This overcomes the difficulty because the damping processes for hydromagnetic waves in hot plasmas are negligible in the present context.

4.3 Reduction of the Streaming Speed

The scattering of the cosmic rays by resonant waves tends to reduce the streaming speed. This may be treated using quasilinear theory.

Evolution of the Streaming Speed

The rate of change of the streaming velocity is given by

$$\frac{dv_{\mathrm{CR}}}{dt} = \frac{1}{2f_0(p)} \int_{-1}^{+1} d\cos\alpha \, \frac{df(p,\alpha)}{dt} \, v\cos\alpha, \tag{4.8}$$

with $df(p,\alpha)/dt$ given by (2.24). In pitch angle scattering only the term involving $D_{\alpha\alpha}$ is retained, as in (4.3), but it is relevant to retain the term involving $D_{\alpha p}$ in order to discuss the effect of the neutral streaming speed. Hence, from (2.24), one inserts $df/dt = (1/\sin\alpha)(\partial/\partial\alpha)\{\sin\alpha[D_{\alpha\alpha}(\partial f \partial\alpha) + D_{\alpha p}(\partial f/\partial p)]\}$ into (4.8). A partial integration over $\cos\alpha$ may be performed trivially and, using (4.3) one finds $\partial f/\partial\alpha = -(v_{\mathrm{CR}}/v)f\sin\alpha$ and $\partial f/\partial p = -bf/p$.

Forward and Backward Propagating Waves

If one evaluates the coefficient $D_{\alpha\alpha}$, ignoring the term proportional to ω in the expression (2.26) for $\Delta\alpha$, the result (3.15) is obtained. Rather than assume that all the waves are generated by the streaming particles themselves, in which case all the waves are in the forward streaming direction, let there be a fraction ζ of the waves propagating in the forward direction, with $1-\zeta$ propagating in the backward direction. The coefficient $D_{\alpha\alpha}$ is independent of ζ, and the for coefficient $D_{\alpha p}$ one finds

$$D_{\alpha p} = -(2\zeta - 1)\frac{\cos\theta}{|\cos\theta|}\frac{v_{\mathrm{A}}p\sin\alpha}{v}D_{\alpha\alpha}. \tag{4.9}$$

On using (4.9), (4.8) reduces to

$$\frac{dv_{\rm CR}}{dt} = \nu_s \left(v_{\rm CR} - (2\zeta - 1) \frac{|\cos\theta|}{\cos\theta} \frac{b}{3} v_{\rm A} \right), \tag{4.10}$$

where the effective scattering frequency given by

$$\nu_s = 3 \int_{-1}^{+1} d\cos\alpha \, \sin^2\alpha \, D_{\alpha\alpha}, \tag{4.11}$$

with $D_{\alpha\alpha}$ given by (3.15).

Scattering by Both Forward and Backward Waves

The waves generated by the streaming particles themselves are propagating in the forward direction, corresponding to $\zeta = 1$ in (4.10). According to (4.10), the scattering then tends to reduce the streaming speed to the neutral streaming speed. However, if the hydromagnetic waves are not generated by the streaming particles, but are due to some other source, one would expect no asymmetry with respect to the forward and backward directions. Thus one would expect $\zeta = \frac{1}{2}$. Equation (4.10) implies that scattering by such waves tends to reduce the streaming speed to zero.

Scattering of Higher Energy Cosmic Rays

It is possible for the higher-energy GCR2 component ($p \gtrsim 10^{15}\,{\rm eV}/c$) to be confined through resonant scattering provided that the resonant waves are generated in some other way. According to (4.2) the required waves have wavelengths $\gtrsim 0.2\,{\rm pc}$. Such waves might be generated through turbulent motions in the ISM, or through stellar winds. There is direct evidence for the presence of turbulence on appropriate scales from observations of the scintillation of pulsars (e.g., Rickett 1990).

Hydromagnetic Waves in the ISM and the Solar Wind

On applying (4.10) with (4.11) to the scattering of galactic cosmic rays, one encounters a practical difficulty in that there is relatively little information on the spectrum of hydromagnetic waves in the ISM. On the other hand, on applying this theory to scattering of fast particles in the solar wind, one has more information on the spectrum of hydromagnetic turbulence (e.g., Zhou and Mattheus 1990). The spectrum of turbulence in the solar wind is a power law, with power-law index approximately equal to 5/3, corresponding to a *Kolmogorov spectrum*:

$$W(k) \propto k^{-5/3}. \tag{4.12}$$

A relatively detailed comparison of theory of resonant scattering and observation of the particles and the waves can be carried out in this case. The comparison (e.g., Lee 1983, 1992), in the sense that the basic theoretical ideas are supported.

4.4 Cross-Field Diffusion

Resonant scattering causes diffusion in pitch angle, which leads to spatial diffusion across field lines. Spatial diffusion across field lines can occur due to two unrelated processes: due to resonant interactions and due to a so-called wandering of field lines (Jokipii 1966). Here a simple model for field-line wandering is presented, and then the general problem of cross-field diffusion is discussed semiquantitatively.

Wandering of Magnetic Field Lines

In the presence of magnetic fluctuations, due to a spectrum of hydromagnetic waves, the direction of a magnetic field line wanders due to the magnetic field, $\delta\boldsymbol{B}$, associated with the fluctuations. Consider two particles that start at the same point at two different times. Each particle sees a different detailed spectrum of fluctuations, and so its guiding center follows a slightly different path. Hence, a large number of particles starting from the same point at different times spread out from the mean direction of the magnetic field as a function of distance, z, along the average direction of the magnetic field, as illustrated in Figure 4.3. This effect is referred to as wandering of magnetic field lines.

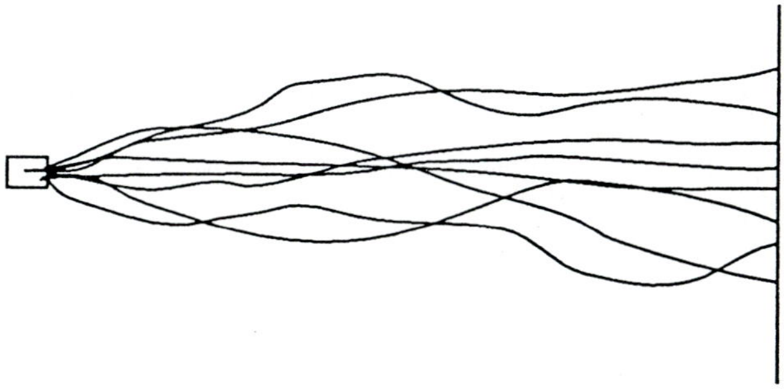

Fig. 4.3. Field line wandering due to a spectrum of magnetic fluctuations is illustrated schematically. The lines represent the paths of the guiding centers of particles starting from the same point at different times.

Fokker-Planck Equation for Field Lines

To treat field line wandering in a quantitative way, one needs a statistical description of the fluctuations. Let the average field be $\boldsymbol{B}_0$. A statistical average over the fluctuations gives

$$\langle\delta\boldsymbol{B}\rangle = 0, \qquad \langle\boldsymbol{B}\rangle = \boldsymbol{B}_0,$$
$$\langle B_i(z)B_j(z+\zeta)\rangle = \langle\delta B_i(z)\delta B_j(z+\zeta)\rangle = S_{ij}(\zeta), \tag{4.13}$$

where $S_{ij}(\zeta)$ is the correlation function. The spectrum of fluctuations can be measured, usually in terms of the spatial Fourier transform of the correlation function:

$$R_{ij}(k) = \int d\zeta \, e^{ik\zeta} S_{ij}(\zeta). \tag{4.14}$$

In the absence of detailed information on the spectrum, one assumes that the fluctuations are gaussian, $S_{ij}(\zeta) \propto \exp(-\zeta^2/2\zeta_0^2)$, where ζ_0 is a correlation length.

An Ensemble of Field Lines

The other ingredient in a theory for field line wandering is a statistical description of the field lines. Let $\mathcal{N}(x,y)$ denote the number of field lines per unit area in the x-y plane. Assuming that the spectrum has cylindrical symmetry about $\boldsymbol{B}_0$, the field line wandering may be described in terms of a spatial diffusion of the field lines in the x-y plane:

$$\frac{\partial \mathcal{N}}{\partial z} = \left(\frac{\partial^2}{\partial x^2} + \frac{\partial^2}{\partial y^2}\right)(D_m \mathcal{N}). \tag{4.15}$$

The spatial diffusion coefficient, D_m, may be evaluated in terms of Fokker-Planck coefficients:

$$D_m = \frac{1}{2}\left\langle \frac{(\Delta x)^2}{\Delta z} \right\rangle = \frac{1}{2}\left\langle \frac{(\Delta y)^2}{\Delta z} \right\rangle. \tag{4.16}$$

Evaluation of the Fokker-Planck Coefficients

The Fokker-Planck coefficients may be evaluated as follows. Starting from the parametric equations for a magnetic field line,

$$\frac{dx}{B_x} = \frac{dy}{B_y} = \frac{dz}{B_z}, \tag{4.17}$$

one integrates with respect to z, with $B_z = B_0$, to find

$$\Delta x = \frac{1}{B_0}\int_0^{\Delta z} dz \, \delta B_x(z), \quad \langle (\Delta x)^2 \rangle = \frac{1}{B_0^2}\int_0^{\Delta z} dz \int_0^{\Delta z} dz' \, \langle \delta B_x(z) \delta B_x(z') \rangle . \tag{4.18}$$

Provided that Δz is much larger than the correlation length ζ_0, the integrals in (4.18) may be rewritten in terms of integrals over $z+z'$ and $z-z'$, and the latter extended to infinity and performed using (4.14). This gives

$$\left\langle \frac{(\Delta x)^2}{\Delta z} \right\rangle = \frac{1}{B_0^2}\int d\zeta \, S_{xx}(\zeta) = \frac{R_{xx}(0)}{B_0^2}. \tag{4.19}$$

Thus the spatial diffusion coefficient (4.16) is determined by the long-wavelength ($k \to 0$) limit of the spectrum of fluctuations.

Cross-Field Diffusion due to Field-Line Wandering

The wandering of field lines leads to cross-field diffusion of particles. Let us write the cross-field diffusion coefficient as $\kappa_\perp$. In the simplest approximation $\kappa_\perp$ is related to D_m, as given by (4.16), by

$$D_\perp = vD_m. \tag{4.20}$$

From (4.19), one may estimate $D_m \sim \zeta_0(\delta B/B_0)^2$, where ζ_0 is the autocorrelation length of the magnetic fluctuations, and $\delta B/B_0$ is their relative amplitude.

In practice, this cross-field diffusion is not a particularly efficient process. This is because the rate of diffusion perpendicular to the field line is partly dependent on the distance the particle has propagated from its source. More efficient cross-field diffusion is required for some purposes, notably for diffusive shock acceleration at oblique shocks. For example, diffusive shock acceleration is impossible at a perpendicular shock in the absence of cross-field diffusion because each particle can cross the shock only once, as the field line to which it is tied crosses the shock. In principle cross-field diffusion allows a particle to encounter the shock more than once even in this case.

Resonant Cross-Field Diffusion

When a particle emits a wave quantum the position of its gyrocenter moves in the direction perpendicular to both the background magnetic field and to the direction of emission. For emission is the x-direction, the gyrocenter moves a distance $|\hbar k_x/qB|$ along the y-axis. As a consequence, resonant scattering by waves with $\theta \neq 0$ always causes some cross-field diffusion. For an axially symmetric distribution of waves, the cross-field diffusion coefficient due to resonant interactions is, cf. (2.25),

$$\kappa_\perp(\boldsymbol{p}) = \sum_{s=-\infty}^{\infty} \int \frac{d^3\boldsymbol{k}}{(2\pi)^3}\, w_M(\boldsymbol{p},\boldsymbol{k},s)\, \frac{\hbar^2 k_\perp^2}{2q^2B^2}\, N_M(\boldsymbol{k}), \tag{4.21}$$

This resonant cross-field diffusion is weak, and ineffective in practice. To see this, first compare (4.21) and the pitch-angle diffusion coefficient. Ignoring angular factors and other factors of order unity, one find $\kappa_\perp/D_{\alpha\alpha} \sim R_\mathrm{g}^2$, where R_g is the gyroradius of the particle. According to (3.23) the spatial diffusion coefficient for parallel diffusion, now written $\kappa_\parallel$ to avoid confusion, is given by $\kappa_\parallel \sim v^2/D_{\alpha\alpha}$. Combing these two results gives

$$\kappa_\perp\kappa_\parallel \sim v^2 R_\mathrm{g}^2. \tag{4.22}$$

The mean free path, λ, for parallel diffusion may be introduced by writing $\kappa_\parallel = v\lambda/3$. Then one finds

$$\kappa_\perp/\kappa_\parallel \sim \epsilon^2, \quad \epsilon = R_\mathrm{g}/\lambda. \tag{4.23}$$

Hence, resonant cross-field diffusion is a small effect for $\epsilon \ll 1$, which is the case for the GCRI component.

A phenomenological quantity introduced to describe cross-field diffusion is the Bohm diffusion coefficient, $D_{\rm B} = R_{\rm g}^2\Omega/3$. This corresponds to scattering causing random displacements $\sim R_{\rm g}$ across the field lines each gyroperiod. Bohm diffusion is not based on a specific theoretical model, but rather on phenomenological descriptions of diffusion in laboratory plasmas. Then one has $\kappa_{\parallel} \sim D_{\rm B}/\epsilon$ and $\kappa_{\perp} \sim \epsilon D_{\rm B}$.

Compound Diffusion

Applied to cosmic rays or to energetic particles in any astrophysical magnetic field, the wandering of field lines implies that particles diffuse slowly across the field lines. In practice the wandering of the field lines itself turns out to be relatively unimportant in causing the cross-field diffusion required in some models, notably in diffusive shock acceleration at oblique shocks. Moreover, cross-field diffusion due to resonant scattering is entirely negligible. However, when field-line wandering is combined with efficient (parallel) diffusion along the field lines, the resulting compound diffusion (Achterberg and Ball 1994) can lead to efficient (perpendicular) diffusion across the field lines.

Compound diffusion may be understood as follows. Parallel diffusion involves a particle whose gyrocenter is moving rapidly back and forth along the actual field lines, with its mean position changing relatively slowly. Consider a particle that starts at a given z. As a result of parallel diffusion, this particle returns to this value of z many times. Due to the field-line wandering, each time the particle returns to this value of z, it is displaced perpendicular to its original position due to the net effect of the deviation of the actual field line from the mean field line along its orbit. In principle, compound diffusion can overcome the major deficiency in field-line diffusion as a cross-field diffusion mechanism, in the sense that it avoids the implication that the particle must travel a long distance along the field line before it wanders significantly across. In compound diffusion, parallel diffusion keeps the particle localized in z, and field line wandering allows it to diffuse across the field lines.

5 Scattering and Acceleration of Electrons

One of the first applications of the theory of resonant scattering was to energetic ions by ion cyclotron waves and electrons by whistlers in the Earth's radiation belts. In this lecture the theory of the scattering of electrons by whistlers is developed and, after a brief review of the structure of the magnetosphere, applied to the Earth's radiation belts. Fermi acceleration is then discussed and applied to prompt relativistic electron component in solar flares.

5.1 Scattering by Whistlers

The threshold condition for an electron to resonate with a hydromagnetic wave requires $v > \Omega/k$ with $k \sim \omega/v_{\rm A}$. Then $\omega < \Omega_{\rm i}$ for hydromagnetic waves requires $v > (\Omega_{\rm e}/\Omega_{\rm i})v_{\rm A}$, that is $v > (m_i/m_e)v_{\rm A} \approx (43)^2 v_{\rm A}$. It follows that only very fast electrons can resonate with hydromagnetic waves. However, it is possible for slower electrons, in the range $43v_{\rm A} \lesssim v \lesssim (43)^2 v_{\rm A}$, to resonate with whistlers.

Whistler Waves

As shown schematically in Figure 3.1, the magnetoacoustic mode joins on continuously to the whistler mode at $\omega \sim \Omega_{\rm i}$. Whistlers exists in the frequency range $\Omega_{\rm i} < \omega < \Omega_{\rm e}$ (in a plasma with $\Omega_{\rm e} < \omega_{\rm p}$). Their name comes from 'whistling atmospherics' detected in the early days of radio as natural audio noise. The noise is generated by a lightning flash in one hemisphere, and the signal is dispersed as it propagates in the whistler mode along the Earth's field lines to the other hemisphere, where it can be heard on a radio receiver as a falling tone.

The properties of whistler waves may be derived using the magnetoionic theory, as discussed in lecture 1. The dispersion relation for whistler waves may be approximated by

$$n_{\rm W}^2 = 1 + \frac{\omega_{\rm p}^2}{\omega(\Omega_{\rm e}|\cos\theta| - \omega)}, \tag{5.1}$$

where $\omega_{\rm p} = (e^2 n_{\rm e}/\varepsilon_0 m_{\rm e})^{1/2}$ is the plasma frequency and $\Omega_{\rm e} = eB/m_{\rm e}$ is the electron cyclotron frequency. For present purposes, only the regime $\omega \ll \Omega_{\rm e}|\cos\theta|$ is considered, and in the plasmas of interest the unit term in (5.1) is then negligible. With these approximations, the properties of whistlers reduce to

$$n_{\rm W}^2 = \frac{\omega_{\rm p}^2}{\omega\Omega_{\rm e}|\cos\theta|}, \qquad e_{\rm W} = \frac{(1, i|\cos\theta|, 0)}{(1+\cos^2\theta)^{1/2}}, \qquad R_{\rm W} = \frac{\omega\Omega_{\rm e}}{2\omega_{\rm p}^2}\frac{1+\cos^2\theta}{|\cos\theta|}. \tag{5.2}$$

Probability of Emission of Whistlers by Electrons

The probability of emission for electrons and whistlers is obtained by inserting (5.2) into (2.14), with $\eta = -1$ for electrons. This gives

$$w_{\rm W}(\boldsymbol{p}, \boldsymbol{k}, s) = \frac{\pi e^2}{4\varepsilon_0}\frac{\Omega_{\rm e}}{\hbar\omega_{\rm p}^2}\frac{v\sin^2\alpha}{|\cos\alpha|}\frac{(1+s|\cos\theta|)^2}{\cos^2\theta}\,\delta(k + s\Omega_{\rm e}/\gamma v\cos\alpha\cos\theta). \tag{5.3}$$

The minimum speed for an electron to resonate with a whistler follows from the δ-function in (5.3), together with the dispersion relation (5.2), with $k = n_{\rm W}\omega/c$. One finds that resonance is possible only for $v > 43v_{\rm A}$. For such electrons, growth is possible provided that the absorption coefficient is negative.

The Absorption Coefficient for Whistlers

The specific form of the absorption coefficient for whistlers due to anisotropic electrons follows from (2.19). After summing over $s = \pm 1$ this gives

$$\gamma_{\mathrm{W}}(\omega) = \frac{\pi^2 k}{2n_{\mathrm{e}}} \int_{-1}^{+1} d\cos\alpha \, \sin^2\alpha \left[p^3 v \left((1+\cos^2\theta) - 2\cos\theta \frac{\cos\alpha}{|\cos\alpha|} \right) \right. \\ \left. \times \left(\frac{\cos\theta}{\sin\alpha} \frac{\partial}{\partial\alpha} - \frac{43 v_{\mathrm{A}}}{v|\cos\alpha|} p \frac{\partial}{\partial p} \right) f(p,\alpha) \right]_{p=p_{\mathrm{R}}}, \tag{5.4}$$

where $p_{\mathrm{R}} = sm\Omega_{\mathrm{e}}/k\cos\alpha\cos\theta$ is the solution of the resonance condition for p, cf. the δ-function in (5.3).

Growth of Whistlers

The conditions for growth of whistlers due to anisotropic distributions of electrons can be inferred from qualitative arguments similar to those given in lecture 3, cf. Figure 3.1. An important difference is that there is only one mode and emission at $s = 1$ always dominates over $s = -1$, cf. (5.3). An immediate consequence is that whistlers can grow due to a compressional-type or positive anisotropy, with $\partial f/\partial \sin\alpha > 0$, but growth is not possible for a rarefactional-type or negative anisotropy, with $\partial f/\partial \sin\alpha < 0$. The waves that grow due to a positive anisotropy satisfy $\cos\theta\cos\alpha/|\cos\theta\cos\alpha| = -1$, so that particles in the forward hemisphere emit waves in the backward direction, and particles in backward hemisphere emit waves in the forward direction.

For a streaming anisotropy, forward-propagating waves resonate with particles in the backward hemisphere at $s = 1$, and this tends to cause the waves to grow. Similarly, backward-propagating waves resonate with particles in the forward hemisphere at $s = 1$, and this also tends to cause the waves to grow. The resonance between forward-propagating waves and particles in the forward hemisphere at $s = -1$, and the resonance between backward-propagating waves and particles in the backward hemisphere at $s = -1$, both tend to cause the waves to damp. The net effect is a tendency to growth, due to the factor $(1+s|\cos\theta|)^2$ in (5.3) being larger in magnitude for $s = 1$ than for $s = -1$. It follows that of the three types of anisotropy under discussion, only a negative anisotropy cannot lead to growth of whistlers.

Pitch-Angle Diffusion due to Whistlers

The pitch-angle diffusion coefficient for electrons resonating with whistlers follows from (2.25) with (2.26), together with (5.2) and (5.3). Assuming that the whistlers are axisymmetric with respect to the direction of $\boldsymbol{B}$, one finds

$$D_{\alpha\alpha} = \frac{4\pi r_0}{\gamma p|\cos\alpha|} 2\pi \int_{-1}^{1} d\cos\theta \, W_{\mathrm{W}}(k_{\mathrm{R}},\theta) \frac{1+\cos^2\theta}{|\cos\theta|}, \tag{5.5}$$

where $W_W(k, \theta)dk$ is the energy density in the whistlers between k and $k + dk$, and where k_R is determined by the δ-function in (5.3).

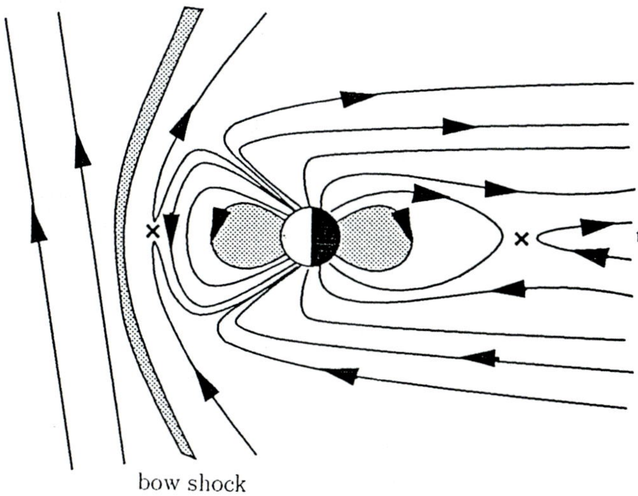

Fig. 5.1. The Earth's magnetosphere is indicated schematically. The Sun is to the left of the figure. The ram pressure of the solar wind draws the magnetic field lines out into a magnetotail in the anti-sunward direction. The bow shock on the sunward side is indicated. The shaded region in the inner magnetosphere is the plasmasphere.

5.2 The Earth's Magnetosphere

The magnetosphere of the Earth (e.g., Akasofu 1977) is the region in which the magnetic field lines are connected to the Earth, cf. Figure 5.1. The magnetosphere forms a cavity in the solar wind, which comes closest to the Earth ($r \gg 10R_E$) on the sunward side. In the solar wind, the dominant energy density is the kinetic energy associated with the flow (at $v_{sw} \sim 500\,\mathrm{km\,s^{-1}}$). The sunward boundary of the magnetosphere is roughly where the ram pressure associated with the flow of the solar wind, $\frac{1}{2}\eta_{sw}v_{sw}^2$, where η_{sw} is the mass density balances the energy density, $B^2/2\mu_0$, in the Earth's magnetic field. A dipolar field falls off as r^{-3}, so that the magnetic pressure falls off $\propto r^{-6}$. The solar wind slows down abruptly at a *bow shock* on the sunward side of the Earth, and then flows around the magnetosphere, which forms a teardrop-like cavity in the solar wind with an extended tail.

The magnetosphere is partly open to the solar wind. This is especially the case when the magnetic field in the solar wind opposes the Earth's magnetic field. Then, as first described by Dungey (1961), magnetic field lines reconnect on the sunward side of the magnetosphere, the flux is transferred to the anti-sunward side, where further reconnection occurs. These reconnections sites are denoted by **x**-points in Figure 5.1.

The Earth's Radiation Belts

There are populations of fast particles trapped in the so-called radiation (or Van Allen) belts (e.g., Roederer 1970; Schultz and Lanzerotti 1974). The dipolar-like field lines of the Earth form magnetic bottles that trap fast particles. In the early days of space research, before 1960, although the existence of trapped distributions of electrons (40 keV–2 MeV) and of trapped ions ($\gtrsim 100\,\mathrm{keV}$) was known, the lifetime of the trapped particles was not known. If the lifetime were determined by inter-particle collisions, called Coulomb interactions, then it would be very long. For example, for a 100 keV electron in a plasma with $n_e = 1\,\mathrm{cm}^{-3}$, the collision time is $\sim 3 \times 10^8\,\mathrm{s}$, that is, about ten years. In 1962 there was an experiment, called STARFISH, in which a nuclear bomb was set off in the magnetosphere, and the decay of the radioactive products was studied. It was found that the life times vary from tens of minutes to a few weeks. Clearly the particles are scattered much more efficiently than they would be through Coulomb interactions alone.

Another observational fact also implies that the lifetime of particles in the radiation belts is relatively short. When a coronal mass ejection hits the Earth's magnetosphere, the outer boundary on the sunward side can be pushed in to a few Earth radii, destroying the radiation belts. However, within a day or so the magnetosphere relaxes back to its initial configuration, and the radiation belts reform. This suggests that the acceleration of the bulk of the particles in the radiation belts occurs within a matter of hours. In a steady state, the lifetime should be of the same order as the acceleration time, suggesting a short lifetime.

Acceleration of Particles in the Radiation Belts

An obvious question is how the particles in the radiation belts are accelerated. A mechanism that seems to explain most of the features of the acceleration can be understood in terms of adiabatic invariants. There are three invariants in an idealized magnetosphere. Let these be denoted (e.g., Roederer 1970; Schultz and Lanzerotti 1974)

$$M = p_\perp^2/B, \quad J = \oint ds\, p_\parallel, \quad \Phi = \oint ds\, A_\phi. \tag{5.6}$$

These may be interpreted as the approximately conserved quantities corresponding to quasi-periodic motions that the particles execute. One may introduce these in terms of action-angle variables, where the angle corresponds to quasi-periodic motion and the action is the conjugate momentum in a hamiltonian sense (e.g., Goldstein 1959). The gyrational motion about the magnetic field lines leads to conservation of the magnetic moment, M, the bounce motion between reflection points in the mirroring motion leads to conservation of J, and the drift motion around the Earth leads to conservation of Φ.

If one writes the radial distance as $r = LR_\mathrm{E}$, where R_E is the radius of the Earth, then for a dipolar field one has $B \propto L^{-3}$, and hence a vector potential $A \propto L^{-2}$. Then, because the lengths in the integrals are both proportional to L, (5.6) gives

$$p_\perp^2 \propto ML^{-3}, \quad p_\parallel \propto JL^{-1}, \quad \Phi \propto L^{-1}. \tag{5.7}$$

Provided that the third adiabatic invariant is conserved, then the particles move on a single L shell and their energy is conserved. However, it turns out that Φ is not conserved, because there are significant fluctuations in the structure of the magnetosphere on time scales in which particles drift around the equator. The fluctuations cause a diffusion in Φ values. According to (5.7), Φ depends only on L, and so diffusion in Φ implies a diffusion in L. Given a source of solar wind particles at the outer boundary of the magnetosphere, such diffusion must cause the particles to diffuse inward. Hence, particles injected from the solar wind at the boundary of the magnetosphere tend to diffuse inward, and as they do so, M and J are conserved.

Diffusive Steady State

As the particles diffuse inward due to violation of the third adiabatic invariant, they become increasingly anisotropic, due to $p_\perp$ increasing $\propto L^{-3/2}$, and $p_\parallel$ increasing $\propto L^{-1}$ as L decreases. Hence the particles develop a positive anisotropy. The number density of the energetic particles also tends to increase as L decreases. The reason for this is that as the particles diffuse inward, the particle distribution function, f, remains constant, so that the number density varies as $n \propto p_\perp^2 p_\parallel f$. Conservation of $M \propto p_\perp^2/B$ and $J \propto p_\parallel L$ implies $p_\perp^2 p_\parallel \propto MJB/L \propto L^{-4}$ for M and J constant. Thus the density tends to build up close to the Earth toward a distribution of particles with number density $\propto L^{-4}$.

Acceleration of Particles

A similar argument implies that the particles gain energy as they diffuse inward. On averaging over pitch angle one finds $\langle p_\perp^2 p_\parallel \rangle \sim p^3 \propto L^{-4}$ at fixed M and J. Hence the kinetic energy of a nonrelativistic particle goes up proportional to $L^{-8/3}$ as it diffuses inward. For example, a 2 keV particle injected from the solar wind at $L = 10$ would have an energy of 150 keV when it reaches $L = 2$. The inward diffusing particles are ultimately scattered into the loss cone, precipitate into the denser regions of the atmosphere and are lost.

Particles are only accelerated by electric fields. In this case the electric field is associated with the motions that violate the third adiabatic invariant. These are convective electric fields, of the form $\boldsymbol{E} = -\boldsymbol{V} \times \boldsymbol{B}$, where $\boldsymbol{V}$ is a fluid motion that does not correspond to corotation with the Earth.

5.3 Scattering of Particles in the Magnetosphere

Historically, resonant scattering of particles in the magnetosphere was the first suggested application of resonant scattering. The first case discussed was the scattering of energetic ions, which was attributed to ion cyclotron waves (Wentzel 1961; Dragt 1961). These ion cyclotron waves are on the rising section of the dispersion curve for Alfvén waves in Figure 3.1. The scattering of energetic electrons was attributed to whistlers (Dungey 1963). Here a simple model is presented for the scattering of the inward-diffusing electrons into the loss cone (Kennel and Petchek 1966), providing the sink for the electrons to balance the source in the solar wind.

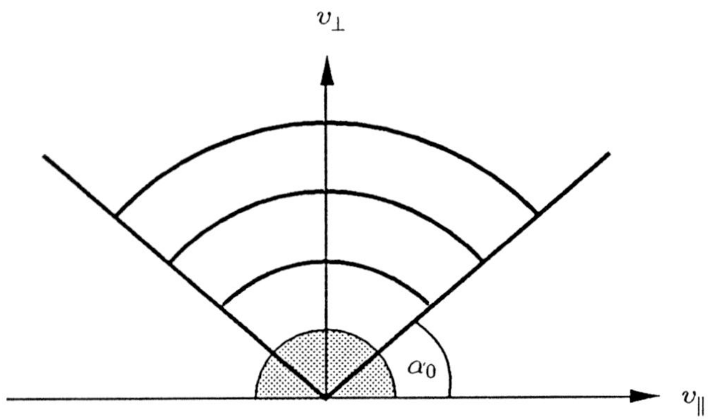

Fig. 5.2. The idealized loss-cone distribution (5.8) is illustrated; α_0 is the loss cone angle, the shaded region is filled by isotropic thermal particles, and the circular arcs denote contours of constant f.

Loss-Cone Driven Growth of Whistlers

Consider an electron distribution with an idealized loss cone of the form

$$f(p,\alpha) = f_0(p)[H(\alpha-\alpha_0) + H(\pi-\alpha_0-\alpha)], \tag{5.8}$$

where $H(x)$ is the step function. That is, (5.8) corresponds to no particles in loss cones $0<\alpha<\alpha_0$ and $\pi/2-\alpha_0<\alpha<\pi/2$, with the distribution independent of α for $\alpha_0<\alpha<\pi/2-\alpha_0$, as illustrated in Figure 5.2. In this case the growth comes from the derivatives of (5.8) with respect of α:

$$\frac{\partial f}{\partial \alpha} = f_0\left[\delta(\alpha-\alpha_0) - \delta(\pi-\alpha_0-\alpha)\right]. \tag{5.9}$$

The first of these δ-functions causes waves at $\theta>\pi/2$ to grow, and the second causes waves at $\theta<\pi/2$ to grow. In either case, the absorption coefficient may be approximated by

$$\gamma_{\mathrm{W}} \approx \frac{n_1}{n_{\mathrm{e}}} \Omega_{\mathrm{e}} \left[\alpha_0^2 - b \left(\frac{43 v_{\mathrm{A}}}{v} \right)^2 \right], \tag{5.10}$$

where n_1 is the number density of the fast particles.

The waves grow provided the growth time is shorter than the effective loss time for the whistlers. The main loss mechanism is simply propagation of the whistlers out of the region where the growth occurs. The distance they need to propagate to escape from the growth region is of order $r = LR_{\mathrm{E}}$. Estimating the group speed v_{g} assuming that whistlers are propagating approximately along the field lines, so that the escape time for the whistlers is

$$T_{\mathrm{W}} \sim \frac{LR_{\mathrm{E}}}{v_{\mathrm{g}}}, \qquad v_{\mathrm{g}} \approx c \left(\frac{\omega_{\mathrm{p}}^2}{\omega \Omega_{\mathrm{e}}} \right)^{1/2} \approx \frac{\omega_{\mathrm{p}}^2}{\Omega_{\mathrm{e}}^2} v, \tag{5.11}$$

where in the final expression the resonance condition and the dispersion relation are used to eliminate ω in favor of the speed v of the resonant electrons. Growth is effective only for $\gamma_{\mathrm{W}} T_{\mathrm{W}} \gg 1$.

Scattering into the Loss Cone

Once the whistlers are generated they scatter particles into the loss cone, so that the loss cone is no longer empty. Suppose that the escape rate for particles in the loss cone is ν_{esc}, with $\nu_{\mathrm{esc}} \approx v/LR_{\mathrm{E}}$. Then pitch-angle diffusion within the loss cone may be described by (3.4) for small α. On balancing the diffusion by the loss term one obtains, writing $D_{\alpha\alpha} = D$,

$$\frac{1}{\alpha} \frac{\partial}{\partial \alpha} \left(\alpha D \frac{\partial f}{\partial \alpha} \right) - \nu_{\mathrm{esc}} f = 0, \tag{5.12}$$

which applies for $\alpha < \alpha_0$. The general solution of (5.12) may be expressed in terms of modified Bessel functions $I_\nu(x)$ of order $\nu = 0$ and argument $x = \alpha(\nu_{\mathrm{esc}}/D)^{1/2}$. On requiring that the distribution function be equal to f_0 at $\alpha = \alpha_0$, one obtains

$$f = f_0 \frac{I_0(\alpha[\nu_{\mathrm{esc}}/D]^{1/2})}{I_0(\alpha_0[\nu_{\mathrm{esc}}/D]^{1/2})}. \tag{5.13}$$

The solution (5.13) is qualitatively different in two limiting cases

$$\begin{aligned} &\text{weak diffusion:} && D \ll \nu_{\mathrm{esc}} \alpha_0^2, \\ &\text{strong diffusion:} && D \gg \nu_{\mathrm{esc}} \alpha_0^2, \end{aligned} \tag{5.14}$$

which correspond to the limits of large and small arguments for the modified Bessel function, with $I_0(x) \approx e^x/(2\pi x)^{1/2}$ for $x \gg 1$ and $I_0(x) = 1 + x^2/4 + \ldots$ for $x \ll 1$. These two cases are illustrated in Figure 5.3. In the case of weak diffusion the loss cone is nearly empty, and in the case of strong diffusion the loss cone is almost full. In effect, in weak diffusion, particles diffuse slowly, and when a particle diffuses into the loss cone it is much more likely to precipitate than to diffuse out again. On the other hand, in the case of strong diffusion,

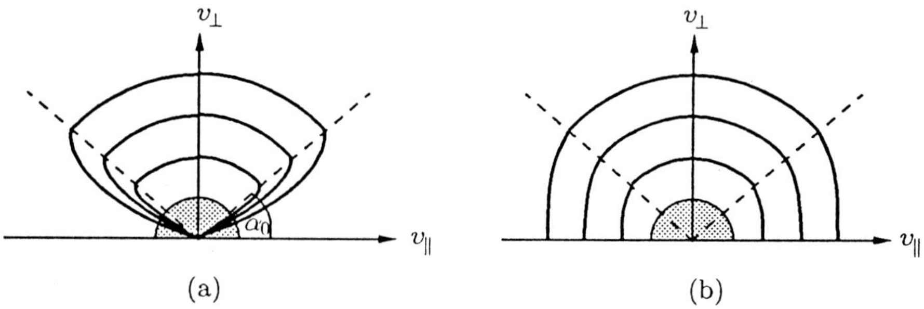

Fig. 5.3. The distribution inside the loss cone is illustrated for (a) weak diffusion and (b) strong diffusion. In weak diffusion particles scattered into the loss cone have a high probability of being lost due to precipitation, so that the loss cone is almost empty. In strong diffusion particles that diffuse into the loss cone have a higher probability of diffusing out again than of precipitating, so that the loss cone is almost full.

particles diffuse in and out of the loss cone on a time scale shorter than the loss time $1/\nu_{\rm esc}$. Then the only electrons that are lost due to precipitation are those so close to the footpoint of the flux tube that they hit the denser regions of the atmosphere is less than a scattering time.

Triggered VLF Emissions

The foregoing theory is satisfactory from a semiquantitative viewpoint in that it accounts reasonably for the properties of the electrons trapped in the radiation belts. The whistler waves are also observed. However, there is a qualitative difference between theory and observation. The theory implies that random phase whistler noise should be generated. Observationally, the precipitating electrons correlate with narrow band whistlers, referred to as discrete VLF emissions.

Discrete VLF emissions have surprising properties, e.g., the review by Matsumoto (1979). They can be triggered by terrestrial radio noise. A Morse code 'dash' (150 ms), but not by a a Morse code 'dot' (50 ms), can trigger them. Also, they appear to be continuously triggered by high harmonics of the AC frequency (50 Hz or 60 Hz in North America) produced as noise by large electricity generating plants. Once triggered, a discrete VLF emission appears on a dynamic spectrum as a narrow line that drifts in frequency. A phenomenological theory for the interpretation of such narrow band, triggered VLF emissions was presented by Helliwell (1967). The theory of discrete VLF emissions remains somewhat controversial.

5.4 Acceleration of Relativistic Electrons

The acceleration of electrons to relativistic energies clearly occurs in synchrotron sources, such as supernova remnants, the lobes of quasars, radio galaxies, and so on. The favored acceleration mechanism is diffusive shock acceleration. This mechanism is not seriously questioned here, but there is a practical difficulty that raises some uncertainty. Although there is observational evidence for diffusive shock acceleration of ions at shocks in the heliosphere, there is no corresponding evidence for diffusive shock acceleration of electrons. As a result, there is no direct evidence on the relative effect of diffusive shock acceleration on electrons and ions. One example of acceleration of electrons to relativistic energies in the heliosphere is in prompt acceleration that produces gamma rays in solar flares. Here, after a brief review of the data, a Fermi-type mechanism is summarized, and then some general comments are made on the wider problem of the acceleration of relativistic electrons.

Acceleration of Electrons in Solar Flares

Solar flares involve acceleration of particles in several different contexts, which can be separated according to the time scale for the acceleration. The primary energy release in a flare appears to go into bulk acceleration of electrons (e.g., Sturrock 1980). Both the time structure and the spatial structure can be determined, but not with corresponding resolutions. This acceleration occurs on a time scale much shorter than a second, with evidence for acceleration on times $\lesssim 10^{-2}\,\mathrm{s}$ (Sturrock et al. 1984). The spatial structure estimated over a much longer time scale suggests energy release in a region $\gtrsim (300\,\mathrm{km})^2$ (de Jager et al. 1987). The resulting electrons are in the range $\sim$ 2–20 keV. Their precipitation into the denser regions of the solar atmosphere produces hard X-ray bursts and optical and UV emission, and some escape to produce type III solar radio bursts (lecture 8). How these electrons are accelerated is not well understood.

On a time scale of about a second, gamma rays are detected from some flares (Chupp 1983). These require ions to be accelerated to $\gtrsim 30\,\mathrm{MeV/nucleon}$, and electrons to be accelerated to relativistic energies. The numbers of such particles and the energy involved in their acceleration are small compared to the bulk energization of the electrons. These particles are fast enough for resonant scattering of ions by hydromagnetic waves and of electrons by whistlers to occur. Before this prompt acceleration was first recognized it was accepted that energetic ions and relativistic electrons are accelerated in association with solar flares, but only on a much longer time scale of tens of minutes to hours (e.g., Wild, Smerd and Weiss 1963). The reduction of at least three orders of magnitude in the acceleration time was thought to place a severe limit on theories of acceleration. However, this seems not to be the case. Several different acceleration mechanisms appear viable. The one favored by the present author (Melrose 1983) is Fermi-type acceleration, and as this mechanism is of much wider interest, it is appropriate to review the present understanding of it here.

Fermi Acceleration

Acceleration of fast particles by MHD turbulence has long been recognized as a possible mechanism for galactic and solar cosmic rays (e.g., Fermi 1949; Davis 1956; Parker 1957). Kulsrud and Ferrari (1971) showed that the interaction between MHD turbulence and fast particles can lead to acceleration through three related processes, including magnetic pumping and the reflection from moving magnetic compressions. The treatment followed here is due to Achterberg (1981), who showed that Fermi acceleration is related to Landau damping of magnetoacoustic waves by fast particles.

Consider the resonant interaction at $s = 0$ between a hydromagnetic wave and a fast particle. When the wave properties (3.7), (3.8) for the hydromagnetic waves are inserted into the expression (2.14) for the probability of emission, one finds that the probability vanishes for $s = 0$ for the Alfvén mode. This is interpreted in terms of magnetic pumping: interaction at $s = 0$ occurs only with waves that involve compressions and rarefactions of the magnetic field, and the Alfvén mode in the approximation (3.7) is purely a shear wave. Hence, only magnetoacoustic waves (and not Alfvén waves) can lead to Fermi acceleration.

The effect of resonant interactions on the particles may be described by (2.23) or (2.24). For $s = 0$ one has $\Delta p_\perp = 0$ in (2.26), and then the only diffusive term that remains in (2.23) gives

$$\frac{df(\boldsymbol{p})}{dt} = \frac{\partial}{\partial p_\parallel}\left[D_{\parallel\parallel}(\boldsymbol{p})\frac{\partial f(\boldsymbol{p})}{\partial p_\parallel}\right]. \tag{5.15}$$

Evaluation of the diffusion coefficient for magnetoacoustic waves gives

$$\begin{aligned} D_{\parallel\parallel}(\boldsymbol{p}) &= \frac{\pi q^2 v_A^2}{\varepsilon_0 c^2}\int \frac{d^3\boldsymbol{k}}{(2\pi)^3}\,\frac{k_\parallel^2 W_m(\boldsymbol{k})}{\omega^2}\,v_\perp^2 J_0'^2\left(\frac{k_\perp v_\perp}{\Omega}\right)\delta(\omega - k_\parallel v_\parallel) \\ &\approx \frac{\pi^2 v_A^2}{B^2/2\mu_0}\,\frac{p_\perp^2 v_\perp^2}{|v_\parallel|^3}\int \frac{dk\,k^3}{(2\pi)^3}\,W_m(k,\theta_R)\left(1 - \frac{v_A^2}{v_\parallel^2}\right), \end{aligned} \tag{5.16}$$

with $\cos\theta_R = v_A/v_\parallel$.

It follows that the resonant interaction at $s = 0$ causes the particles to diffuse in one dimension. In the absence of resonant scattering the particle distribution would become anisotropic, with $\langle p_\parallel^2\rangle \gg \langle p_\perp^2\rangle$. This tends to suppress the energy exchange, and the acceleration stops.

Effect of Resonant Scattering

If resonant scattering is effective in keeping the distribution approximately isotropic, then the energy exchange continues. Assuming the particle distribution remains approximately isotropic, the evolution of the average (over pitch angle) distribution function is given by

$$\frac{df_0(p)}{dt} = \frac{1}{p^2}\frac{\partial}{\partial p}\left[p^2 D_{pp}(p)\frac{\partial f_0(p)}{\partial p}\right],$$

$$D_{pp}(p) = \tfrac{1}{2}\int_{-1}^{1} d\cos\alpha\,\cos^2\alpha\, D_{\|\|}(\boldsymbol{p}). \tag{5.17}$$

On making the small gyroradius approximation, as in (3.11) but now for $s = 0$, the integrals in (5.16) and (5.17) may be evaluated and one finds

$$D_{pp}(p) \approx \frac{\pi\bar{\omega}v_A W_m}{8(B^2/2\mu_0)}\frac{p^2}{v}\ln\frac{v}{v_A}, \tag{5.18}$$

where W_m is the energy density in the magnetoacoustic waves and $\bar{\omega}$ is their mean frequency.

The mean acceleration rate implied by (5.17) is

$$\left\langle\frac{d\varepsilon}{dt}\right\rangle = \frac{1}{p^2}\frac{\partial}{\partial p}[p^2 v D_{pp}(p)]. \tag{5.19}$$

On inserting (5.18), to within a factor of order unity, one finds

$$\left\langle\frac{d\varepsilon}{dt}\right\rangle \approx 4\bar{\omega}v_A p\left(\frac{\delta B}{B}\right)^2, \tag{5.20}$$

with $(\delta B)^2/\mu_0 = W_m$, which is the well-known form for Fermi acceleration.

Application of Fermi Acceleration to Prompt Electrons

Now consider whether the Fermi mechanism can plausibly account for the prompt acceleration of electrons to relativistic energies in solar flares. The specific question asked is whether Fermi acceleration can produce relativistic electrons in less than about a second.

The acceleration rate (5.20) implies that an electron can be accelerated to relativistic energies in an acceleration time

$$t_{acc} = \left(\frac{\pi\bar{\omega}}{4}\frac{W_M}{B^2/2\mu_0}\frac{v_A}{c}\right)^{-1}. \tag{5.21}$$

By way of illustration, consider the favorable parameters: $2\pi/\bar{\omega} \sim 10^{-2}$ s from the shortest time scale observed in the energy release, $v_A/c \sim 1/30$, and $W_M \sim B^2/2\mu_0$. Then (5.20) gives $t_{acc} \sim 6\times10^{-2}$ s, so that it appears that the acceleration can be fast enough in principle.

An important requirement is that efficient resonant acceleration occurs. This requires that the resonant waves grow fast enough. They take several growth times to reach a level where they are effective in scattering the electrons. This conditions is satisfied for resonant whistlers for $\gamma_W \gtrsim \bar{\omega}$, with γ_W given by (5.10). With $\alpha_0^2 \sim 1$, this requires that the ratio of the number densities of relativistic to thermal electrons satisfy $n_1/n_e \gtrsim \bar{\omega}/\Omega_e \sim 10^{-6}$, which is a modest constraint. One can conclude that simple theory allows such very rapid acceleration without requiring any extreme choices of parameters.

The Pre-Acceleration Problem

One difficulty with prompt acceleration in solar flares is common to many contexts where relativistic electrons are accelerated. This is the injection problem: how are the ions pre-accelerated to $v > v_A$ and the electrons to $v > 43v_A$ so that resonant acceleration becomes possible? One can assume that the bulk energization provides the fast electrons for further acceleration, but it is not clear how the ions are pre-accelerated.

In the case of diffusive shock acceleration in synchrotron sources, the problem is with the electrons rather than the ions. Shock waves with Alfvénic Mach number M_A have shock speed $v_1 = M_A v_A$ in the frame in which the plasma ahead of the shock is at rest. An ion that is stationary in this frame before the shock arrives, has a speed $> v_A$ relative to the post-shock plasma after the shock passes. Thus, a thermal ion that leaks across the shock would find itself with $v > v_A$ on the other side of the shock. Thus it may not be necessary to appeal to any specific pre-acceleration mechanism for the ions, although opinions differ on this point. The situation with electrons is different because they need to exceed $43v_A$ to be effectively pre-accelerated. It is not known how this is achieved.

The inadequate understanding of pre-acceleration leads to an uncertainty in the rate that thermal particles become suprathermal. In the theory of diffusive shock acceleration, this rate, which is the injection rate of new fast particles at the shock, is a free parameter.

Possible Pre-Acceleration Mechanisms for Electrons

The following are possible pre-acceleration mechanisms for the electrons.

(1) Electron acceleration occurs only for strong shocks with $M_A > 43v_A$. There is no direct evidence for this, but there is also no evidence for diffusive shock acceleration of electrons at weaker shocks in the solar system.

(2) There is a specific wave-particle pre-acceleration mechanism. One that has been suggested involves wave that are generated by streaming ions ahead of the shock being Landau damped by the electrons, and so pre-heating them (Cargill and Papadopoulos 1988).

(3) The electrons are pre-accelerated through shock drift acceleration. There is evidence for shock drift acceleration of electrons for some shocks in the solar system that generate type II solar radio bursts (e.g., Leroy and Mangeney 1984).

(4) The plasma β (the ratio of gas pressure to magnetic pressure) is of order unity, implying that the thermal speed of ions is $\sim v_A$ and the thermal speed of electrons is $\sim 43v_A$. The particles in the thermal tails then have speeds in excess of the required thresholds, and no pre-acceleration is required.

Which, if any, of these mechanisms operates in practice is unclear.

6 Gyromagnetic Absorption and Cyclotron Maser Emission

Gyromagnetic emission is the generic name for emission due to a particle spiraling in a magnetic field. Gyromagnetic emission from nonrelativistic electrons is called cyclotron emission; it is concentrated at lines near harmonics $s = 1, 2, 3\ldots$ of the cyclotron frequency. Gyromagnetic emission from mildly relativistic electrons is called gyrosynchrotron emission, and gyromagnetic emission from ultrarelativistic electrons is called synchrotron emission.

6.1 Exact Results for Gyromagnetic Emission in Vacuo

There is a well-known exact result for the power radiated by a single particle due to gyromagnetic emission in vacuo. The general formalism adopted here may used to rederive and generalize this result.

The Emissivity in the Magnetoionic Modes

The probability for gyromagnetic emission is written down in a general form in (2.14). Applied to the emission of magnetoionic waves by electrons, (2.14) gives

$$w_M(\boldsymbol{p}, s, \omega, \theta) = \frac{(2\pi c)^3}{\hbar\omega^3 n_M^2 \partial(\omega n_M)/\partial\omega}\,\eta_M(s, \omega, \theta)$$

$$\eta_M(s, \omega, \theta) = \frac{n_M r_0 m_e c\omega^2 \beta^2 \sin^2\alpha}{14\pi(1+T_M^2)}\,\delta[\omega(1 - n_M\beta\cos\theta\cos\alpha) - s\Omega]$$
$$\times\left|\frac{L_M\sin\theta + T_M(\cos\theta - n_M\beta\cos\alpha)}{n_M\beta\sin\alpha\sin\theta}\,J_s + J_s'\right|^2, \qquad (6.1)$$

with $\beta_\parallel = v_\parallel/c$, $\beta_\perp = v_\perp/c$, $k_\parallel = (n_M\omega/c)\cos\theta$, and where $r_0 = e^2/4\pi\varepsilon_0 m_e c^2$ is the classical radius of the electron. The argument of the Bessel functions is $x_M = (\omega/\Omega)n_M\beta\sin\alpha\sin\theta$. The quantity $\eta_M(s, \omega, \theta)$ is the *emissivity* at the sth harmonic in the mode M; it is defined as the power emitted per unit frequency and per unit solid angle (about the wave normal direction).

The probability (6.1) is used in three separate ways here. One is to calculate the power radiated in vacuo exactly. The second is relevant to applications to nonrelativistic electrons, when the probability is simplified by retaining only the leading terms in the power series expansions of the Bessel functions. The third is to synchrotron emission, when the Bessel functions have argument nearly equal to their order ($1 - x_M \ll 1$) and are approximated by Airy integrals.

Polarization for Gyromagnetic Emission in Vacuo

The power radiated in vacuo by a single particle in gyromagnetic emission may be derived from (6.1) as follows. First note that in vacuo ($n_M = 1$) there are two transverse states of transverse polarization ($L_M = 0$). The polarization, described by the axial ratio, T, is determined by the emission process. The emission at each harmonic is completely polarized, with emission at the sth harmonic having

$$T = \frac{\cos\theta - \beta_\parallel}{\beta_\perp \sin\theta} \frac{J_s(sx)}{J_s'(sx)}. \tag{6.2}$$

Here the power emitted is evaluated separately in two linearly polarized components, polarized along the major axes of the ellipse.

The total power emitted, P, is evaluated by multiplying the probability (6.1) by $\hbar\omega$, summing over $s \geq 1$ and integrating over $d^3\boldsymbol{k}/(2\pi)^3$:

$$P = \sum_{s=1}^{\infty} \int_{-1}^{1} d\cos\theta \, \frac{r_0 m_e c s^2 \Omega_e^2 \beta_\perp^2}{\gamma^2 (1 - \beta_\parallel \cos\theta)^3} \left\{ \left[\frac{z}{x} J_s(sx)\right]^2 + [J_s'(sx)]^2 \right\}, \tag{6.3}$$

with

$$z = \frac{\cos\theta - \beta_\parallel}{1 - \beta_\parallel \cos\theta}, \qquad x = \frac{\beta_\perp \sin\theta}{1 - \beta_\parallel \cos\theta}. \tag{6.4}$$

Separating into the linearly polarized components corresponds to separating the terms involving J_s^2 and $J_s'^2$ in (6.3). These correspond to polarization along the projection of $\boldsymbol{B}$ on the plane orthogonal to $\boldsymbol{k}$ ($T = \infty$), and along the direction $\boldsymbol{B} \times \boldsymbol{k}$ ($T = 0$), respectively. The powers in these two components are denoted $P^\parallel$ and $P^\perp$, respectively.

The Angular Integral

The $\cos\theta$-integral in (6.3) is performed after making a change of variables to

$$\cos\theta' = \frac{\cos\theta - \beta_\parallel}{1 - \beta_\parallel \cos\theta}, \qquad \beta' \sin\theta' = \frac{\beta_\perp \sin\theta}{1 - \beta_\parallel \cos\theta}, \qquad \beta' = \frac{\beta_\perp}{(1 - \beta_\parallel^2)^{1/2}}. \tag{6.5}$$

One finds

$$P^{\parallel,\perp} = r_0 m_e c \Omega_e^2 \beta'^2 (1 - \beta'^2) \sum_{s=1}^{\infty} s^2 \int_{-1}^{1} d\cos\theta' \, (1 + \beta_\parallel \cos\theta') \, F_s^{\parallel,\perp},$$

$$F_s^\parallel = \frac{\cos\theta'}{\beta' \sin\theta'} J_s(s\beta' \sin\theta'), \qquad F_s^\perp = J_s'(s\beta' \sin\theta'). \tag{6.6}$$

The integrals in (6.6) reduce to standard integrals for the Bessel functions. Evaluating them gives

$$P^{\|,\perp} = 2r_0 m_e c\Omega_e^2 \beta'^2(1-\beta'^2)\sum_{s=1}^{\infty} s^2 G_s^{\|,\perp},$$

$$G_s^{\|} = \frac{1}{\beta'^2}\int_0^{\beta'} dy\,\frac{J_{2s}(2sy)}{y} - \frac{1}{\beta'^3}\int_0^{\beta'} dy\,J_{2s}(2sy),$$

$$G_s^{\perp} = \frac{1}{s\beta'}J'_{2s}(2s\beta') - \frac{1}{\beta'^2}\int_0^{\beta'}\frac{dy}{y}J_{2s}(2sy) + \frac{1}{\beta'}\int_0^{\beta'} dy\,J_{2s}(2sy). \quad (6.7)$$

The Sum over Harmonic Number

The final step is to perform the sum over s. The relevant sums are Kapteyn series. The following series were first evaluated explicitly for the present purpose by Schott (1912):

$$\sum_{s=1} s^2 J_{2s}(2sy) = \frac{y^2(1+y^2)}{2(1-y^2)^4}, \quad \sum_{s=1} sJ'_{2s}(2sy) = \frac{y}{2(1-y^2)^2}, \quad (6.8)$$

The final result is

$$P = P^{\|} + P^{\perp} = \frac{2r_0\Omega_e^2 p_\perp^2}{3m_e c} = \frac{2}{3}r_0 m_e c\Omega_e^2\gamma^2\beta_\perp^2,$$

$$\frac{P^{\|}-P^{\perp}}{P^{\|}+P^{\perp}} = -\frac{2+\beta_\perp^2-2\beta_\|^2}{4(1-\beta_\|^2)} = -\frac{2m_e^2c^2+3p_\perp^2}{4(m_e^2c^2+p_\perp^2)}. \quad (6.9)$$

The results (6.7) and (6.9) are used in lecture 7 in treating synchrotron radiation. In the ultrarelativistic limit $p_\perp^2 \gg m^2c^2$, (6.3) implies 75% polarization in the perpendicular direction, that is, with electric vector orthogonal to the direction of $\boldsymbol{B}$. This is indicative of the high degree of polarization of synchrotron emission

6.2 Cyclotron Absorption

Emission and absorption by thermal electrons is important in producing cyclotron lines in radio emission from the solar atmosphere, in optical and infrared emission from hot spots on some accreting white dwarf stars (cataclysmic variables), and in X-rays from hot spots on some neutron stars (notably Herculis X1).

Average over a Maxwellian Distribution

A thermal distribution of electrons is described by the Maxwellian distribution function

$$f(p) = \frac{n_e\, e^{-v^2/2V_e^2}}{(2\pi)^{3/2}m_e^3V_e^3}, \quad (6.11)$$

with $T_e = m_e V_e^2$, written in terms of $\beta_0 = V_e/c$ in the following.

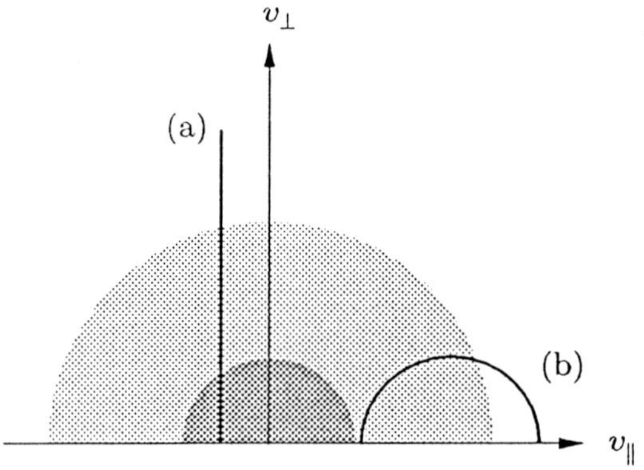

Fig. 6.1. Examples of resonance ellipses in the non-relativistic region of velocity space with the region occupied by thermal electrons shown darkly shaded with nonthermal electrons in the lightly shaded region. Curve (a) is relevant to thermal emission and absorption (when the nonthermal electrons are ignored, and curve (b) is relevant to cyclotron maser emission.

The absorption coefficient at the sth harmonic is found by setting $\gamma = 1$ in (6.1), retaining only the leading term in the power series expansion of each Bessel function, $J_s(z) \approx (z/2)^s/s!$, inserting the distribution function (6.11) into the expression (2.19) for the absorption coefficient, and evaluating the integrals in (2.19). This gives

$$\gamma_M(s,\omega,\theta) = \frac{\pi^{1/2}\omega_{\rm p}^2 A_M(s,\omega,\theta)}{2\omega n_M \beta_0 |\cos\theta|\partial(\omega n_M)/\partial\omega}\, e^{-(\omega - s\Omega_{\rm e})^2/2\omega^2 n_M^2 \beta_0^2 \cos^2\theta}. \tag{6.12}$$

The detailed form of $A_M(s,\omega,\theta)$ involves modified Bessel functions, and simplifies in the small gyroradius limit, $[n_M\beta_0(\omega/\Omega_{\rm e})\sin\theta]^2 \ll 1$, to

$$A_M(s,\omega,\theta) \approx \frac{(\frac{1}{2}\lambda_M)^{s-1}}{2s!(1+T_M^2)}\left[\frac{\omega}{\Omega_{\rm e}}(L_M\cos\theta - T_M\sin\theta)\tan\theta + sT_M\sec\theta + s\right]^2,$$
$$\lambda_M = [n_M\beta_0(\omega/\Omega_{\rm e})\sin\theta]^2. \tag{6.13}$$

It is convenient to denote the magnetoionic modes by writing $M \to \sigma = \pm 1$ with $\sigma = +1$ for the o-mode and $\sigma = -1$ for the x-mode. Then, in the high-frequency limit $\omega^2 \gg \omega_{\rm p}^2$, (6.13) simplifies further to

$$A_\sigma(s,\omega,\theta) \approx \frac{s^2}{4s!}(\tfrac{1}{2}\lambda_\sigma)^{s-1}(1 - \sigma|\cos\theta|)^2. \tag{6.14}$$

The approximation to the magnetoionic modes actually corresponds to assuming that they are approximately circularly polarized, and this breaks down for $|\cos\theta| \lesssim \Omega_{\rm e}/2\omega_{\rm p}$, where they become nearly linearly polarized.

The Line Width

The absorption is confined to relatively narrow lines around each harmonic, with the x-mode more strongly absorbed than the o-mode. The width, $(\Delta\omega)_s$, of the line at the sth harmonic, $(\Delta\omega)_s$, is determined by the Doppler broadening due to the thermal motions. From the exponential dependence in (6.12), one estimates

$$(\Delta\omega)_s/s\Omega_e = n_M\beta_0|\cos\theta|. \tag{6.15}$$

Gyromagnetic Absorption in the Solar Corona

Gyromagnetic absorption by thermal electrons can be important in preventing escape of radio emission generated at $\omega < 2\Omega_e$ or $\omega < 3\Omega_e$ from the solar corona. Specifically, suppose that the magnetic field decreases with radial distance r from the center of the Sun as $dB/dr = -B/L_B$; then the optical depth for absorption at the sth harmonic may be estimated from the product of the maximum value of the absorption coefficient (6.12) times the light propagation time over a distance in which $s\Omega_e$ changes by $(\Delta\omega)_s$. This gives an estimate for the optical depth, $\tau_{\sigma s}$, for cyclotron absorption at the sth harmonic:

$$\tau_{\sigma s} = \sqrt{2}n_e\beta_0|\cos\theta|L_B\gamma_{Ms}/c. \tag{6.16}$$

For $n_e = 10^{15}\,\mathrm{m}^{-3}$, $T_e = 3\times10^6\,\mathrm{K}$, $B = 10^{-2}\,\mathrm{T}$, $L_B = 10^7\,\mathrm{m}$, $\theta = 45°$, one finds that $\tau_{\sigma s}$ is very large for both modes at $s = 2$, and is large for the x-mode at $s = 3$. Hence one expects no radiation to escape through the second harmonic layer $\omega = 2\Omega_e$.

6.3 Cyclotron Maser Emission

Electron cyclotron absorption can be negative leading to electron cyclotron maser emission (ECME). There are several different versions of ECME (e.g., Melrose 1986). The most favorable version, proposed by Wu and Lee (1979), involves negative absorption of x-mode waves at the fundamental $s = 1$ near $\theta = \pi/2$ in a plasma with $\omega_p \ll \Omega_e$. The required source of free energy is a distribution of electrons with $\partial f/\partial p_\perp > 0$ at small $p_\perp$.

The Semirelativistic Approximation

The case of nearly perpendicular propagation, $\theta \approx \pi/2$, is of special significance because of the form of the resonance condition defined by the δ-function in (6.1). The strictly nonrelativistic approximation involves replacing the Lorentz factor γ by unity, and this corresponds to assuming that the ellipse is approximated by a vertical line, cf. Figures 2.3 and 6.1. In contrast, for $\theta = \pi/2$ the resonance ellipse reduces to a circle centered on the origin of velocity space and with radius $\beta = (1-\omega^2/s^2\Omega_e^2)^{1/2}$. A vertical line is a poor approximation to this circle when its radius is small. A *semirelativistic* approximation to the resonance condition is

$$\omega[1 - n_M(\omega, \theta)\beta_\parallel \cos\theta] - s\Omega_e(1 - \tfrac{1}{2}\beta_\perp^2 - \tfrac{1}{2}\beta_\parallel^2) = 0. \quad (6.17)$$

This corresponds to a resonance circle with its center displaced from the origin, cf. curve (b) in Figure 6.1. The circle drawn is chosen so that it lies just outside the region where the thermal electrons are located so that thermal gyromagnetic absorption may be neglected. In evaluating the gyromagnetic absorption coefficient (2.19), the nonrelativistic approximation is made everywhere except in the δ-function, given by (6.17).

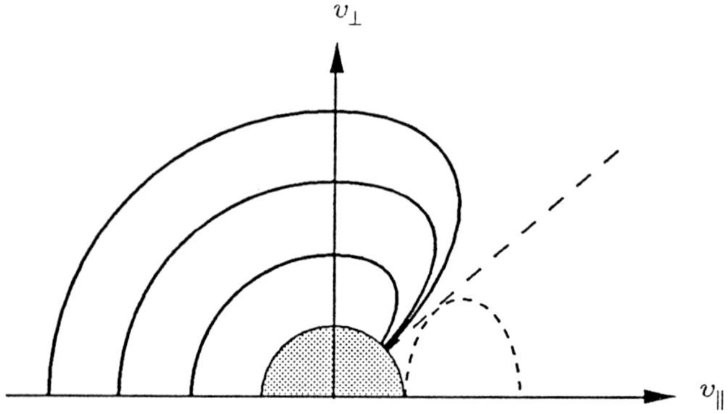

Fig. 6.2. A loss-cone distribution with loss-cone angle indicated by the dashed line. The solid lines denoted contours of constant f. The resonance ellipse illustrated by the dashed curve favors maser action.

Loss-Cone Driven Growth

There are two important requirements that need to be satisfied for ECME to be a possible source of escaping radiation. One is that there be a distribution of electrons with available free energy in the appropriate form, and the other is that the resulting radiation be able to escape directly from the plasma.

The source of free energy requires a feature in the distribution function with $\partial f/\partial p_\perp > 0$ at small $p_\perp$. One such distribution is the loss-cone distribution illustrated in Figure 6.2. Loss cones form naturally in a magnetic trap where electrons are confined due to the magnetic mirror effect. This requirement, $\partial f/\partial p_\perp > 0$ at small $p_\perp$, is satisfied for magnetically trapped electrons under a wide variety of circumstances. However, the back reaction to the maser emission causes electrons to be scattered into the loss cone, and if the maser operates this scattering is very efficient. Hence, for the maser to operate effectively, one requires a process that continuously reestablishes the loss cone. This is usually attributed to acceleration of the electrons in the planetary magnetosphere or the stellar corona. Then ECME can operate only while the acceleration continues to provide a fresh source of precipitating electrons.

Emission in the x-mode or the o-mode

A pre-condition for the radiation to escape is that, as the plasma frequency decreases away from the source, the waves do not encounter any stop band. Inspection of Figure 1.1 shows that only the o-mode and the x-mode satisfy this requirement.

The x-mode at given s has a larger growth rate than the o-mode. This is because the polarization of the x-mode is such that its electric vector rotates in the same sense as the electrons gyrate, whereas the electric vector in the o-mode rotates in the opposite sense. As a consequence, the coupling between electrons and x-mode waves is much stronger than the coupling between electrons and o-mode waves.

Emission at $s = 1$ in the x-mode must be above the cutoff frequency, $\omega = \omega_x$ as given by (1.30), which reduces to

$$\omega_x \approx \Omega_e + \omega_p^2/\Omega_e \tag{6.18}$$

for $\omega_p \ll \frac{1}{2}\Omega_e$. Thus emission at $s = 1$ in the x-mode is possible only in a plasma with $\omega_p \ll \Omega_e$.

The condition $\omega > \omega_x$ excludes emission at $\theta = \pi/2$, because for $\cos\theta = 0$, the resonance condition reduces to $\omega = \Omega_e/\gamma < \Omega_e < \omega_x$. The value of $\cos\theta$ determines the center of the resonance ellipse and, as illustrated in Figure 6.2, the center needs to be close to the origin, $|\cos\theta| \ll 1$, but not at the origin, $\cos\theta \neq 0$. It follows that ECME occurs at small but nonzero $|\cos\theta|$.

Integration Around the Resonance Ellipse

The evaluation of the absorption coefficient may be simplified by integrating around the resonance ellipse. The parameters (2.12) of the ellipse may be approximated by those of a circle:

$$\frac{v_c}{c} = \frac{k_\parallel c}{s\Omega_e}, \quad \frac{v_c}{c} = \left[\left(\frac{v_c}{c}\right)^2 - \frac{2(\omega - s\Omega_e)}{s\Omega_e}\right]^{1/2}. \tag{6.19}$$

The integration around the circle involves changing variables from $v_\parallel$, $v_\perp$ to polar variables v', ϕ' relative to the center of the circle:

$$v_\parallel = v_c - v'\cos\phi', \quad v_\perp = v'\sin\phi', \quad \delta(\omega - s\Omega - k_\parallel v_\parallel) = \frac{c^2}{s\Omega_e v_R}\delta(v' - v_R). \tag{6.20}$$

The expression (2.19) for the absorption coefficient, with the general form (2.14) for the probability of emission, then reduces to

$$\gamma_M(\boldsymbol{k}) = -\sum_{s=-\infty}^{\infty} \frac{16\pi^3 m_e^4 c^4 r_0 v_R R_M(\boldsymbol{k})}{s\Omega_e\omega} \int_0^\pi d\phi' \times \left\{ |\boldsymbol{e}_M^*(\boldsymbol{k}) \cdot \boldsymbol{V}(\boldsymbol{k},\boldsymbol{p},s)|^2 \left[\frac{s\Omega}{v_\perp}\frac{\partial}{\partial p_\perp} + k_\parallel \frac{\partial}{\partial p_\parallel} f(\boldsymbol{p})\right]\right\}_{v'=v_R}. \tag{6.21}$$

Approximations to the Growth Rate

Some useful approximate analytic forms for the absorption coefficient (6.21) for ECME have been derived (e.g., Hewitt, Melrose and Rönnmark 1982). However, in most cases the evaluation needs to be performed numerically (e.g., the review by Wu 1986), or by particle-in-cell techniques (e.g., Winglee, Dulk and Pritchett 1988).

An analytic estimate of the growth rate for ECME at $s = 1$ in the x-mode for a loss-cone distribution, with electrons with a typical speed $\langle\beta\rangle c$ and loss-cone angle α_0, is (Hewitt, Melrose and Rönnmark 1982)

$$|\gamma_x| \approx \frac{\pi\omega_\mathrm{p}^2}{\Omega_\mathrm{e}} \frac{n_1}{n_\mathrm{e}} \frac{1}{\langle\beta\rangle^2 \sin\alpha_0}, \tag{6.22}$$

where n_1 is the number density that would be required to fill the loss cone. In practice this is a large growth rate, and the maser should saturate quickly. As a result, the actual value of the growth rate is relatively unimportant. Thus, ECME is an efficient process when the conditions for it to occur are satisfied.

6.4 Applications of Cyclotron Maser Emission

Electron cyclotron maser emission (ECME) is the favored emission mechanism for certain planetary radio emission, for solar spike bursts, and for very bright radio emission from some flare stars.

AKR

The Earth is a spectacular radio source. However the frequency of the radiation is below the ionospheric cutoff frequency, and hence it cannot be detected directly from the ground. This radiation, called the auroral kilometric radiation (AKR), was not studied in detail until the early 1970s (e.g., Gurnett 1974). AKR correlates with "inverted-V" events, which are a particular class of auroral electron precipitations with energies $\approx$ 2 to 10 keV. Inverted-V electrons are thought to be accelerated by parallel (to $\boldsymbol{B}$) electric fields (e.g., Mozer et al. 1980; Boström et al. 1988).

The theory of ECME accounts naturally for the polarization of AKR, which is almost entirely in the x-mode. Another success was the prediction (Melrose 1976) that the condition $\omega_\mathrm{p} \ll \Omega_\mathrm{e}$ must be satisfied in the source region. AKR occurs in localized density cavities, where ω_p decreases from the surrounding values by a large factor (Benson, Calvert and Klumpar 1980). The density cavities are attributed to the parallel electric field depopulating the magnetosphere in these localized regions.

The observed pitch-angle distributions of inverted-V electrons are such that, when used to calculate the growth rate, they imply that ECME should occur (Omidi and Gurnett 1982; Melrose, Rönnmark and Hewitt 1982). This provides strong support for the interpretation in terms of ECME.

DAM

Jupiter is a strong, sporadic radio source at decametric (DAM) wavelengths (e.g., Warwick 1967; Carr and Gulkis 1969). DAM correlates with the position of the innermost Galilean satellite Io, especially near the highest frequency ($\approx 40\,\mathrm{MHz}$) at which DAM occurs.

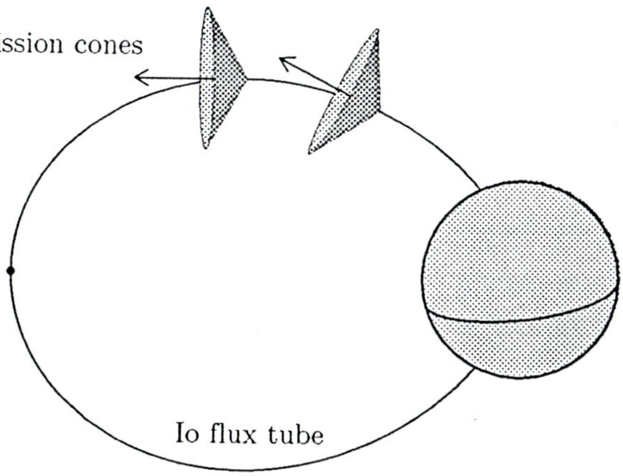

Fig. 6.3. The emission pattern for DAM is modeled in terms of emission on the surface on a cone whose apex lies on a flux tube that passes through the orbit of Io, and with the frequency of emission determined by the cyclotron frequency at the point of emission.

Since the 1960s it has been thought that DAM is due to some form of ECME from mildly relativistic electrons ($\approx$ few times $10^2\,\mathrm{keV}$) with the source located on the magnetic field lines that pass through the orbit of Io (e.g., Goldreich and Lynden-Bell 1969). Strong evidence in support of the interpretation emerged once the loss-cone-driven version of the theory was formulated (Wu and Lee 1979). A surprising inference from the early data concerns the angular emission pattern: it was inferred that DAM is emitted locally on the surface of a hollow cone with opening angle $\approx 80°$ with a thickness of $\approx 1°$ (Dulk 1967). This seemingly bizarre pattern is just what is predicted for loss-cone-driven ECME.

The fly-bys of Jupiter by the Pioneer and Voyager spacecraft provided detailed information relevant to DAM (e.g., Dessler 1983). The observed emission pattern is consistent with emission into hollow cones, with apex of the cone on the magnetic field lines that pass through the orbit of Io, and with emission at the local cyclotron frequency (Goldstein and Goertz 1983). This pattern is illustrated schematically in Figure 6.3. The equator of Jupiter is nearly in the ecliptic plane, and radiation can be seen from the Earth only if it is emitted nearly in this plane.

DAM is polarized in the sense of the x-mode, and is consistent with the predictions of the ECME theory (Melrose and Dulk 1993).

Solar Spike Bursts

Spike bursts are one special class of solar radio burst that occur at decimetric wavelengths (frequencies $\sim 1\,\mathrm{GHz}$) in association with solar flares (e.g., Benz 1986, 1993). They have a short rise time, $\sim 10\,\mathrm{ms}$, that indicates a small source size and hence a brightness temperature considerable greater than for other solar bursts. They can also be also highly circularly polarized. These features favor ECME (Holman, Eichler and Kundu 1980; Melrose and Dulk 1982; Vlahos 1987), rather than plasma emission (lecture 8). It is thought that spike burst are produced close to the site of electron acceleration in solar flares, with the maser driven by accelerated electrons propagating away from their source toward increasing B.

Bright Emission from Flare Stars

The radio emission from flare stars can be separated into two components (e.g., Dulk 1985). One has a maximum brightness temperature, $T_\mathrm{b} \lesssim 10^{10}\,\mathrm{K}$, and is attributed to gyrosynchrotron emission. The other has $T_\mathrm{b} \gg 10^{10}\,\mathrm{K}$ and so requires a coherent emission mechanism. Electron cyclotron maser emission is the favored mechanism, although plasma emission cannot be ruled out (Kuijpers 1985).

6.5 Difficulties with Cyclotron Maser Emission

Electron cyclotron maser emission is now an accepted coherent emission mechanism in plasma astrophysics. However, there remain difficulties with all the suggested applications.

Sources of Free Energy for AKR

A maser requires a pump, which provides free energy that is converted into radiation. A pump for ECME is required to produce the feature in the distribution function with $\partial f/\partial p_\perp > 0$ that drives the instability. A loss-cone anisotropy due to electrons precipitating from a source in a planetary magnetosphere or stellar corona appears to be a favorable mechanism. However, although this has been the accepted mechanism for AKR, more recent studies of the detailed form of the distribution cast some doubt on this.

As illustrated schematically in Figure 6.4, several different sources of free energy for AKR are apparent in the measured distributions: a loss-cone anisotropy, a 'hole', and 'trapped' electrons (Louarn et al. 1990; Roux et al. 1993). The 'hole' is in a region of velocity space below a precipitating beam of electrons. The trapped distribution is in a 'forbidden' region of velocity space, which is inaccessible to precipitating electrons under steady-state conditions. Nevertheless, a peak in the distribution is located in just this region. Which of these features, if any, is the actual source of free energy for ECME is unclear. Present observations cannot resolve this uncertainty because the growth time of the maser

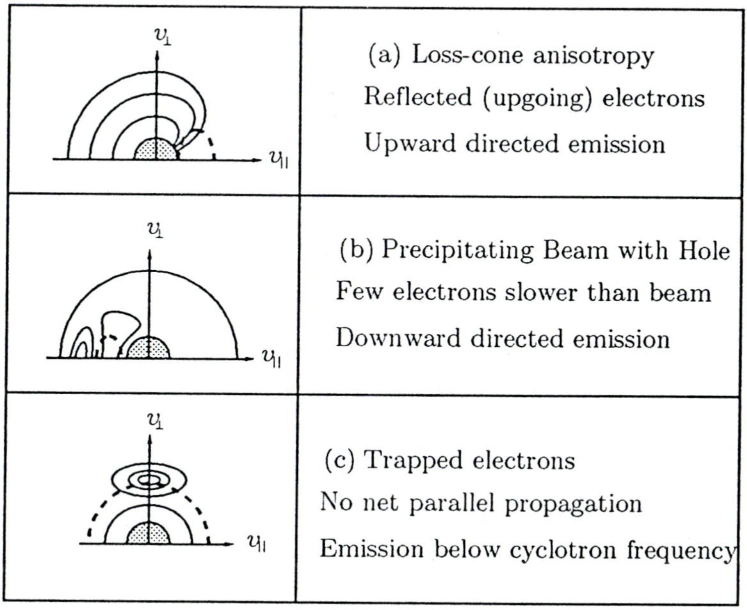

Fig. 6.4. Three sources of free energy identified in AKR. The solid curves are contours of the electron distribution function, thermal electrons are confined to the shaded region, and the dashed curves denote the most favorable resonance ellipses. [After Roux et al. (1993)]

is much shorter than the time required to determine the distribution function. Consequently, the measured distribution function is relaxed, in the quasilinear sense, and the specific features that cause the waves to grow should no longer be present. Only particle data with higher time resolution can make it possible to identify unambiguously the particular feature that drives the ECME.

Fine Structures

Dynamic spectra of AKR and DAM show narrowband structures, with bandwidth so narrow that they seem to preclude an explanation in terms of any maser mechanism. This is because maser theory applies in the random phase approximation, which requires that the bandwidth of the growing waves exceed the growth rate. As a consequence, these fine structures cannot be explained in terms of the version of ECME discussed here. There is an alternative (reactive) version of ECME that allows faster growth than the weak-turbulence version of the mechanism (e.g., Melrose 1986), and involves growth of a phase-coherent disturbance due to self-bunching of the particle distribution. However, the observed fine structures are similar in appearance to the discrete VLF emissions in the whistler mode, and it may be that a model with some features in common with Helliwell's (1967) model for discrete VLF emissions is appropriate.

Other kinds of structure that cannot be explained as a natural consequence of ECME include reported harmonic structure in AKR and spike bursts, and a weak o-mode component in AKR. In a maser the fastest growing mode should dominate virtually completely, exhausting the available free energy before more slowly growing modes build up to a significant level. If the fastest growing mode is the x-mode at $s = 1$, then one expects effectively no emission at other harmonics or in the o-mode. Explanation for weak components at harmonics or in the o-mode should be sought in terms of propagation effects, such as reflection off overdense regions, or due to nonlinearities and inhomogeneities in the plasma.

Cyclotron Absorption of Spike Bursts

Perhaps the most serious unsolved problem remains in the application of ECME to solar spike bursts: cyclotron absorption at the second-harmonic layer should prevent it from escaping from the solar corona (Melrose and Dulk 1982). Suggestions as to how this difficulty might be overcome include the following four. 1) The escaping ECME is at $s \geq 2$, and is generated above the second harmonic layer, or maybe in the o-mode, for which the absorption is weaker than for the x-mode. 2) The maser produces z mode waves and the escaping radiation results from coalesce of z mode waves with themselves (Melrose 1991). 3) There are 'windows' through which the radiation can escape. The absorption coefficient at $s = 2$ varies as $\sin^2\theta$, cf. (6.14) with (6.13), providing a window at small $\sin\theta$. Another window near $\theta = \pi/2$ also exists for the o-mode (Robinson 1989). 4) Radiation incident on the absorbing layer from below with $\omega < 2\Omega_e$ can modify the distribution function of the electrons there, allowing re-emission at $\omega > 2\Omega_e$ (McKean, Winglee and Dulk 1989).

The quantitative interpretation of solar spike bursts is strongly dependent on what assumptions are made concerning how the radiation escapes. If none of these or other explanations for the escape is found satisfactory then the ECME theory for spike bursts may have to be discarded.

7 Synchrotron Emission

Synchrotron emission is gyromagnetic emission from highly relativistic particles. In this lecture, after a summary of the theory of synchrotron emission (e.g., Ginzburg and Syrovatskii 1965; Bekefi 1966; Rybicki and Lightman 1979; Melrose 1980a; Shu 1991), several aspects of the application to astrophysical sources are discussed, emphasizing the interpretation of spectra and of the polarization.

7.1 Qualitative Properties of Synchrotron Emission

The application of gyromagnetic emission to ultrarelativistic particles includes some subtleties (Ginzburg and Syrovatskii 1965). An important preliminary point applies to any emission by a relativistic particle.

Emission into a Forward Cone

A simple argument based on Lorentz transforming from the laboratory frame, K, to the frame in which the radiating particle is instantaneously at rest, K_0, leads to the following two general results for ultrarelativistic particles ($\gamma \gg 1$):

1) All angles of emission in K_0, with the exception of a range of order γ^{-1} about the direction $-\boldsymbol{v}$, transform into a forward cone with half-angle of order γ^{-1} about the direction $\boldsymbol{v}$ in K.
2) All but a fraction γ^{-1} of the power emitted in K is confined to a forward cone with half-angle of order γ^{-1} about the direction $\boldsymbol{v}$.

For gyromagnetic emission 2) implies that all but a fraction γ^{-1} of the power emitted by a particle with pitch angle α is confined to angles θ satisfying

$$\theta = \alpha + O(\gamma^{-1}). \tag{7.1}$$

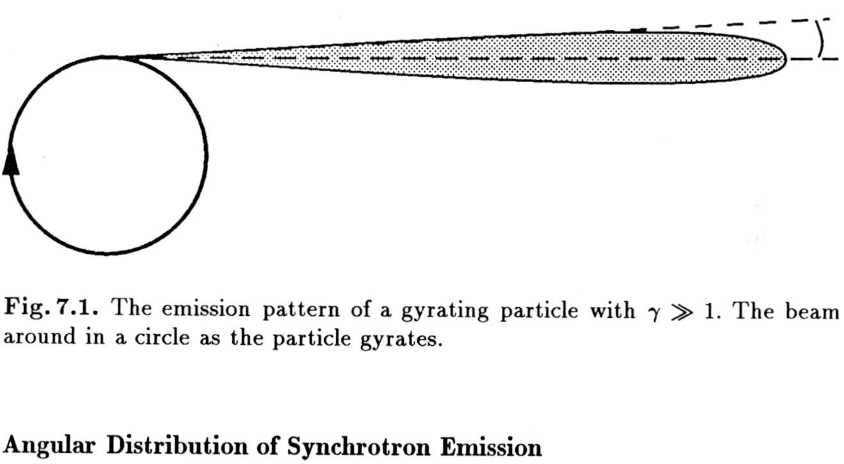

Fig. 7.1. The emission pattern of a gyrating particle with $\gamma \gg 1$. The beam sweeps around in a circle as the particle gyrates.

Angular Distribution of Synchrotron Emission

For gyromagnetic emission emission by a relativistic particle in vacuo, the exact result for the power radiated (6.9) gives

$$P \approx \frac{2m_e r_0 c\Omega_e^2}{3}\gamma^2 \sin^2\theta. \tag{7.2}$$

The resonance condition implied by the δ-function in (6.1) simplifies to

$$\omega \approx s\Omega_e/\gamma \sin^2\theta. \tag{7.3}$$

The typical frequency emitted cannot be inferred from (7.3) alone because the value of s is undetermined.

Synchrotron Emission as a Sequence of Pulses

A further semiquantitative argument enables one to estimate the typical frequency in synchrotron emission by considering the temporal distribution of the radiation received by an observer. First note that because of the highly anisotropic angular distribution of synchrotron emission, significant emission from a particle with pitch angle α is seen by an observer only in the narrow range of angles about $\theta = \alpha$. One pulse of radiation is received each revolution of the particle, with this pulse corresponding to the range of angles of order γ^{-1} during which the emission cone of the particle sweeps across the observer, as illustrated in Figure 7.1. Consider the case $\alpha = \pi/2$, which corresponds to the particle moving in a circle. The particle performs one gyration in a time $2\pi\gamma/\Omega_e$. The time interval per gyration in which it is traveling towards the observer is a fraction of order γ^{-1} of this, that is, a time interval $\approx 2\pi/\Omega_e$. Consider the pulse received when a particle with speed βc moves directly towards the observer for a time Δt. Suppose that the particle is traveling in the x direction and that it starts radiating at $t = 0$ when it is at $x = 0$ and stops radiating at time $t = \Delta t$ when it is at $x = \beta c \Delta t$. (This starting and stopping of emission in the direction of the observer simulates the sweeping of the beam illustrated in Figure 7.1 across the direction to the observer.) The length of the wave train received by the observer is then $(1 - \beta)c\Delta t \approx c\Delta t/2\gamma^2$, and the duration of the pulse received per gyration by the observer is $\Delta t_{\rm rec} \approx \pi/\gamma^2\Omega_e$. A pulse of duration Δt contains Fourier components $\omega \lesssim 1/\Delta t$. Hence one expects synchrotron radiation to contain frequencies

$$\omega \lesssim \pi^{-1}\gamma^2\Omega_e \sin\theta, \tag{7.4}$$

where the factor $\sin\theta$ is included by making a Lorentz transformation from the frame where the particle has pitch angle $\pi/2$ to the frame where its pitch angle is θ. It then follows from (7.3) that the important harmonic numbers involved are $s \lesssim \pi^{-1}(\gamma\sin\theta)^3$. Thus, for $(\gamma\sin\theta)^3 \gg 1$, the emission is dominated by high harmonics and one is justified in treating s as a continuous variable.

Derivation of Synchrotron Formulas

There are several different ways of deriving the formulas for synchrotron emission, and all involve making Airy integral approximations to Bessel functions. In the approach described here, the Airy integral approximation is made at the last possible stage.

The synchrotron formulas may be derived by following the procedure outlined in lecture 6 to derive the power emitted by a single particle. Instead of integrating over angles of emission, one integrates over the pitch-angle distribution of the radiating electrons. According to (7.1) these integrals are equivalent to lowest order in an expansion in $1/\gamma$. In the counterpart of (6.7) one assumes that s is a continuous variable, and the integral over s is reexpressed as an integral over ω using (7.3). Then, noting that the argument of the resulting Bessel functions is $2s\beta'$, with β' given by (6.5), the Airy integral approximation gives

$$J_{2s}(2s\beta') \approx \frac{K_{1/3}(R)}{\pi\sqrt{3}\gamma\sin\theta}, \qquad R = \frac{2s}{3\gamma^3\sin^3\theta} = \frac{\omega}{\omega_c}, \qquad \omega_c = \tfrac{3}{2}\Omega_e\gamma^2\sin\theta. \tag{7.5}$$

The volume emissivity in synchrotron emission is then found by multiplying the resulting expression by the pitch-angle distribution, $\phi(\theta)$, evaluated at the angle of emission, and dividing by 4π steradians.

7.2 Synchrotron Formulas

The procedure outlined above gives the emissivity in synchrotron emission for a single particle. The integral over a power-law distribution leads to a power-law frequency spectrum.

The Synchrotron Emissivity

The emissivity in synchrotron emission reduces to

$$\eta^{\|,\perp}(\omega,\theta) = \frac{\sqrt{3}}{8\pi^2} m_e r_0 c\Omega_e\, \delta(\theta-\alpha)\, F^{\|,\perp}(\omega,\theta),$$

$$F^{\|,\perp}(\omega,\theta) = \frac{\omega}{\omega_c}\int_{\omega/\omega_c}^{\infty} dt\, K_{5/3}(t) \mp (\omega/\omega_c)K_{2/3}(\omega/\omega_c). \tag{7.6}$$

The δ-function expresses the condition (7.1). The power radiated per unit frequency is found by integrating (7.6) over solid angle and summing over the two states of polarization:

$$P(\omega) = \frac{\sqrt{3}}{2\pi} m_e r_0 c\Omega_e \sin\theta\, F(\omega/\omega_c),$$

$$F(R) = \tfrac{1}{2}[F^{\|}(R) + F^{\perp}(R)] = R\int_R^{\infty} dt\, K_{5/3}(t). \tag{7.7}$$

Expansions of the function $F(R)$ for small and large arguments are given by

$$F(R) = \begin{cases} \dfrac{4\pi}{\sqrt{3}\Gamma(1/3)}\left(\dfrac{R}{2}\right)^{1/3}\left[1 - \tfrac{1}{2}\Gamma(1/3)\left(\dfrac{R}{2}\right)^{2/3} + \cdots\right], \\ \left(\dfrac{\pi R}{2}\right)^{1/2} e^{-R}\left[1 + \dfrac{55}{72R} + \cdots\right], \end{cases} \tag{7.8}$$

for $R \ll 1$ and $R \gg 1$, respectively. In between these limiting cases, there is a maximum at $F(0.29) = 0.92$. The function $F(R)$ is plotted in Figure 7.2. A simple analytic approximation to it is $F(R) \approx 1.8\, R^{0.3}\, e^{-R}$.

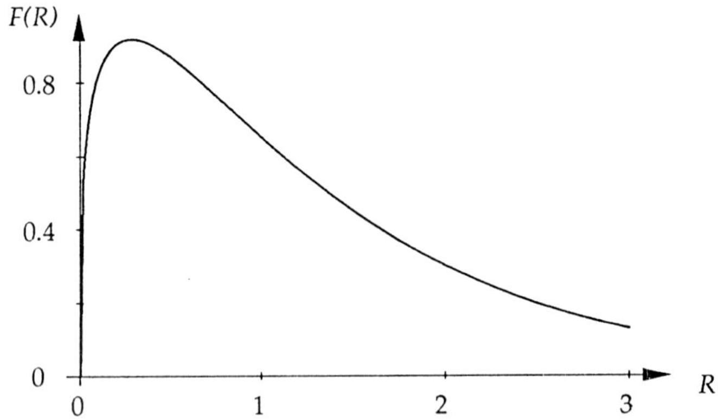

Fig. 7.2. The function $F(R)$ defined by (7.7).

Volume Emissivity for a Power-Law Distribution

In most synchrotron sources the frequency spectrum, at least at higher frequencies, has a power-law form $I(\omega) \propto \omega^{-\alpha}$, where α is the *spectral index*. This type of spectrum is interpreted in terms of emission from relativistic electrons with a power-law energy distribution.

The emission due to a distribution $f(p, \alpha) = f(p)\phi(\alpha)$ of electrons is found by integrating the emissivity (7.6) over the distribution of electrons; this gives the emissivity per unit volume, referred to above as the emission coefficient $J(\omega, \theta)$. For a power-law distribution in energy it is convenient to introduce the *energy spectrum* $N(\varepsilon)$, with $\varepsilon = pc$ for ultrarelativistic particles, by writing

$$4\pi \int_0^\infty dp\, p^2 f(p) = \int_0^\infty d\varepsilon\, N(\varepsilon), \quad N(\varepsilon) = \frac{4\pi}{c^3}\, \varepsilon^2 f(\varepsilon/c). \tag{7.9}$$

The energy spectrum is assumed to be a power law of the form

$$N(\varepsilon) = \begin{cases} K\varepsilon^{-a} & \text{for } \varepsilon_1 \le \varepsilon \le \varepsilon_2, \\ 0 & \text{otherwise.} \end{cases} \tag{7.10}$$

In integrating the emissivity over this energy spectrum, one sets $\varepsilon_1 \to 0$, $\varepsilon_2 \to \infty$ (numerical integration is required for other limits). The volume emissivity reduces to

$$J^{\|,\perp}(\omega, \theta) = \frac{K(mc^2)^{-a+1}}{16\pi^2 c} \frac{\sqrt{3} q^2 \Omega_e \sin\theta\, \phi(\theta)}{4\pi\varepsilon_0}\, j^{\|,\perp}(a) \left(\frac{2\omega}{3\Omega_e \sin\theta}\right)^{-(a-1)/2},$$

$$j^{\|}(a) = \frac{2/3}{a+1} 2^{(a-3)/2}\, \Gamma\left(\frac{3a+7}{12}\right) \Gamma\left(\frac{3a-1}{12}\right), \quad j^{\perp}(a) = \frac{a+5/3}{2/3}\, j^{\|}(a). \tag{7.11}$$

It follows that the spectral index a of the particle energy spectrum, and the spectral index α of the synchrotron emission are related by $\alpha = \frac{1}{2}(a-1)$, or $a = 2\alpha + 1$.

The Degree of Polarization

The degree of linear polarization of the synchrotron emission is given by

$$r_\ell = -\frac{a+1}{a+7/3} = -\frac{\alpha+1}{\alpha+5/3}. \tag{7.12}$$

The relatively high degree of polarization is a characteristic signature of synchrotron emission. For a value of $a = 3$, which is typical for many sources, (7.12) implies $r_\ell = -3/4$, that is a degree of linear polarization of 75% in the direction orthogonal to the projection of the magnetic field $\boldsymbol{B}$ in the source on the plane of the sky. This prediction is based on the seemingly unrealistic assumption that the magnetic field in the source is uniform. In practice one expects the orientation of $\boldsymbol{B}$ to vary along the line of sight and across the source, and this would reduce the degree of polarization observed; the degree of polarization would be zero for a source in which the orientation of the magnetic field is random. Nevertheless many synchrotron sources show relatively high degrees of polarization, for example, $r_\ell \gtrsim 30\%$ and a few have $r_\ell \gtrsim 70\%$.

Synchrotron Absorption

To every emission process there is a corresponding absorption process, implying the existence of synchrotron absorption, sometimes called synchrotron self-absorption.

The absorption coefficients for the two linear polarizations are given by

$$\gamma^{\|,\perp}(\omega,\theta) = -\frac{(2\pi)^4 c^3}{\omega^2}\int_{-1}^{1} d\cos\alpha\,\phi(\alpha)\int_0^\infty d\varepsilon\,\varepsilon^2\,\eta^{\|,\perp}(\omega,\theta)\,\frac{d}{d\varepsilon}\left[\frac{N(\varepsilon)}{\varepsilon^2}\right]. \tag{7.13}$$

The absorption coefficient (7.13) cannot be negative, as may be shown by partially integrating (Wild, Smerd and Weiss 1963). Thus, except under extreme conditions that are not considered here, there can be no synchrotron maser emission. For the power-law distribution (7.10), with the limits $\varepsilon_1 \to 0$, $\varepsilon_2 \to \infty$, (7.13) gives

$$\begin{aligned}\mu^{\|,\perp}(\omega,\theta) = \frac{(2\pi)^3 c}{\omega^2}\,\frac{K(mc^2)^{-a}}{16\pi^2c^2}\,\frac{\sqrt{3}q^2\Omega_0\sin\theta\phi(\theta)}{4\pi\varepsilon_0}\\ \times (a+2)j^{\|,\perp}(a+1)\left(\frac{2\omega}{3\Omega_0\sin\theta}\right)^{-a/2},\end{aligned} \tag{7.14}$$

with $j^{\|,\perp}(a)$ given by (7.11).

When a source is optically thick to synchrotron radiation it is said to be self absorbed. The spectrum in the optically thick region is determined by the ratio of the emission coefficient (7.11) to the absorption coefficient (7.14). These are $\propto \omega^{-(a-1)/2}$ and $\propto \omega^{-2-a/2}$, respectively. This implies $I(\omega) \propto \omega^{5/2}$ for a self-absorbed source. The $\omega^{5/2}$-spectrum may be understood by noting the intensity for a source with brightness temperature T_b is $I(\omega) \propto \omega^2 T_b$, and that for a self-absorbed source the effective temperature is approximately equal to the typical energy of the particles radiating at ω. Using (7.4), this energy is $\varepsilon \propto \omega^{1/2}$, implying $T_b \propto \omega^{1/2}$.

7.3 Synchrotron Spectra

The main features of an observed synchrotron spectrum are its shape, usually a power law or combination of power laws, any observed low-frequency turnover, and its polarization.

Power-Law Spectra

In the simplest case, the frequency distribution is a power law over the range of interest, $F_\omega \propto \omega^{-\alpha}$, and the interpretation is that this is due to a power-law distribution of particles, $N(\varepsilon) \propto \varepsilon^{-a}$. The power-law indices are related by, cf. (7.11),

$$\alpha = \tfrac{1}{2}(a-1), \qquad a = 2\alpha + 1. \tag{7.15}$$

Synchrotron sources with moderately steep spectra are usually power laws. A wide range of values of α is found, with $\alpha = 1$ implying $a = 3$ being a value often chosen for illustrative purposes.

Flat spectrum sources, $\alpha \approx 0$, are relatively common. Setting $\alpha = 0$ in (7.15) implies $a = 1$, which is flatter than can be explained by any plausible acceleration mechanism. For very energetic sources, such as AGN, flat spectra are usually attributed to a self-absorbed source whose peak optical depth varies with the size of the source in such a way as to produce a flat spectrum (e.g., Blandford, Netzer and Woltjer 1990, p. 232). There are other synchrotron sources with flat spectra for which this model seems inappropriate. However, there is no other accepted explanation for flat spectra.

Low-Frequency Turnovers

The usual interpretation of a low-frequency turnover in synchrotron source is in terms of synchrotron self absorption. However, there are several other possible types of turnover, as illustrated schematically in Figure 7.3. The low-frequency dependences in Figure 7.3 may be understood as follows. 1) External free-free absorption involves an optical depth $\tau_{\mathrm{ff}} \propto \omega^{-2}$, so that $\exp(-\tau_{\mathrm{ff}})$ cuts off very sharply with decreasing ω. 2) The Razin effect is due to suppression of synchrotron emission in the presence of a plasma, as explained below. 3) The $\omega^{5/2}$ dependence for a self-absorbed source, is derived and interpreted following (7.14). 4) Internal free-free absorption results from the emissivity $\propto \omega^{-\alpha}$ being balanced by free-free absorption $\propto \omega^{-2}$. 5) The derivation of the $\propto \omega^{1-\alpha/2}$ dependence

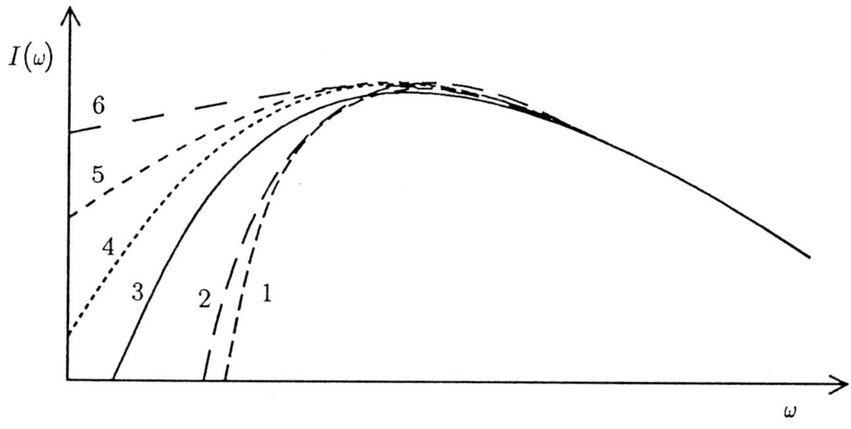

Fig. 7.3. Low-frequency turnovers in synchrotron spectra (log-log plot): 1) external free-free absorption, 2) Razin suppression, 3) synchrotron self absorption, $\propto \omega^{5/2}$, 4) internal free-free absorption, $\propto \omega^{2-\alpha}$, 5) induced Compton scattering, $\propto \omega^{1-\alpha/2}$, 6) low-energy cutoff, $\propto \omega^{1/3}$.

for induced Compton scattering is outlined following (7.32) below. 6) The $\omega^{1/3}$-dependence is characteristic of the emission of all relativistic electrons at sufficiently low frequencies, cf. (7.8).

The Razin Effect

The Razin effect is *not* an absorption effect, but is merely due to the presence of a plasma suppressing emission. One extreme example of this effect can be imagined by considering emission at some natural frequency (say the 21 cm line of hydrogen) in plasmas of increasing density. Provided the plasma frequency is much less than the natural frequency, its effect is negligible. However, as the plasma frequency approaches the natural frequency, the refractive index $n = (1-\omega_\mathrm{p}^2/\omega^2)^{1/2}$ becomes significantly less than unity. The emissivity varies as a power of the refractive index ($\propto n^3$ for the magnetic dipole transition involved in 21 cm emission), and so the power radiated goes down. A collection of atoms radiates less due to the presence of the medium. Once the plasma frequency exceeds the natural frequency, the radiation ceases, and an atom in the excited state is unable to make the radiative transition.

Qualitatively, the Razin effect may be understood by applying this suppression to emission at the sth harmonic with $s \gg 1$. For electric 2^s-multipole emission, the emissivity depends on the refractive index as $\propto n^{2s+1}$. With $n = 1 - \omega_\mathrm{p}^2/2\omega^2$, one has $n^{2s} \approx \exp(-s\omega_\mathrm{p}^2/\omega^2)$. This correctly implies an exponential cutoff with decreasing ω, as occurs in the Razin effect. However, the details of this argument are not appropriate because emission by ultrarelativistic particles cannot be treated as multipole emission. In fact, the important way that the refractive index appears is through β in vacuo being replaced by $n\beta$ in a plasma, cf. (6.1). Then one has

$$n\beta \approx \left(1-\frac{\omega_{\rm p}^2}{2\omega^2}\right)\left(1-\frac{1}{2\gamma^2}\right)\approx 1-\frac{1}{2\gamma^2}\left(1+\frac{\gamma^2\omega_{\rm p}^2}{2\omega^2}\right). \tag{7.16}$$

It follows that, if one writes $n\beta \approx 1 - 1/2\gamma_{\rm eff}^2$, then for $\omega \ll \gamma\omega_{\rm p}$, the particle appears to be less relativistic, $\gamma_{\rm eff} \ll \gamma$. As a consequence the emission by the particle is suppressed by an exponential factor as $\omega < \gamma\omega_{\rm p}$ decreases. The frequency at which this suppression occurs follows by writing $\omega = \gamma\omega_{\rm p} = \gamma^2 \Omega_{\rm e} \sin\theta$ and eliminating γ. This gives the Razin-Tsytovich frequency, $\omega_{\rm RT} = \omega_{\rm p}^2/\Omega_{\rm e}\sin\theta$.

The Magnetic Field in Self-Absorbed Sources

There is a major uncertainty in determining the magnetic field, B. The characteristic frequency of synchrotron emission, $\propto \varepsilon^2 B$, is determined by a combination of the energy, ε, of the emitting particle and the magnetic field strength B. These two parameters cannot be inferred separately without further information.

One additional observational feature that can allow one to determine B is when there is a low-frequency turnover. If a peak in the frequency spectrum exists, the observations provide two additional parameters: the frequency, $\omega_{\rm peak}$, and the brightness temperature, $T_{\rm peak}$, at the peak.

Assuming that the turnover is due to synchrotron absorption, $T_{\rm peak}$ must be of the order of the energy, $m_{\rm e}c^2(\gamma_{\rm peak}-1)$, of the electrons whose mean frequency of emission occurs at $\omega_{\rm peak}$. Hence one has

$$\omega_{\rm peak} \sim 0.5\Omega_{\rm e}\gamma_{\rm peak}^2, \qquad T_{\rm peak} \sim m_{\rm e}c^2\gamma_{\rm peak}. \tag{7.17}$$

Hence, measurement of $\omega_{\rm peak}$ and $T_{\rm peak}$ allows one to estimate B and $\gamma_{\rm peak}$ separately.

The Minimum Energy Argument

In the absence of a turnover, the observed intensity of the emission at given ω determines only the quantity $K(mc^2)^{-a+1}B^{(a+1)/2}$, cf. (7.11). Thus one cannot separately determine B or the number density n_1 or the energy density $W_{\rm par}$, in the energy spectrum (7.10):

$$n_1 = \frac{K}{a-1}\left(\varepsilon_1^{1-a}-\varepsilon_2^{1-a}\right), \qquad W_{\rm par} = \frac{K}{a-2}\left(\varepsilon_1^{2-a}-\varepsilon_2^{2-a}\right). \tag{7.18}$$

The *equipartition* argument corresponds to assuming equality of the energy densities of the particles and the magnetic field. Writing

$$W_{\rm par} = \xi W_m, \qquad W_m = B^2/2\mu_0, \tag{7.19}$$

the factor ξ remains a major uncertainty. If one assumes that only the energy density of the radiating electrons is included in the assumed equality, then one has $\xi = 1$. Alternatively, one assumes equipartition between the magnetic field and the sum of the energies in all energetic particles, and say only 3% of these are electrons with energy density $W_{\rm par}$; then one has $\xi = 0.03$.

There is a theoretical argument in support of (7.19). The argument is a relatively weak one: that the total energy density $W_{\rm par} + B^2/2\mu_0$ be a minimum for the observed synchrotron emission. On expressing the particle energy density in terms of B using (7.11) at fixed ω, one finds $K \propto B^{-(a+1)/2}$ for a fixed synchrotron spectrum, and then (7.18) and (7.11) imply $W_{\rm par} \propto B^{-3/2}$. Minimizing $W_{\rm par} + B^2/2\mu_0$ with respect to B then gives (7.19) with $\xi = 4/3$.

Polarization of Sources

The polarization of synchrotron emission is intrinsically high, with $r_l = -(a+1)/(a+7/3)$, which gives $r_l = -0.75$ for $a = 3$. The polarization is linear in the direction orthogonal to the projection on the plane of the sky of the magnetic field lines in the source. There is a small circularly polarized component of synchrotron radiation, which appears in next order in the expansion in $1/\gamma$, and is normally too small to be measured accurately in synchrotron sources.

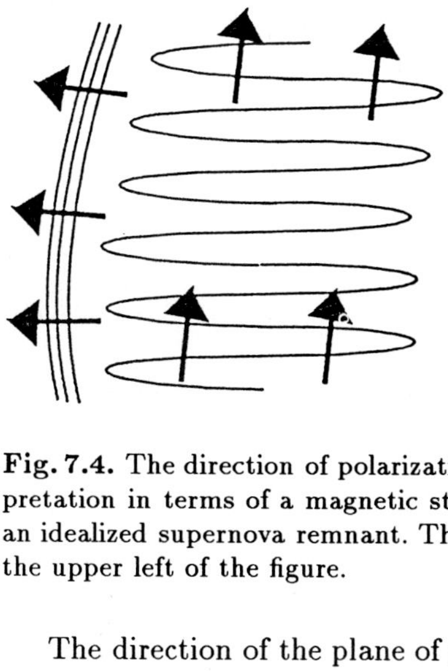

Fig. 7.4. The direction of polarization, shown by solid arrows, and its suggested interpretation in terms of a magnetic structure is indicated for a portion near the edge of an idealized supernova remnant. The shock is propagating from the lower right toward the upper left of the figure.

The direction of the plane of polarization provides information on the structure of the magnetic field in a resolved source. A notable example is for some supernova remnants, as illustrated schematically in Figure 7.4. The polarization implies that the magnetic field lines are nearly tangential at a sharp edge of a remnant, and are nearly radial just inside this boundary. A plausible explanation is that a shock driven by the ejecta from the supernova compresses the interstellar magnetic field in a thin layer where it is nearly tangential to the flow. Acceleration of electrons occurs at the shock, or perhaps more likely in subshocks embedded in a larger shock transition. These electrons cannot escape upstream, accounting for the sharp edge. Behind the shock the pressure of the energetic particles is substantial and exerts a pressure on the gas tending to accelerate it. As a consequence, there is a pressure gradient in which a light fluid (the gas of energetic particles) is accelerating a dense fluid (the swept up interstellar plasma). Such a situation is well-known to be unstable to the Rayleigh–Taylor instability, leading to the light fluid forming fingers that penetrate into the dense fluid, and the dense fluid forming fingers that penetrate into the light fluid. Thus this instability plausibly leads to fingers of each gas penetrating through the other, dragging along the magnetic field lines into a nearly radial direction, as illustrated in Figure 7.4.

Faraday Rotation

The polarization of synchrotron emission can be modified by Faraday rotation either in the source or in the interstellar medium. (Faraday rotation in the Earth's ionosphere can also be significant at lower frequencies, but is well enough understood for it to be taken into account in analyzing the radio data.)

Faraday depolarization results from differential Faraday rotation within the source itself. To see how this occurs consider two rays, one originating from the back and the other from the middle of the slab in an idealized slab model with a uniform magnetic field. If Faraday rotation causes one complete rotation (through 180°) for the ray from the back of the source as it crosses the source, then the plane of polarization rotates through 90° for the ray from the middle of the source. The sum of these two rays gives net polarization of zero, and one finds that a similar cancellation of the polarization occurs for all rays in this case. The condition for Faraday depolarization is $\Delta k L \gtrsim 1$, with

$$\Delta k = \frac{\omega_{\rm p}^2 \Omega_{\rm e} \cos\theta}{\omega^2 c}, \tag{7.20}$$

is the difference in wave number between the two magnetoionic modes at fixed ω, θ, and where L is the thickness of the source along the line of sight.

The Rotation Measure

Faraday rotation in the interstellar medium produces a characteristic signature: according to (7.20), the rate of rotation with distance along the ray path is inversely proportional to the square of the frequency. It is found that when the plane of linear polarization is measured at several different frequencies, the angle, ψ say, satisfies this predicated relation well. This enables one to remove the effect of Faraday rotation in the interstellar medium and to find the intrinsic plane of polarization at the source. The amount of Faraday rotation also enables one to estimate a parameter called the *rotation measure*, denoted RM, that depends only on the properties of the interstellar medium along the ray path:

$$\psi = \frac{e r_0 c}{2\pi m_{\rm e} \omega^2} \,\mathrm{RM}, \qquad \mathrm{RM} = \int ds\, n_{\rm e} B \cos\theta. \tag{7.21}$$

When combined with another parameter called the dispersion measure, DM = $\int ds\, n_e$, the rotation measure can be used to estimate the strength of the interstellar magnetic field. Both RM and DM can be measured for pulsars, whose radiation has a linearly polarized component that is subject to Faraday rotation. The dispersion results from the group speed of waves in a plasma depending on frequency, $v_g = nc \approx (1 - \frac{1}{2}\omega_p^2/\omega^2)c$, causing the arrival time of pulses to vary with frequency. Measurement of the delay enables one to estimate the integral of $\omega_p^2 \propto n_e$ along the ray path, that is, enables one to estimate DM. A typical value for the interstellar magnetic field is $\approx 0.5\,\mathrm{nT} = 5\,\mu\mathrm{G}$.

7.4 Evolution of Synchrotron Spectra

The spectrum of a synchrotron source evolves due to synchrotron losses, and also due to adiabatic expansion, and acceleration of particles (Kardashev 1962), neither of which are included here.

Synchrotron Losses

The rate of energy loss by an electron due to synchrotron is equal to the power radiated, P, cf. (6.9). Hence one has

$$\dot{\varepsilon} = -b(\alpha)\varepsilon^2, \qquad b(\alpha) = \frac{2r_0\Omega_e^2 \sin^2\alpha}{3m_e c^3}. \tag{7.22}$$

For the present let us assume that the pitch angle distribution is isotropic and consider the loss averaged over pitch angle. With $\langle\sin^2\alpha\rangle = 2/3$, (7.22) gives

$$b = \langle b(\alpha)\rangle = \frac{32\pi r_0^2}{9m_e^2c^3} W_m, \qquad W_m = \frac{B^2}{2\mu_0}. \tag{7.23}$$

Evolution of the Spectrum due to Synchrotron Losses

The evolution of a distribution of electrons subject to synchrotron losses is described by

$$\frac{\partial N(\varepsilon,t)}{\partial t} = -\frac{\partial}{\partial\varepsilon}[\dot{\varepsilon}N(\varepsilon,t)] + Q(\varepsilon,t) - \nu_{\rm esc}N(\varepsilon,t), \tag{7.24}$$

where the three terms on the right hand side describe the synchrotron losses, injection of electrons and escape of electrons (escape rate $\nu_{\rm esc}$), respectively. Equation (7.24) may be integrated explicitly by first Laplace transforming.

In the special case where the source term, $Q(\varepsilon) = A\varepsilon^{-a}$, is a power law, and the injection occurs at a constant rate (independent of time), the solution is

$$N(\varepsilon,t) = \frac{Ae^{-\nu_{\rm esc}/b\varepsilon}}{b\varepsilon^2}\int_{\varepsilon}^{\varepsilon/(1-b\varepsilon t)} d\varepsilon'\, e^{\nu_{\rm esc}/b\varepsilon'}\varepsilon'^{-a}, \tag{7.25}$$

with the upper limit replaced by ∞ for $b\varepsilon t > 1$.

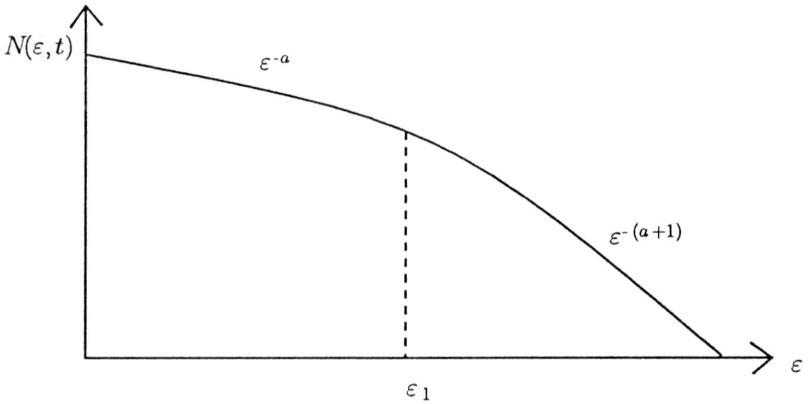

Fig. 7.5. The solution (7.26) is plotted (log–log plot) schematically, showing the increase in the power-law index a to $a+1$ at $\varepsilon_1 = 1/bt$.

In the special case where there is no escape, that is, $\nu_{\text{esc}} = 0$, (7.25) simplifies to

$$N(\varepsilon, t) = \frac{A\varepsilon^{-a}}{(a-1)b\varepsilon}\begin{cases} 1-(1-b\varepsilon t)^{a-1} & \text{for } b\varepsilon t \le 1, \\ 1 & \text{for } b\varepsilon t > 1. \end{cases} \tag{7.26}$$

This solution is illustrated in Figure 7.5. Synchrotron losses cause the power-law spectrum $\propto \varepsilon^{-a}$ to

steepen at higher energies to $\propto \varepsilon^{-(a+1)}$. The bend occurs at an energy $\varepsilon = 1/bt$ which decreases with increasing time. This implies a steepening of the synchrotron spectrum such that α increases by 0.5.

The energy at which the increase in slope occurs corresponds to electrons whose synchrotron half life, which is the time for the energy of an electron to decrease by a factor of one half, is approximately equal to the time elapsed. Solving (7.22) for ε at time t, given $\varepsilon = \varepsilon_0$ at $t = 0$, one finds

$$1/\varepsilon = 1/\varepsilon_0 + bt. \tag{7.27}$$

Hence, one identifies the half lifetime as

$$t_{1/2} = 1/b\varepsilon = (3c/2r_0)\left(\Omega_e \sin^2\theta\right)^{-2}\left(\varepsilon/m_e c^2\right)^{-1}. \tag{7.28}$$

7.5 Inverse and Induced Compton Scattering

Compton scattering has several important implications for synchrotron sources. Inverse Compton scattering can be a source of very high energy photons, and can limit the brightness temperature of synchrotron sources, and induced Compton scattering can modify the spectrum at low frequencies.

Inverse Compton Scattering

Formally, Compton scattering is the scattering electrons by photons, but the term is also used in a generic sense to include all the consequences of the interaction between photons and free electrons. These include the scattering of radiation by electrons, called Thomson scattering for nonrelativistic electrons and inverse Compton scattering when the electrons are ultrarelativistic.

In a synchrotron source there are relativistic electrons and photons present, and the relativistic electrons must scatter the photons. Inverse Compton scattering is analogous to synchrotron emission, and may be treated by making minor modifications to the formulas for synchrotron emission (averaged over angles of emission). Specifically, if one replaces the cyclotron frequency Ω_e in the synchrotron formulas by the frequency ω of the unscattered photons, and the magnetic energy density $B^2/2\mu_0$ by the energy density W_{ph} in these photons, then the synchrotron formulas transform into formulas for inverse Compton scattering. Ignoring factors of order unity, the typical frequency of synchrotron emission due to particles with Lorentz factor γ is $\omega \approx \Omega_e\gamma^2$, and if the same electrons scatter these photons, then the resulting inverse Compton scattered photons have frequency

$$\omega' \approx \omega\gamma^2 \approx \Omega_e\gamma^4. \tag{7.29}$$

For some sources, notably the Crab Nebula, the highest frequency emission is thought to be due to this process, that is, due to inverse Compton scattering of synchrotron photons. This is referred to as the synchrotron-self-Compton process.

The Inverse Compton Catastrophe

There is an important limitation imposed by inverse Compton scattering on the brightness temperature in a synchrotron source. That such a limit must exist can be seen by the following argument. The ratio of the power radiated in inverse Compton scattering, P_C, to that in synchrotron emission, P_S, is determined by the ratio of the energy densities in the photons and the magnetic field, that is, one has $P_C/P_S = W_{\mathrm{ph}}/W_m$. Note, however, that W_{ph} consists of contributions from synchrotron photons, $\omega \approx \Omega_e\gamma^2$, these photons scattered once, $\omega' \approx \Omega_e\gamma^4$, these photons scattered again, $\omega'' \approx \Omega_e\gamma^6$, and so on. This is a geometric series which may be summed,

$$W_{\mathrm{ph}} = W_{\mathrm{ph}0}\left[1 + \frac{W_{\mathrm{ph}0}}{W_m} + \left(\frac{W_{\mathrm{ph}0}}{W_m}\right)^2 + \cdots\right] = \frac{W_{\mathrm{ph}0}}{1 - W_{\mathrm{ph}0}/W_m}, \tag{7.30}$$

where $W_{\mathrm{ph}0}$ is the energy density in the photons after the first scattering. For $W_{\mathrm{ph}0} = W_m$ a *Compton catastrophe* occurs in which the power radiated diverges. The value at which this occurs is relatively insensitive to the details of the source, and corresponds to a brightness temperature for the synchrotron emission $T_b \approx 10^{12}$ K (Kellerman and Pauliny-Toth 1969). There is no direct evidence for this, in that no catastrophe has been observed, but there is strong indirect evidence in that no synchrotron sources are observed with T_b significantly $> 10^{12}$ K, whereas there are many sources with $T_b \lesssim 10^{12}$ K. It is reasonable to conclude that T_b for synchrotron sources is limited to $\lesssim 10^{12}$ K due to inverse Compton scattering by the relativistic electrons.

Induced Compton Scattering

The low-frequency turnover in Figure 7.3 due to induced Compton scattering may be understood as follows. *Induced Compton scattering* is related to Compton scattering in essentially the same way as absorption is related to emission for any emission process. This effect is important when the optical depth for Thomson scattering is large. The optical depth is

$$\tau_0 = \sigma_{\rm T} n_{\rm e} L, \qquad \sigma_{\rm T} = \frac{8\pi r_0^2}{3}, \tag{7.31}$$

where $\sigma_{\rm T}$ is the Thomson cross section.

Induced Compton scattering is included in the Kompaneets equation, which describes the effect of Compton scattering by thermal electrons on isotropic radiation, described by its occupation number, $N(\omega)$:

$$\frac{dN(\omega)}{dt} = \frac{\sigma_{\rm T} n_{\rm e} \hbar}{m_{\rm e} c} \left(\frac{1}{\omega^2} \frac{d}{d\omega} \omega^4 \right) \left\{ T_{\rm e} \frac{dN(\omega)}{d\omega} + N(\omega) + [N(\omega)]^2 \right\}. \tag{7.32}$$

The terms on the right hand side of (7.32) describe the effects of the Doppler spread due to the thermal motion of the scattering electrons, the quantum recoil and induced Compton scattering, respectively. The quantum recoil and induced Compton scattering do not depend on the temperature of the electrons, nor on other details of their distribution function, provided they are nonrelativistic and isotropic.

Induced Compton scattering involves the derivative, with respect to frequency, of $\omega^2 N(\omega) \propto I(\omega)/\omega$. This derivative implies that the effect is large where the slope of the frequency spectrum changes rapidly. Induced Compton scattering tends to distort a spectrum by pumping photons from higher to slightly lower frequencies across a bend in the spectrum. The frequency spectrum below the turnover illustrated in Figure 7.3 for induced Compton scattering is found by balancing the term involving $[N(\omega)]^2$ in (7.32) with a source term of the form of the power law at higher frequencies $\propto \omega^{-(\alpha+3)}$, where $I(\omega) \propto \omega^3 N(\omega)$ is used.

8 Plasma Emission

Solar radio bursts are due to an emission mechanism that is quite different from the emission mechanisms that operate in other radio sources. The emission is at the fundamental or second harmonic of the plasma frequency, and it is polarized in the sense of the o-mode, unlike gyromagnetic emission which favors the x-mode. The brightness temperatures can be high ($T_b > 10^{10}\,$K) so that the emission process must be coherent. The relevant emission processes are referred to as plasma emission.

8.1 Solar Radio Bursts

Solar radio emission was discovered in the early 1940s. The radio bursts discussed here occur in the range $30 \lesssim \omega/2\pi \lesssim 300\,$MHz in the solar corona and, for type III bursts, at much lower frequencies in the solar wind. The earlier data and their various interpretations were reviewed by Wild, Smerd and Weiss (1963), Wild and Smerd (1971), Melrose (1980b), McLean and Labrum (1985), Goldman and Smith (1986), Benz (1993).

Types III, II & I Bursts

The bursts are classified according to their appearances on dynamic spectra.

Type III bursts : Type III bursts are rapidly drifting, from high to low frequencies. They are interpreted as streams of electrons propagating outward through the solar corona at speeds $\sim 0.1c$. Type III bursts occur in groups associated with solar flares. Often the electron streams propagate through the solar corona into the solar wind, to beyond the orbit of the Earth. Such type III bursts in the solar wind have been studied in detail, and the information from spacecraft on such bursts complements the ground-based information on burst from the solar corona.

Some type III bursts are associated with type I–III storms. Storms anticorrelate with flares. Such storm type III bursts start at much lower frequencies than flare-associated type III bursts, indicating a source much higher in the corona. The energy of the electrons also tends to be lower (a few keV, rather than a few tens of keV).

Type II bursts : Type II bursts drift from high to low frequency much more slowly than type III bursts. They are associated with shock waves produced by a solar flare. These shock waves can also be detected as Moreton waves, seen in optical line emission from the chromosphere as the shock propagates away from the flare site. Sometimes type II emission continues as the shock propagates into the solar wind, but it fades much more quickly than type III emission. Not all shock waves generate type II bursts, and in the solar wind only a small fraction of all observed shocks have associated type II emission.

Type I bursts : Type I emission occurs in storms that can involve continuum and burst emission. Storms can last for days to weeks. The bursts are of short duration and can drift either to higher or lower frequency. Type I emission occurs above a minimum frequency $\gtrsim 50\,\mathrm{MHz}$, which tends to vary slowly with time. In type I–III storms the type III bursts occur below this frequency, sometimes with the type I bursts forming chains (Hanasz 1966) out of which the type III bursts appear to grow. It is thought that the electrons that generate both the type I and the type III bursts are accelerated as a result of reconnection in large loops connecting different active regions.

Harmonic Structure

Type III bursts and type II bursts show harmonic structure: emission occurs at a fundamental frequency and its second harmonic. The frequency ratio is usually slightly less than two. There is some evidence for higher harmonics, but if they exist at all, higher harmonic are so rare that they should be treated as special cases. The ratio of the intensity at the two harmonics is often close to unity. This is especially the case for type III bursts in the solar wind, after an initial phase in which the fundamental dominates. Hence, a theory needs to account for two, and only two harmonics of typically similar intensity.

In contrast, there is no evidence for harmonic structure in type I emission. Type I emission appears to be entirely at the fundamental.

Polarization

All three types of burst are usually polarized in the sense of the o-mode of magnetoionic theory. Fundamental type III and type II bursts can have a relatively high polarization, up to $\sim 70\%$. However, the polarization can also be low or undetectable, especially for bursts originating near the limb. It appears that initially highly polarized radio emission can be depolarized as it propagates through the solar corona, and that this depolarization effect is much stronger for sources near the solar limb. Second harmonic emission is relatively weakly polarized or unpolarized.

Type I emission can be completely polarized in the sense of the o-mode, especially for sources near the central meridian. As with type III bursts and type II bursts, the degree of polarization tends to decrease as the storm center moves toward the solar limb. The degree of polarization can maintain a characteristic intermediate value for hours (Zlobec 1975).

Brightness Temperature

To calculate T_b for a radio burst requires an estimate of the area of the burst, as well as of its intensity. Radioheliograph data on source sizes led to a relatively wide range of values, from faint bursts around the brightness of the background corona, $T_\mathrm{b} \sim 10^6\,\mathrm{K}$, to bright bursts, $T_\mathrm{b} \sim 10^{10}$–$10^{12}\,\mathrm{K}$. Type III bursts in the solar wind are brighter, with a characteristic maximum $T_\mathrm{b} \sim 10^{14}$–$10^{15}\,\mathrm{K}$.

The apparent sizes of sources in the corona are not thought to give a realistic estimate of the actual source sizes. Several different observations imply that the apparent sources are scatter images of the actual sources at much greater heights than the actual source. This was apparent in early attempts at modeling radio bursts, where standard models for the corona were used with the density arbitrarily increased by a factor ten or greater to fit the radio data. Radioheliograph observations show sources at a given frequency all about the same size and about the same height. Thus, for example, the fundamental and harmonic sources at a given time appear to come from widely different heights. Correspondingly, fundamental and harmonic sources at a given frequency, and hence at different times when the electron stream is at widely separated heights in the corona, appear to come from the same height.

Models for propagation through an inhomogeneous corona to account for these scatter images encounter a fundamental difficulty. The scattering, described by ray tracing through an ensemble of inhomogeneous coronas, implies that both the source area, A, and the solid angle, $\Delta\Omega$, to which the radiation is approximately confined, increase with increasing scattering. However, Liouville's theorem requires that the product $A\,\Delta\Omega$ remain constant (Melrose and Dulk 1988). This inconsistency can only be resolved by arguing that emission from the apparent source has only a small filling factor. An implication is that that if this filling factor is ignored in calculating $T_{\rm b}$, the resulting value underestimates the true value of $T_{\rm b}$.

8.2 Qualitative Discussion of Plasma Emission

Plasma emission is any indirect emission process in which (a) the exciting agency generates plasma turbulence which cannot escape directly from the plasma, and (b) this turbulence leads to escaping radiation through some secondary process. Plasma emission occurs in solar radio bursts, of which there are several types with the most familiar being type III bursts due to beams of fast electrons.

Outline of the Theory

The first detailed theory for plasma emission from the solar corona was for type III bursts by Ginzburg and Zheleznyakov (1958). At the time the theory was proposed it was accepted that type III bursts involve emission at the fundamental ($\omega = \omega_{\rm p}$) and second harmonic ($\omega = 2\omega_{\rm p}$), and that the emission is excited by a stream of electrons. A variant of their theory is outlined schematically in Figure 8.1. It consists of three stages: 1) generation of Langmuir turbulence through a streaming instability, 2) production of fundamental plasma emission by scattering of Langmuir waves into transverse waves by plasma particles, and 3) production of second harmonic emission through coalescence of two Langmuir waves to form a transverse wave. Since the theory was originally proposed, the details of each of these stages has been updated several times as ideas on the underlying plasma theory evolved. There is now a vast literature on the theory

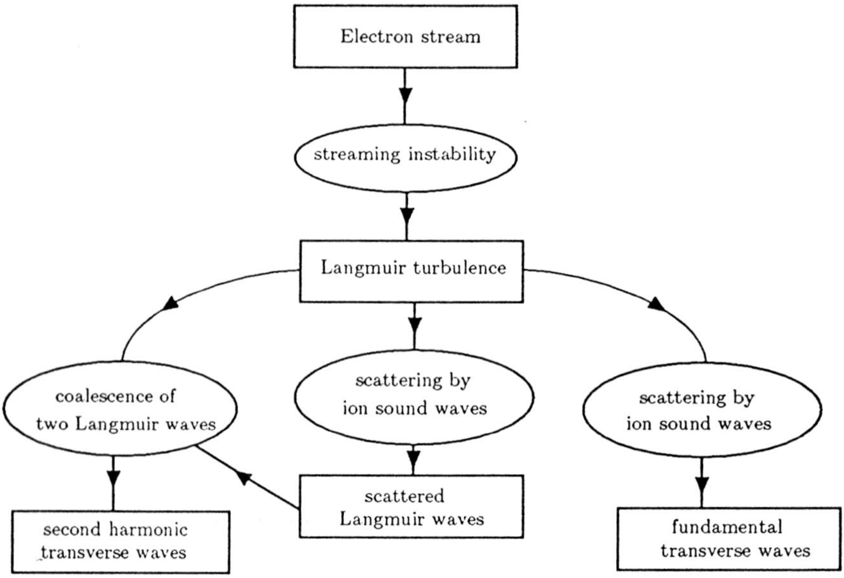

Fig. 8.1. Flow diagram for a variant of the theory of Ginzburg and Zheleznyakov (1958) for the generation of plasma emission. In other variants the processes indicated that involve ion sound waves are replaced by other nonlinear plasma processes.

of solar radio bursts, and there is a wide variety of detailed ideas on the specific processes that are important.

Three-Wave Interactions

In the variant of the theory of plasma emission illustrated schematically in Figure 8.1 the scattering is attributed to ion sound waves, also called ion acoustic waves. These scattering processes may be described using weak turbulence theory.

Qualitatively, the relevant processes involve three-wave interactions in which two waves beat to generate a third wave. Let the three initial waves be described by frequencies ω_1, ω_2, ω_3 and wave vectors $\boldsymbol{k}_1$, $\boldsymbol{k}_2$, $\boldsymbol{k}_3$. Then the specific beat process $1+2 \rightarrow 3$, in which waves 1 and 2 beat to form wave 3, satisfies the beat (or Manley-Rowe) conditions

$$\omega_1 + \omega_2 = \omega_3, \quad \boldsymbol{k}_1 + \boldsymbol{k}_2 = \boldsymbol{k}_3. \tag{8.1}$$

These beat conditions play the role of resonance conditions for three-wave interactions, in the sense that they appear as the arguments of δ-functions in the probabilities for the three-wave interaction.

One process that can lead to fundamental plasma emission is the beat L+S $\rightarrow$ T, where L refers to a Langmuir wave, S to an ion sound wave and T to a transverse wave. The process corresponding to scattering of Langmuir waves is the beat L + S → L′, where L′ refers to the scattered Langmuir wave. These are wave coalescence processes. The inverse of a coalescence is a decay process. The decay processes L → T + S and L → L′ + S are qualitatively similar to the corresponding coalescence processes, and lead to a slight downshift, rather than a slight upshift, in frequency. Second harmonic emission results from the coalescence process L + L′ → T.

Dispersion Relations

Each wave must satisfy the relevant dispersion relations. For Langmuir waves (1.28), ion sound waves (1.29) and transverse waves (1.26), the dispersion relations are, respectively,

$$\omega_{\rm L}(\boldsymbol{k}) \approx \omega_{\rm p} + 3k^2V_{\rm e}^2/2\omega_{\rm p}, \quad \omega_{\rm S}(\boldsymbol{k}) \approx kv_s, \quad \omega_{\rm T}(\boldsymbol{k}) = (\omega_{\rm p}^2 + k^2c^2)^{1/2}, \tag{8.2}$$

where $V_{\rm e} = (T_{\rm e}/m_{\rm e})^{1/2}$ is the thermal speed of electrons and $v_s \approx V_{\rm e}/43$ is the ion sound speed. In practice the dispersion relations can place severe restrictions on when specific three-wave interactions are allowed.

Effect of a Magnetic Field

When the effect of a magnetic field is included there is a richer variety of possible coalescence and decay processes, including, for example, upper hybrid or z-mode waves in place of the L-waves, lower hybrid or electron acoustic waves in place of the S-waves, and o-mode or x-mode waves in place of the T-waves.

An important consequence of the inclusion of the magnetic field is that the theory implies that fundamental emission should be completely polarized in the sense of the o-mode. To see this one needs to compare the frequency $\omega_{\rm L}$ with the cutoff frequency, $\omega_{\rm x}$ (1.30) for the x-mode. The change in frequency in fundamental emission is negligible here, and hence for $\omega_{\rm L} < \omega_{\rm x}$ only emission in the o-mode is possible. One has

$$\omega_{\rm L} - \omega_{\rm x} \approx \frac{3k^2V_{\rm e}^2}{2\omega_{\rm p}} - \frac{\Omega_{\rm e}}{2} \approx \frac{\omega_{\rm p}}{2}\left(\frac{3V_{\rm e}^2}{v_\phi^2} - \frac{\Omega_{\rm e}}{\omega_{\rm p}}\right), \tag{8.3}$$

where the phase speed, v_ϕ, of the Langmuir waves is of order the streaming speed for the bump-in-tail instability, cf. Figure 2.2. For plausible parameters, one has $3V_{\rm e}^2/v_\phi^2 < 0.03$ and $\Omega_{\rm e}/\omega_{\rm p} > 0.1$, so that the emission should be 100% in the o-mode. This is the case for some type I bursts, but it seems that type III and type II bursts are always significantly depolarized.

8.3 Weak Turbulence Theory

In weak turbulence theory an expansion of the basic equations is made in powers of the amplitude of the waves and the waves are treated in the random phase approximation. To treat three-wave interactions, the linear theory outlined in lecture 2 needs to be extended to the lowest order of nonlinearity.

Kinetic Equation for Three-Wave Interactions

Extending the theory in detail involves some tedious calculations. However, the most important result for present purposes is the form of the kinetic equation for three-wave interactions. Kinetic equations are derived by a simple generalization of the argument based on the Einstein coefficients (lecture 2). The important ingredient in the argument is that the rates of transition when waves in mode M are emitted and absorbed is $\propto [1+N_M(\boldsymbol{k})]$ and $\propto N_M(\boldsymbol{k})$, respectively. Hence the rates for the three-wave processes $\mathrm{L}+\mathrm{S}\to M$ and $\mathrm{L}\to M+\mathrm{S}$ are proportional to $[1+N_M(\boldsymbol{k})]N_\mathrm{L}(\boldsymbol{k}')N_\mathrm{S}(\boldsymbol{k}'')$ and $N_M(\boldsymbol{k})[1+N_\mathrm{L}(\boldsymbol{k}')][1+N_\mathrm{S}(\boldsymbol{k}'')]$, respectively. The kinetic equation for waves in the mode M involves the difference between these. This difference contains a term $N_M(\boldsymbol{k})$ that is independent of $N_\mathrm{L}(\boldsymbol{k}')$ and $N_\mathrm{S}(\boldsymbol{k}'')$. This term describes photon splitting, which is intrinsically quantum mechanical in the sense that it depends intrinsically on the value of Planck's constant. This term is ignored here. Photon splitting of gamma rays needs to be taken into account in pulsars with superstrong magnetic fields ($B > 0.2B_\mathrm{c}$, cf. lecture 9).

It is convenient to combine the kinetic equation derived in this way for the wave in the mode M due to a three-wave process $\mathrm{L}+\mathrm{S}\leftrightarrow M$ (upper sign), with that for the process $\mathrm{L}\leftrightarrow M+\mathrm{S}$ (lower sign), with $M=\mathrm{T}$ or L' the cases of interest here. One finds

$$\begin{aligned}\frac{dN_M(\boldsymbol{k})}{dt}=\int\frac{d^3\boldsymbol{k}'}{(2\pi)^3}\frac{d^3\boldsymbol{k}''}{(2\pi)^3}\,&u^{\pm}_{M\mathrm{LS}}(\boldsymbol{k},\boldsymbol{k}',\boldsymbol{k}'')\\&\times\left\{N_\mathrm{L}(\boldsymbol{k}')N_\mathrm{S}(\boldsymbol{k}'')-N_M(\boldsymbol{k})\left[N_\mathrm{S}(\boldsymbol{k}'')\pm N_\mathrm{L}(\boldsymbol{k}')\right]\right\}.\end{aligned}\tag{8.4}$$

A similar argument gives the kinetic equation for the generation of the second harmonic due to the process $\mathrm{L}+\mathrm{L}'\to\mathrm{T}$:

$$\begin{aligned}\frac{dN_\mathrm{T}(\boldsymbol{k})}{dt}=\int\frac{d^3\boldsymbol{k}'}{(2\pi)^3}\frac{d^3\boldsymbol{k}''}{(2\pi)^3}\,&u^{\pm}_{\mathrm{TLL}}(\boldsymbol{k},\boldsymbol{k}',\boldsymbol{k}'')\\&\times\left\{N_\mathrm{L}(\boldsymbol{k}')N_{\mathrm{L}'}(\boldsymbol{k}'')-N_\mathrm{T}(\boldsymbol{k})\left[N_\mathrm{L}(\boldsymbol{k}')+N_{\mathrm{L}'}(\boldsymbol{k}'')\right]\right\}.\end{aligned}\tag{8.5}$$

Probabilities for Three-Wave Processes

The actual form of the probabilities $u^{\pm}_{M\mathrm{LS}}$ and u_{TLL} is relatively unimportant in the following discussion. Nevertheless, it is appropriate to write them down and comment on them.

The two probabilities $u^{\pm}_{M\mathrm{LS}}$ differ only in the sign in the beat conditions $\omega_{\mathrm{L}} \pm \omega_{\mathrm{S}} = \omega_M$, $\boldsymbol{k}_{\mathrm{L}} \pm \boldsymbol{k}_{\mathrm{S}} = \boldsymbol{k}_M$, which appear as the arguments of δ-functions in the following expression:

$$
\begin{aligned}
u^{\pm}_{MP\mathrm{S}}(\boldsymbol{k}, \boldsymbol{k}', \boldsymbol{k}'') = {} & \frac{e^2 \hbar \omega_{\mathrm{p}}^2}{2\varepsilon_0 m_{\mathrm{e}}^2 V_{\mathrm{e}}^2} \frac{\omega_{\mathrm{S}}(\boldsymbol{k}'') R_M(\boldsymbol{k}) R_P(\boldsymbol{k}')}{\omega_M(\boldsymbol{k}) \omega_P(\boldsymbol{k}')} |\boldsymbol{e}^*_M(\boldsymbol{k}) \cdot \boldsymbol{e}_P(\boldsymbol{k}')|^2 \\
& \times (2\pi)^4 \delta^3(\boldsymbol{k} - \boldsymbol{k}' \mp \boldsymbol{k}'')\, \delta\big(\omega_M(\boldsymbol{k}) - \omega_P(\boldsymbol{k}') \mp \omega_{\mathrm{S}}(\boldsymbol{k}'')\big),
\end{aligned} \tag{8.6}
$$

with $P = \mathrm{L}$ for Langmuir waves. The Manley-Rowe condition (8.1) plays the role of a resonance condition and is contained in the δ-functions in (8.6).

For transverse waves the factor involving the polarization vectors is summed over the two final states of polarization or averaged over the initial states of polarization. The various possible cases then give

$$
|\boldsymbol{e}^*_M(\boldsymbol{k}) \cdot \boldsymbol{e}_P(\boldsymbol{k}')|^2 = \begin{cases} |\boldsymbol{\kappa}_{\mathrm{L}} \cdot \boldsymbol{\kappa}_{\mathrm{L}'}|^2 & M = \mathrm{L},\ P = \mathrm{L}', \\ |\boldsymbol{\kappa}_{\mathrm{L}} \times \boldsymbol{\kappa}_{\mathrm{T}}|^2 & M = \mathrm{L},\ P = \mathrm{T}, \\ \frac{1}{2}|\boldsymbol{\kappa}_{\mathrm{T}} \times \boldsymbol{\kappa}_{\mathrm{L}}|^2 & M = \mathrm{T},\ P = \mathrm{L}, \\ \frac{1}{2}(1 + |\boldsymbol{\kappa}_{\mathrm{L}} \cdot \boldsymbol{\kappa}_{\mathrm{T}}|^2) & M = \mathrm{T},\ P = \mathrm{T}'. \end{cases} \tag{8.7}
$$

The probability for the process $\mathrm{L} + \mathrm{L}' \to \mathrm{T}$ of second harmonic plasma emission, after summing over the two states of polarization of the transverse waves, is

$$
\begin{aligned}
u_{\mathrm{TLL}'}(\boldsymbol{k}_{\mathrm{T}}, \boldsymbol{k}_{\mathrm{L}}, \boldsymbol{k}_{\mathrm{L}'}) = {} & \frac{e^2 \hbar}{32\varepsilon_0 m_{\mathrm{e}}^2 \omega_{\mathrm{p}}} \frac{(k_{\mathrm{L}}^2 - k_{\mathrm{L}'}^2)^2}{k_{\mathrm{T}}^2} |\boldsymbol{\kappa}_{\mathrm{L}} \times \boldsymbol{\kappa}_{\mathrm{L}'}|^2 \\
& \times (2\pi)^4 \delta^3(\boldsymbol{k}_{\mathrm{T}} - \boldsymbol{k}_{\mathrm{L}} - \boldsymbol{k}_{\mathrm{L}'})\, \delta\big(\omega_{\mathrm{T}}(\boldsymbol{k}_{\mathrm{T}}) - \omega_{\mathrm{L}}(\boldsymbol{k}_{\mathrm{L}}) - \omega_{\mathrm{L}}(\boldsymbol{k}_{\mathrm{L}'})\big).
\end{aligned} \tag{8.8}
$$

The Beat Conditions

For the three-wave interaction to occur the beat conditions (8.1) must be satisfied. The vector sum of the wavevectors is illustrated in Figure 8.2a. The frequency of Langmuir waves or of transverse waves is much greater than that of the ion sound waves, so that the initial wave (M or P) and final wave (P or M) must have approximately the same frequency. For $\mathrm{L} \pm \mathrm{S} \to \mathrm{L}'$, the condition $\boldsymbol{k} = \boldsymbol{k}' \pm \boldsymbol{k}''$ imposes a somewhat subtle kinematic restriction that is not discussed here. For $\mathrm{L} \pm \mathrm{S} \to \mathrm{T}$, the transverse wave has frequency close to that of the Langmuir wave, and so it leads to fundamental plasma emission. The transverse wave has wavenumber k very much smaller than the wavenumber k' of Langmuir waves; one then requires that the ion sound wave has $\boldsymbol{k}'' \approx \mp \boldsymbol{k}'$.

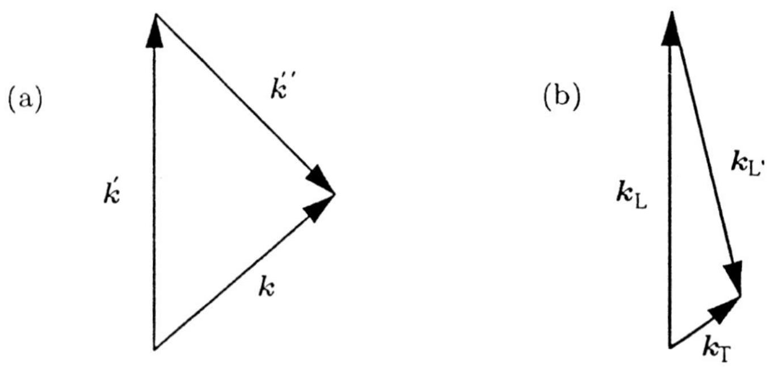

Fig. 8.2. (a) The vector sum (8.1) in an arbitrary case. (b) Coalescence of two Langmuir waves in the head-on case.

The Head-on Approximation

Langmuir waves generated directly by the streaming electrons have wavenumber much greater than that of second harmonic transverse waves. To see this, note that with $\omega \approx 2\omega_{\rm p}$, the dispersion relation (8.2) for T-waves implies $k_{\rm T}^2 = 3\omega_{\rm p}^2/c^2$, and that one has $k_{\rm T}^2/k_{\rm L}^2 \approx 3v_\phi^2/c^2$, where the phase speed v_ϕ of the Langmuir wave is determined by the streaming speed. Provided one has $v_\phi \ll c/\sqrt{3}$, the beat condition implied by the δ^3-function in the probability (8.8), requires $\boldsymbol{k}_{\rm L'} \approx -\boldsymbol{k}_{\rm L}$. This corresponds to the two Langmuir waves meeting head on, as illustrated in Figure 8.2b. The streaming instability only generates Langmuir waves in the forward direction. Production of the second harmonic can proceed only when nonlinear processes, such as $\rm L \pm S \to L'$, build up a nonthermal distribution of Langmuir waves in the backward direction.

The probability (8.8) simplifies in the head-on approximation, when the following apply in (8.8):

$$\delta\left[\omega_{\rm T}(\boldsymbol{k}_{\rm T}) - \omega_{\rm L}(\boldsymbol{k}_{\rm L})\right] \approx (1/\sqrt{3}c)\delta(k_{\rm T} - \sqrt{3}\omega_{\rm p}/c),$$

$$\frac{(k_{\rm L}^2 - k_{\rm L'}^2)^2}{k_{\rm T}^2}|\boldsymbol{\kappa}_{\rm L} \times \boldsymbol{\kappa}_{\rm L'}|^2 \approx (3\omega_{\rm p}^2/c^2)|\boldsymbol{\kappa}_{\rm T} \cdot \boldsymbol{\kappa}_{\rm L}|^2|\boldsymbol{\kappa}_{\rm T} \times \boldsymbol{\kappa}_{\rm L}|^2. \tag{8.9}$$

The angular dependence implied by (8.9) is quadrupolar, as is seen by introducing the angle $\psi = \arccos(\boldsymbol{\kappa}_{\rm T} \cdot \boldsymbol{\kappa}_{\rm L})$ and noting that the angular dependence has the quadrupolar form $\sin^2\psi\cos^2\psi$.

Saturation of Three-Wave Interactions

A quantitative treatment of plasma emission based on the foregoing equations is cumbersome even when simplifying assumptions are made. However, there is one limiting case that may be treated simply and which may be understood in terms of simple physical arguments. This is the limit in which the three-wave interactions saturate.

The processes $\mathrm{L} \pm \mathrm{S} \to M$ with $N_\mathrm{S} \gg N_\mathrm{L}$ build up the level of waves in the modes $M = \mathrm{T}$ or $M = \mathrm{L}'$. Initially one has $N_M \ll N_\mathrm{L}$, and the term $N_\mathrm{L} N_\mathrm{S}$ dominates the other two terms in the braces in (8.4). As N_M increases the other two terms increase and the magnitude of the quantity in braces in (8.4) decreases. The three-wave interaction is said to saturate when the quantity in braces in (8.4) approaches zero and the interaction ceases to cause N_M to increase. Thus saturation occurs for

$$N_M = N_\mathrm{L} N_\mathrm{S} / (N_\mathrm{S} \pm N_\mathrm{L}) \approx N_\mathrm{L}. \tag{8.10}$$

In particular, for fundamental plasma emission due to $\mathrm{L}\pm\mathrm{S} \to \mathrm{T}$ with $N_\mathrm{S} \gg N_\mathrm{L}$, the process saturates at $N_\mathrm{T} = N_\mathrm{L}$. Due to the near equality of the frequencies of the fundamental transverse and Langmuir waves ($\omega_\mathrm{T} \approx \omega_\mathrm{L} \gg \omega_\mathrm{S}$), $N_\mathrm{T} = N_\mathrm{L}$ corresponds to $T_\mathrm{T} = T_\mathrm{L}$. Hence the fundamental saturates at an intrinsic brightness temperature determined by the effective temperature of the Langmuir waves. Thus, despite the highly nonthermal nature of these processes, this saturation level is what one might expect on a thermodynamic-type argument.

A similar argument based on (8.9) implies that the second harmonic also saturates at $N_\mathrm{T} = N_\mathrm{L}$ for $N_\mathrm{L} \gg N_{\mathrm{L}'}$. In this case, $\omega_\mathrm{T} \approx 2\omega_\mathrm{L}$ implies a saturation level $T_\mathrm{T} = 2T_\mathrm{L}$. It follows that such a saturation model leads naturally to harmonics of similar intensity, as is often observed.

Saturation of the Bump-in-Tail Instability

The saturation model implies that the intrinsic brightness temperature of plasma emission is determined by the effective temperature of the Langmuir waves. A model in which the streaming instability also saturates allows one to estimate T_L in terms of the parameters of the stream.

The bump-in-tail instability is amenable to a relatively simple treatment in a one-dimensional model. Here 'one-dimensional' refers to a three-dimensional theory in which the only Langmuir waves considered have $\boldsymbol{k}$ along the direction of the beam and the particle distribution function is integrated over the two momentum component orthogonal to the beam direction. The instability saturates when the distribution function becomes sufficiently modified by quasilinear relaxation as to reduce the growth rate substantially. Semiquantitatively, saturation occurs when the energy density in the Langmuir waves becomes a significant fraction of the initial energy density in the electron beam.

A rough estimate of the effective temperature of the Langmuir waves at saturation may be made as follows. For simplicity assume that the saturation level of the Langmuir waves is approximately constant, $W(v_\phi) = W_0$ say, for waves in a range $v_b - \Delta v < v_\phi < v_b$, and zero otherwise, where v_b is the beam speed. If the energy density of these waves is a fraction ζ of the initial energy density in the beam, then one has

$$W_0 = \zeta n_1 m v_b^2 / 2\Delta v. \tag{8.11}$$

Further assume that the Langmuir waves are confined to a range $\Delta\Omega$ of solid angle. The range of solid angles $\Delta\Omega$ can be estimated only in terms of a three-dimensional theory for the instability. One simple estimate is to write $\Delta\Omega = \pi(\Delta\theta)^2$,with $\Delta\theta = \Delta v / v_\mathrm{b}$. Then the effective temperature T_L and W_0 are related by, for $\Delta v \ll v_\mathrm{b}$,

$$T_\mathrm{L} = \frac{(2\pi)^3 v_\mathrm{b}^4}{\omega_\mathrm{p}^3 \Delta\Omega} W_0 = \frac{\zeta (2\pi)^2 n_1 m v_\mathrm{b}^8}{\omega_\mathrm{p}^3 (\Delta v)^3}. \tag{8.12}$$

8.4 Discussion of Plasma Emission

There is no example in astrophysics or space physics where one can have confidence in quantitative treatments of a coherent emission process. The examples for which we have the most detailed information on the radiation and on the properties of the particles that generate it are for AKR (lecture 6), for type III bursts in the solar wind, and for plasma emission from electrons accelerated at the bow shocks of the Earth and Jupiter.

Langmuir Waves and Type III Bursts

When type III bursts in the solar wind were first probed directly by spacecraft, an obvious test of the theory was to confirm that the bursts are indeed generated by streams of electrons that generate Langmuir waves. The streams of electrons were found to correlate with type III bursts, but at first no Langmuir waves were found. The Langmuir waves were later recognized (Lin et al. 1981) as being distributed very inhomogeneously. They occur is localized regions where their energy density is large. However, the filling factor for these clumps of Langmuir waves is small, so that the average energy density in the Langmuir waves is only a small fraction of the energy density in the streaming electrons.

The recognition that the Langmuir waves are in clumps solved one problem, but raised several more. The problem that was solved arises from the energetics when the bump-in-tail instability saturates: the stream loses so much energy that it would stop almost immediately. The low mean energy density in the observed clumpy Langmuir waves is consistent with the stream propagating with little loss of energy. One question raised by this observation is whether these clumps are associated with strong-turbulence effects, as discussed below. Another problem is why the wave growth is apparently restricted to localized regions (e.g., Melrose and Goldman 1987; Robinson 1993). A further question is whether the clumpiness invalidates the use of quasilinear theory, which presupposes a spatially uniform distribution of Langmuir turbulence. Under relatively mild restrictions, a clumpy distribution has essentially the same quasilinear effect on the electrons stream as a corresponding uniform distribution of Langmuir waves with the same mean energy density (Melrose and Cramer 1989).

Weak-Turbulence Saturation Model for Type III Bursts

The weak-turbulence saturation model for type III bursts that is outlined above accounts naturally for the approximately equal intensity of the two harmonics in many type III bursts. The clumpiness of the Langmuir waves does not affect the result provided that the emission from individual clumps saturates, or provided that the radiation saturates after encountering a large number of clumps.

Saturation of the bump-in-tail instability may occur in the clumps. There is some indirect evidence that this is the case. Numerical calculations of the evolution of the electron distribution function as the stream propagates, subject to quasilinear relaxation in the one-dimensional approximation, produces a distribution function that is similar to that observed (Grognard 1985). Indeed, if the observed distribution function for type III electrons is used as l into the code, the output, which is the self-consistent solution, is essentially unchanged from the input. This is surprising in view of the number of simplifying assumption in the one-dimensional quasilinear model.

It may be concluded that there is some evidence in support of the weak-turbulence saturation model, but the evidence is far from compelling. Even if this model is accepted, at best it leads to only a semiquantitative theory for the average emission in type III bursts.

Strong Turbulence

An alternative to these weak turbulence effects are *strong turbulence* effects (e.g., Nicholson 1983; Goldman 1984). Strong turbulence concepts that have been applied to plasma emission include parametric instabilities, modulational instabilities and Langmuir collapse. Specific strong turbulence effects are relevant only when the energy density of the Langmuir turbulence exceeds an appropriate threshold value.

There is an important qualitative distinction between the strong-turbulence form of a three-wave interaction (a parametric instability) and the weak-turbulence form. In weak turbulence theory a three-wave process causes the turbulence to evolve only when there are two excited wave distributions that beat together to generate the third wave distribution. In strong-turbulence theory, one excited wave distribution acts as a pump that generates the other two wave distributions. A modulational instability is similar to a parametric instability in that only one wave distribution (the pump) is required, but, from a kinematic viewpoint, the evolution is regarded as a four-wave interaction rather than as a three-wave interaction. One such instability (the oscillating two stream instability) was proposed and explored in connection with the evolution of the Langmuir turbulence in type III streams (e.g., Goldman 1983). However, the strong-turbulence effect now thought most relevant for type III bursts is Langmuir collapse.

Langmuir Collapse

Langmuir collapse is related to self-focusing of laser light in nonlinear optics: a uniform distribution of turbulence breaks up into localized regions in which the energy density in the turbulence increases to very high values on a very short timescale. These regions, sometimes called cavitons, reach a terminal size $\sim 20\lambda_{De}$ and the distribution of electric amplitudes can be estimated (Robinson and Newman 1990)

The suggested relevance to type III bursts involves a preliminary stage in which the Langmuir turbulence evolves due to weak-turbulence effects, which systematically reduce the frequency of the waves. This is closely analogous to induced Compton scattering, cf. (7.32), in the sense that radiation is driven from higher to lower frequencies. As a result, the Langmuir turbulence is driven toward small k, where it is said to form a condensate. The resulting small-k condensate of Langmuir turbulence is subject to collapse. The observational data do not support the suggestion that the observed clumps are the direct result of such collapse. However, there is some observational support for strong-turbulence effects playing a significant role. Thus, it seems that both strong and weak turbulence effects need to be taken into account.

9 Pulsars

Since their discovery in 1967, theoretical modeling of the magnetospheres of pulsars has been a major challenge. Recently two books have been published on the theory, and they present remarkably different viewpoints as to the status of our current understanding. Michel (1991), in commenting on his earlier review (Michel 1982) remarked that when he finished writing it "the entire theoretical foundation of [the] standard model had collapsed." In contrast, Beskin, Gurevich and Istomin (1993) stated that "the physical picture of the basic processes in the pulsar magnetosphere seems on the whole to be clear." My opinion will emerge during this lecture.

9.1 Electromagnetic Fields Around a Rotating Neutron Star

Pulsars are rotating neutron stars. They emit radio waves in beams, which are detected as pulses as the beam sweeps across the Earth once per rotation period of the neutron star. The central problem in the theory of pulsars is to determine the distribution of particles that constitute the magnetosphere of the neutron star, and to show how this accounts for the observed radio and other emissions.

Rotating Neutron Star in Vacuo

Consider a magnetized star, of radius R_*, rotating with angular velocity $\boldsymbol{\Omega}$. Inside the star, which is assumed to be a good conductor, there is a *corotation electric field*, $\boldsymbol{E} = -(\boldsymbol{\Omega} \times \boldsymbol{x}) \times \boldsymbol{B}$, which is such as to cause an electric drift, $\boldsymbol{E} \times \boldsymbol{B}/B^2$, that is equal to the rotation velocity $\boldsymbol{\Omega} \times \boldsymbol{x}$.

The simplest model for the fields exterior to the star is to assume that the star is surrounded by a vacuum. Then there is a discontinuity in $\boldsymbol{E}$ at the surface of the star, implying a surface charge on the star. Outside the star there is an electric field with a component, $E_{\parallel}$, along the magnetic field lines. Assuming a magnetic dipole at the center of the star, the surface charge has a quadrupolar distribution and produces an exterior electric field with component parallel to the magnetic field

$$E_{\parallel} = R_* \Omega B_* (R_*/r)^4 \cos^3\theta, \tag{9.1}$$

where B_* is the polar magnetic field. Taking the values $\Omega = 10^1\,\mathrm{s}^{-1}$, $B_* = 10^8\,\mathrm{T}$ and $R_* = 10\,\mathrm{km} = 10^4\,\mathrm{m}$, one finds $E_{\parallel} \approx 10^{13}\,\mathrm{V\,m}^{-1}$ at the surface of a typical pulsar. This electric field is large enough to rip particles off the surface and accelerate them to relativistic energies. This process should continue until there is enough charge around the pulsar to shield out $E_{\parallel}$. Hence, one expects the star to have a magnetosphere populated with particles. The vacuum model is then inappropriate because the plasma tends to shield out $E_{\parallel}$ and to set up a corotation electric field.

Goldreich-Julian Density

The opposite extreme to the vacuum model is a rotating magnetosphere with the corotating electric field set up everywhere. This electric field has a nonzero divergence, requiring a local charge density. In planetary magnetospheres and stellar coronas, this charge density requires only a very small excess of charges of one sign over the other. However, in a pulsar magnetosphere this charge density, which is required to shield out $E_{\parallel}$, is substantial.

The charge density for rigid rotation is

$$\rho = \varepsilon_0 \mathrm{div}\,\boldsymbol{E} = \varepsilon_0\left(-2\boldsymbol{\Omega}\cdot\boldsymbol{B} + \boldsymbol{\Omega}\times\boldsymbol{x}\cdot\mathrm{curl}\,\boldsymbol{B}\right). \tag{9.2}$$

The second term inside the braces is of order $|\boldsymbol{\Omega}\times\boldsymbol{x}|^2/c^2$ times the first term, and so is negligible for $r \ll \Omega/c$. The number density determined by setting $\rho = en_{\mathrm{GJ}}$ in (9.2) is called the *Goldreich-Julian* density (Goldreich and Julian 1969):

$$n_{\mathrm{GJ}} = -\frac{2\varepsilon_0\boldsymbol{\Omega}\cdot\boldsymbol{B}}{e}. \tag{9.3}$$

For a dipolar magnetic field $\boldsymbol{\Omega}\cdot\boldsymbol{B}$, and hence n_{GJ}, has the same sign in a region around the northern pole and a corresponding region around the southern pole, and the opposite sign in an equatorial region. The sign of n_{GJ} reverses on surfaces where $\boldsymbol{B}$ is perpendicular to the rotation axis, as illustrated in Figure 9.1.

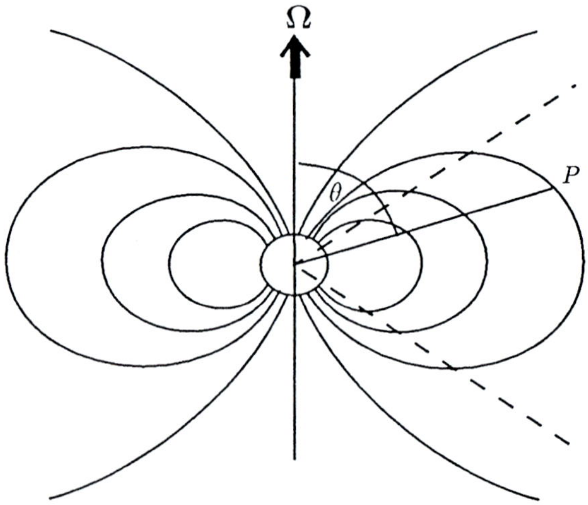

Fig. 9.1. On a dipolar magnetic field line, the point P satisfies $r = r_0 \sin^2\theta$, where the point P_0 is at $r = r_0$. For an aligned rotator, the locus of points corresponding to $\boldsymbol{\Omega}\cdot\boldsymbol{B} = 0$ that separates regions of opposite sign of the charge density corresponds to $|\cos\theta| = 1/3$, as indicated by the dashed lines.

The Polar Caps

The region of corotation cannot extend beyond a surface, called the *light cylinder*, at which the corotation speed would equal the speed of light, as illustrated in Figure 9.2. In a model in which the rotation and magnetic axes are aligned, this corresponds to a radius (in the equatorial plane)

$$r_{\rm lc} = c/\Omega. \tag{9.4}$$

The polar cap angle, $\theta_{\rm cap}$, defining the radius of the polar cap on the surface of the star, may be estimated using the equation in polar coordinates for a dipolar field line:

$$R_* = r_{\rm lc}\sin^2\theta_{\rm cap}, \qquad \theta_{\rm cap} = \arcsin(R_*/r_{\rm lc})^{1/2} \approx (\Omega R_*/c)^{1/2}. \tag{9.5}$$

For a typical slowly rotating pulsar, $r_{\rm lc}$ is of order 10^8 m, which is about $10^4\,R_*$, implying $\theta_{\rm cap} \approx 10^{-2} \approx 0.5°$. For the Crab pulsar, rotating with a period of 33 ms, implying $\Omega = 2\times10^2\,{\rm s}^{-1}$, the light cylinder is of order a hundred stellar radii, and the polar cap angle is of order 5°. For millisecond pulsars, the light cylinder is of order ten stellar radii.

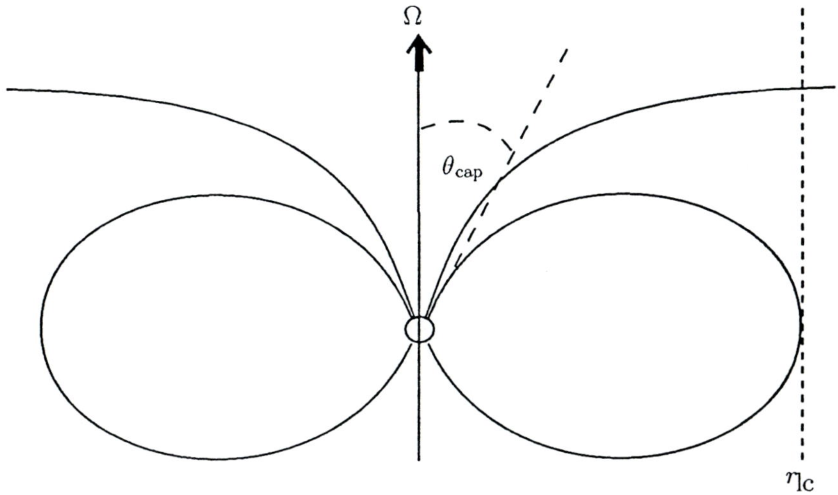

Fig. 9.2. The polar cap is defined by the set of field lines that extend to beyond the light cylinder.

Current Closure

Particles on field lines in the polar cap are accelerated outward by the parallel electric field, and apparently must escape from the pulsar. Particles of the same sign escape from both polar caps. Particles of opposite signs are trapped in corotating regions nearer the equatorial plane.

There are some obvious problems with the model even at this stage, and it is not clear how these problems are to be resolved. 1) The escaping charges from both polar caps seem to imply a net current and hence a net loss of charge. This is unacceptable because the star would charge up rapidly, and set up a Coulomb field until this is strong enough to prevent further loss of charge. If the pulsar is not to charge up, there must be a return current, but how this is set up remains a problem with most models. 2) Alternatively, it has been suggested that the pulsar does become charged (e.g., Michel 1991). Such a model appears not to radiate and is not discussed further here. 3) A question arises concerning the obvious difference between the two signs of $\boldsymbol{\Omega} \cdot \boldsymbol{B}$. It has been remarked that pulsars should come in two sexes, one corresponding to electrons being ripped off the polar cap, and the other corresponding to ions being ripped of the polar cap (Ruderman 1981). However, there is no observational evidence for two such distinct classes of pulsars. 4) Another difficulty is that $n_{\rm GJ}$ reverses sign along some field line, cf. Figure 9.1. The charges cannot adjust to allow this to occur without the appearance of outer gaps (Holloway 1973), which are regions where the charge density is inadequate to shield out $E_{\|}$.

9.2 Gaps

Although there is no completely consistent model for a pulsar magnetosphere, it is possible to identify several ingredients that are likely to be incorporated in an acceptable model. One of these is the development of vacuum gaps which discharge through the production of electron–positron pairs.

The Ruderman and Sutherland (1975) Model

The first detailed model that incorporates a vacuum gap was proposed by Ruderman and Sutherland (1975).

Binding of Ions

The Ruderman and Sutherland model includes an assumption that the binding energy of ions on the surface of a neutron star is so strong that they cannot be ripped off by the available potential. This assumption was consistent with known results in 1975, but present-day results invalidate the assumption (e.g., Neuhauser, Koonin and Langanke 1987). Nevertheless, let us accept this assumption for the sake of discussion. An implication is that if the sign of $\boldsymbol{\Omega} \cdot \boldsymbol{B}$ is negative, so that (9.2) requires a positive charge density over the polar cap, then it is not possible to provide the required charge from the stellar surface due to the strong binding of the ions. (This model applies only to neutron stars with negative $\boldsymbol{\Omega} \cdot \boldsymbol{B}$, and those for which this quantity is positive could not be pulsars.)

The Polar Gap

In the Ruderman and Sutherland model a vacuum gap, called a polar gap, forms above the polar zones, where there is a large unshielded parallel electric field. The positrons are accelerated upward and the electrons are accelerated downward. These primary particles emit gamma rays leading to secondary production of pairs (Sturrock 1971), and the region above the polar cap becomes sufficiently well populated by the pair plasma to shield out $E_{\parallel}$. The pair plasma then flows out along the field lines in the polar cap. This pair plasma contains both primary positrons, with a typical energy $\varepsilon \approx e\Delta\Phi$, where $\Delta\Phi$ is potential drop across the polar cap region, and lower energy pairs. The model is illustrated schematically in Figure 9.3.

The model illustrated in Figure 9.3, in which the vacuum gap occurs immediately above the stellar surface in the polar cap, effectively implies that the field lines inside the polar cap are not rotating and those outside the polar cap are corotating. Although this is unrealistic in detail, it is thought to provide a reasonable estimate of the expected potential drop that becomes available. The potential may be identified as $\Phi = \boldsymbol{\Omega} \times \boldsymbol{x} \cdot \boldsymbol{A}$, where $\boldsymbol{A}$ is the the vector potential. One has $\boldsymbol{A} = \mathrm{curl}\,(\boldsymbol{\mu}/r)$ for a magnetic dipole $\boldsymbol{\mu}$. The potential difference between the pole ($\theta = 0$) and the edge of the polar cap ($\theta = \theta_{\rm cap}$) is then given by

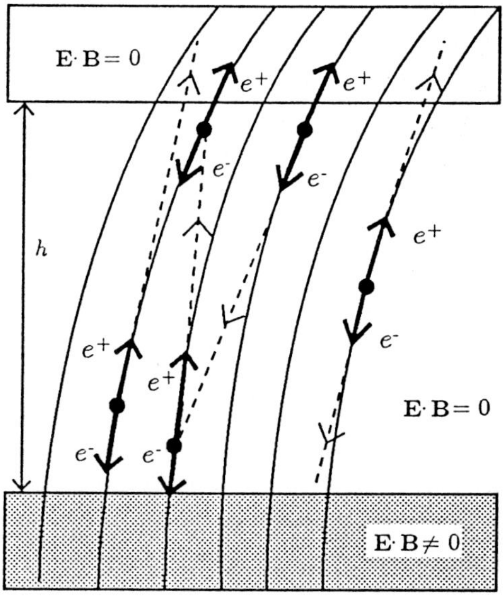

Fig. 9.3. The vacuum gap of height h is illustrated for the model of Ruderman and Sutherland (1975). The parallel electric field in the region between the stellar surface (shaded region) and the overlying magnetosphere accelerates electrons (e^-) downward and positrons upward (e^+). Curvature photons, indicated by dashed lines, generate secondary pairs.

$$\Delta\Phi = \frac{\Omega|\boldsymbol{\mu}|\sin^2\theta_{\text{cap}}}{R_*} = \frac{\Omega^2 R_*^3 B_*}{2c}. \tag{9.6}$$

With $\Omega = 10^1\,\mathrm{s}^{-1}$, $R_* = 2\times10^4\,\mathrm{m}$, $B_* = 10^8\,\mathrm{T}$ and $c = 3\times10^8\,\mathrm{m\,s}^{-1}$, (9.6) gives $\Delta\Phi \approx 10^{14}\,\mathrm{V}$. The potential is larger for young pulsars, notably the Crab and Vela pulsars, which are more rapidly rotating.

Self-Consistent Models

The Ruderman and Sutherland model, although no longer considered viable because ions are too weakly bound to the stellar surface, is one example of an attempt to solve the electrodynamics of a pulsar magnetosphere in a self-consistent manner. On the one hand, the corotation electric field cannot be present everywhere, specifically, not on field lines that extend beyond the light cylinder where it would imply motion at greater than c. On the other hand, the parallel electric field cannot be shielded everywhere because then there would be no adequate source of plasma to provide the shielding. A self-consistent model requires localized (in space or time) regions with large parallel electric field that provides a source of pairs to maintain the corotation electric field throughout most of the magnetosphere. In addition, the current implied by the outflow of plasma on open field lines must be balanced by a return current.

There are basically three types of model for the source region of the pair plasma: polar cap models, space-charge limited models, which are discussed below, and outer gap models (e.g., Cheng, Ho and Ruderman 1986). The model of Beskin et al. (1983) involves an assumed double layer at the surface of the star where the required pairs are accelerated; this is a polar-cap model, similar to that of Ruderman and Sutherland (1975), but the physical basis for the formation of the double layer is unclear.

Space-Charge Limited Models

There are models based on the assumption that charges can be freely ripped off the surface of the star. There are several different detailed models (e.g., Arons 1983, 1992; Mestel 1993). Such models also lead to large unshielded parallel electric fields, but for a different reason to polar-cap models.

In these models the basic assumption is that particles drawn from the surface of the star supply the number density (9.3) immediately above the surface of the star. As in the polar cap regions these charges flow freely out along the field lines, where there is no other obvious source of particles. Let their flow velocity along the field lines be $u_{\|}$, so that there is a current density $J = en_{\mathrm{GJ}}u_{\|}$. As the cross-sectional area ($\propto 1/B$) of a magnetic flux tube changes, the current density, $J = en_{\mathrm{GJ}}u_{\|}$, must change to satisfy $J \propto B$. Once $u_{\|}$ becomes relativistic, these requirements become $n_{\mathrm{GJ}} \propto B_z$, where the z-axis is along the rotation axis, and $n_{\mathrm{GJ}} \propto B$. These requirements are incompatible in general, leading to a parallel electric field building up.

In the absence of a parallel electric field, $u_{\|}$ is constant, and the change in the cross section ($\propto B^{-1}$) of a given flux tube implies that n_{GJ}/B is a constant. In the absence of any source of charge, one must have $\operatorname{div} \boldsymbol{J} = 0$, and $J \propto B$ is inconsistent with this. Thus, in the absence of an additional source of plasma, a parallel electric field must build up, and this field can act in a similar way to a vacuum gap in producing a pair plasma.

9.3 The Pair Plasma

There are several possible sources for pairs in a pulsar magnetosphere: 1) in a sufficiently strong electric field in vacuo, pairs are produced spontaneously, 2) a single photon propagating across a magnetic field can decay into pair if its energy exceeds a threshold of about twice the rest energy of the electron, 3) a photon-photon interaction can lead to a pair if the center-of-mass energy of each photon exceeds the rest energy of the electron, and 4) other photon-matter interactions can produce pairs if the photon energy exceeds twice the rest energy of the electron. Of these, the single-photon decay is the most important in standard pulsar models.

Emission of Gamma Rays by Primary Particles

The primary particles emit gamma rays due to several possible processes. One is curvature emission: the electron or positron propagates along curved field lines, and due to this curvature the particle experiences a centripetal acceleration toward the center of curvature of the field line, and this causes it to radiate. In addition, the particle may not be created in its lowest Landau orbital and, as discussed below, it then emits gamma rays as it relaxes to its ground state. Under appropriate conditions these gamma rays can themselves decay into pairs, so that one has an avalanche resulting in a plasma composed of relativistic electrons and positrons. The condition for this cascade process to operate is assumed to be the condition for a neutron star to be a pulsar. When the rotation becomes too slow to satisfy this condition, the pulsar dies.

One-Photon Pair Production

The decay mechanism for a gamma ray with $\varepsilon_{\rm ph} > 2m_{\rm e}c^2$, which corresponds to $\varepsilon_{\rm ph} > 1\,{\rm MeV}$, depends intrinsically on the presence of a superstrong magnetic field. By ***superstrong*** is meant a magnetic field that is a significant fraction of the so called critical magnetic field

$$B_{\rm crit} = m_{\rm e}^2c^2/e\hbar, \tag{9.7}$$

which corresponds to the cyclotron energy $\hbar\Omega_{\rm e}$ being equal to the rest energy, $m_{\rm e}c^2$. Numerically one has $B_{\rm crit} = 4.4 \times 10^9\,{\rm T}$, and the magnetic field at the poles of neutron stars can exceed about ten percent of this value.

There are several processes which are strictly forbidden in the absence of a superstrong magnetic field and which are allowed in its presence. One of these is the decay of a gamma ray into a pair. The probability per unit time of decay of a gamma ray with frequency $\omega \gg 2m_{\rm e}c^2/\hbar$ into a pair due to propagation at an angle θ to the magnetic field lines is

$$\alpha_\perp \approx \frac{\alpha}{\omega}\frac{m_{\rm e}^2c^4}{\hbar^2}\frac{B\sin\theta}{B_{\rm crit}}T(\chi), \quad T(\chi) \approx 0.46\exp\left(-\frac{4}{3\chi}\right), \quad \chi = \frac{\hbar\omega}{2m_{\rm e}c^2}\frac{B\sin\theta}{B_{\rm crit}}, \tag{9.8}$$

with $\alpha = e^2/4\pi\varepsilon_0\hbar c \approx 1/137$ the fine structure constant. The result (9.8) applies for $\chi \ll 1$. For relevant parameters this decay occurs quickly except for θ very close to zero.

There are two processes that should be mentioned, but which are not discussed in detail here. One is the formation of bound pairs (positronium) rather than free pairs (e.g., Shabad and Usov 1984; Shabad 1993), which needs to be taken into account for $B \gtrsim 0.1B_{\rm crit}$. The other is photon splitting, in which one gamma ray decays into two gamma rays, which needs to be taken into account for $B \gtrsim 0.2B_{\rm crit}$. Both these processes affect the efficiency of the production of pairs.

Curvature Emission of Gamma Rays

Let R_c be the radius of curvature of the field lines. Curvature emission, which is somewhat analogous to synchrotron emission in that both are due to emission by relativistic particles moving in an arc of a circle, has a characteristic frequency $(c/R_c)\gamma^3$, with $R_c \approx 3R_*$ near the surface of the star. This gives $(c/R_c)\gamma^3 \approx 10^4\gamma^3\,\mathrm{s}^{-1}$. Expressing the frequency ω as a photon energy $\varepsilon_{\rm ph} = \hbar\omega$ one has $\varepsilon_{\rm ph} \approx 10^{-11}\gamma^3\,\mathrm{eV}$. Provided that a gamma ray produced by curvature emission has $\varepsilon_{\rm ph} > 2m_ec^2$, it is energetically possible for it to decay into a pair. Such decay leads to secondary pair production. In this way, one initial primary particle, with $\gamma \approx 10^6$, can produce a large number of secondary pairs.

Relaxation to the Lowest Landau Level

In a magnetic field the perpendicular motion of particles is quantized, and the energy is given by (2.9), viz.

$$\varepsilon_n(p_\parallel) = (m^2c^4 + p_\parallel^2c^2 + 2neB\hbar c^2)^{1/2}. \tag{9.9}$$

The states with different n are referred to as the Landau levels, with $n = 0$ the ground state. In lecture 2 this quantization is used only as a calculational tool, and the classical limit is taken by expanding in $\hbar$. In a pulsar magnetosphere the typical spacing between Landau levels, which is the cyclotron energy $\hbar\Omega_e = m_ec^2(B/B_{\rm crit})$, is substantial.

Primary electrons or positrons generated in high Landau levels, $n \gg B_{\rm crit}/B$, emit gamma rays through synchrotron radiation. These gamma rays may act as a source of secondary pairs provided their energy exceeds the threshold for pair production, $n \gtrsim (B_{\rm crit}/B)^2$.

One-Dimensional Plasma

A further important feature of the pair plasma, at least in the inner part of a pulsar magnetosphere, is that it is one dimensional. That is, the electrons and positrons have no motion around the field lines. The relaxation to the ground state occurs through a sequence of emission processes. For $n \gg B_{\rm crit}/B$, when the perpendicular motion is ultrarelativistic, the emission is essentially the same as synchrotron emission in the non-quantum limit. Once the perpendicular motion becomes nonrelativistic ($n \lesssim B_{\rm crit}/B$), the particle relaxes toward its ground state by jumping in a stepwise process from state n to $n-1$ to $n-2$, and so on. The rate per unit time that such transitions occur (for $p_\parallel = 0$) is

$$R_{n\to n-1} = n\alpha\,\frac{4m_ec^2}{3\hbar}\left(\frac{B}{B_{\rm crit}}\right)^2. \tag{9.10}$$

The lifetime for the decay to the ground state is of order the inverse of the transition rate for the slowest transition, which is the transition from the first excited state to the ground state. This gives a lifetime of order $3/B^2$ s, where B

is in tesla. For the magnetic field near a pulsar one has $B \approx 0.1 B_{\rm crit} = 4 \times 10^8$ T, and the lifetime for the decay is then extremely short $\sim 2 \times 10^{-17}$ s. One expects all electrons and positrons to be in their ground state, corresponding to one-dimensional motion.

An interesting formal question concerns the force required to keep the particles moving along the curved magnetic field line. The force is due to the curvature drift motion, which causes the particles to drift perpendicular to $\boldsymbol{B}$ with a velocity $\boldsymbol{v}_{\rm cd}$. The curvature drift speed is of magnitude

$$v_{\rm cd} = \frac{v^2 \gamma}{\Omega_{\rm e} R_{\rm c}}, \tag{9.11}$$

where v is the parallel velocity of the particle and $R_{\rm c}$ is the radius of curvature of the field line. The direction of the curvature drift (opposite for opposite signs of the charge) is such that the Lorentz force $q\boldsymbol{v}_{\rm cd} \times \boldsymbol{B}$ provides the centripetal acceleration, toward the center of curvature of the field line, needed to make the particle follow the curved path.

The Pulsar Wind

Pulsars are observed to be slowing down. Granted that the moment of inertia, I, is reasonably well known, the observed $\dot{\Omega}$ implies an angular momentum loss at a rate $I\dot{\Omega}$. This angular momentum must be carried off either by escaping radiation or by escaping particles. The angular momentum carried off by the observed radiation is far too small to account for the spin down. A magnet rotating in a vacuum emits magnetic dipole radiation with a frequency equal to that of the rotation, but this frequency is well below the plasma frequency of the surrounding plasma so that the radiation cannot escape. A flux of Alfvén waves has been suggested (e.g., Beskin et al. 1993), but no mechanism has been proposed for generating the required flux of Alfvén waves. If electromagnetic radiation and a flux of Alfvén waves are excluded, the only remaining possibility is a flux of kinetic energy, that is, a wind.

It is now widely accepted that pulsars lose most of their energy through a relativistic wind. However, there are serious difficulties in understanding how the wind forms and how the escaping energy and angular momentum is transferred from the electromagnetic field (in a Poynting flux) to outflowing particles. A simple model that would allow this involves the current closing by flowing across field lines near the light cylinder, with the cross-field motions associated with radiation reaction to the emission of gamma rays. However, this model is not consistent with the observations because it would imply most of the power going into gamma rays (e.g., Mestel et al. 1985; Mestel 1993). This difficulty is avoided in the model of Beskin et al. (1993) by having the current close across field lines in a shock wave near the light cylinder. In its simplest form this model simply by-passes the question of how the energy and angular momentum are lost.

Inner Region of the Crab Nebula

There is observational support from the Crab Nebula for the presence of a pulsar wind. The site of the acceleration of the relativistic electrons that power the synchrotron nebula is thought to be a shock that forms some ten percent of the distance from the central pulsar to the edge of the nebula (e.g., Kennel and Coroniti 1984a,b). Inside this region there seems to be a deficiency in the synchrotron emission, and just outside this region "wisps" of activity that propagate outward and vary on a time scale of about a month are observed. It is believed that the whole nebula is powered by the pulsar: a nebula powered in this way is called a plerion. The shock is where the ram pressure of the wind balances the pressure due to the relativistic electrons in the nebula.

9.4 Radio Emission Mechanisms

The radio emission process for pulsars is not known. Several different emission mechanisms are under consideration, and it is unclear which (if any) of these will eventually become the accepted mechanism.

Specific Emission Mechanisms

Various radio emission mechanisms proposed for pulsars include the following.

Emission by bunches : An early theory for the radio emission is coherent curvature emission by bunches. The basic idea is the familiar one that N particles in a volume less than a cubic wavelength radiate like a macrocharge $Q = Ne$, and because the power radiated is proportional to Q^2, the power is N^2 times the power from an individual particle. There are seemingly insurmountable difficulties with this theory (Melrose 1981). For example, when one takes into account the highly anisotropic nature of curvature emission by relativistic particles, the bunch really needs to be a pancake with its normal within an angle $1/\gamma$ of the direction of the magnetic field. An obvious difficulty is to identify a mechanism that allows such an exotic bunch to form; none of the suggested mechanisms works. Moreover, even if such a bunch did form, the bending of the field lines would cause its normal to deviate to more than $1/\gamma$ away from the direction of the magnetic field in a very short time. For these and other reasons, coherent curvature emission by bunches is unacceptable.

Relativistic plasma emission : An alternative mechanism favored in the 1970s is based on a relativistic version of plasma emission (lecture 8). This involves a streaming instability, which generates waves that are analogous to Langmuir waves in that they cannot escape from the plasma. Nonlinear processes in the plasma partially convert these waves into escaping radiation. This theory requires a detailed discussion of the wave properties in a relativistic, streaming one-dimensional pair plasma (e.g., Arons and Barnard 1986; Beskin et al. 1993). Such a plasma supports a Langmuir-like mode and two high-frequency modes that are somewhat analogous to the o-mode and x-mode.

The streaming instability could be due to the high energy beam of positrons moving through the pair plasma, to a relative motion of the electrons and positrons in the pair plasma, or to a less obvious types of relative motion. However, the growth rate for these instabilities is too small: the pair plasma leaves the magnetosphere before it has given up significant energy. A larger growth rate can result if the generation of the pair plasma fluctuates in time, producing a sequence of beams, with the faster particles in a following beam overtaking the slower particles in a preceding beam (Ursov and Usov 1988).

Relativistic plasma emission is perhaps the most widely favored pulsar radio emission mechanism, and development of the theory continues (e.g., Asseo, Pelletier and Sol 1990).

Maser curvature emission : Curvature emission is like synchrotron radiation in that, in the simplest case, the absorption coefficient cannot be negative (Blandford 1975; Melrose 1978), so that maser emission cannot occur. However, the proof is invalid when the curvature drift (9.11) is included (Zheleznyakov and Shaposhnikov 1979; Luo and Melrose 1992). Maser curvature emission can occur due to at least two effects: the curvature drift across field lines and a twist of the field lines, corresponding to curved field lines that are not confined to a plane.

A form of curvature-associated instability proposed by Beskin et al. (1993) exists in the limit $B \to \infty$, when the drift speed (9.11) vanishes. However, the claim that this instability is spurious (Nambu 1989; Machabeli 1991) has not been refuted.

Cyclotron instability : An instability that involves electrons (or positrons) having a cyclotron transition through the anomalous cyclotron resonance $s = -1$ in (2.8) (Kazbegi, Machabeli and Melikidze 1991). This leads to emission near the cyclotron frequency. The waves that grow in this instability have refractive index > 1 and so cannot escape directly. To produce escaping radiation these waves need to be converted into the high-frequency modes through a plasma-emission type process. For this process to produce radiation in the radio range, the instability must occur at large distances from the star, where B is sufficiently small.

Free electron maser emission : A form of linear acceleration emission in which the relativistic particles are accelerated by an oscillating electric field can lead to maser emission (Melrose 1978). The characteristic frequency of the emission is $\omega \sim \omega_0\gamma^2$, where ω_0 is the greater of the typical frequency of the oscillating electric field, or the typical wavenumber times c. A detailed treatment of this mechanism (Rowe 1992a,b) shows that it exists in two regimes. One corresponds to a form of relativistic plasma emission, in which the energy in the emitted waves comes primarily from the energy in the oscillating electric field. The other corresponds to a form of free electron maser emission, in which the oscillating field acts as a wiggler field.

Millisecond Pulsars

The emission from millisecond pulsars is remarkably similar to that from ordinary pulsars, in the sense that one could not identify a millisecond pulsar from its pulse profile alone. A constraint that could be applied to all emission mechanisms is that they be capable of accounting for similar emission from ordinary pulsars and from millisecond pulsars. A parameter that is markedly different for the two classes of pulsars is B. The emission mechanisms that are sensitive to B are the cyclotron mechanism and curvature-drift induced maser emission. This constraint would argue against these mechanisms. However, the remaining mechanisms depend on B only in the combination $B\Omega \propto n_{\rm GJ}$, which is not so different for the two classes of pulsars.

Which is the Most Plausible Mechanism?

In my opinion, emission by bunches is unacceptable, and should not be considered further. The two mechanisms that are sensitive to B would require that a different emission mechanism operate in millisecond pulsars, and this is intrinsically implausible. Although some form of relativistic plasma emission is perhaps the most plausible emission mechanism, free electron maser emission and maser curvature emission due to a twisted magnetic field have not been ruled out.

Returning to the views expressed by Michel (1991) and Beskin et al. (1993), in my opinion there are several important aspects of the physics of pulsar magnetospheres that are not adequately understood. Besides the radio emission mechanism, these include the location of the primary acceleration region, the return current path, and the transfer of energy and angular momentum to a pulsar wind. The problem of formulating a self-consistent model for a pulsar magnetosphere is far from solved.

Acknowledgements

It is a pleasure to thank Arnold Benz and Thierry Courvoisier for their efficient organization of the winter school, and for the excellent way the winter school was run, and to thank the Swiss Society of Astrophysics and Astronomy for their hospitality in Les Diablerets. I also thank Stephen Hardy, Andrew Melatos, Mick Pope, Mike Wheatland and Andrew Willes for their assistance in proof reading the notes, and Andrew Willes for Figure 6.3.

References

Achterberg, A. (1981): "On the nature of small amplitude Fermi acceleration" *Astron. Astrophys.* **97**, 259–264

Achterberg, A. (1990): "Particle acceleration near astrophysical shocks" in (eds) Brinkmann, W., Fabian, A.C., Giovannelli, F., *Physical Processes in Hot Cosmic Plasmas* (Kluwer: Dordrecht) pp. 67–80

Achterberg, A., Ball, L. (1994): "Particle acceleration at superluminal quasi-perpendicular shocks" *Astron. Astrophys.* **284**, 687–704

Akasofu, S.-I. (1977): *Physics of Magnetospheric Substorms* (D. Reidel: Dordrecht)

Akhiezer, A.I., Akhiezer, I.A., Polovin, R.V., Sitenko, A.G. Stepanov, K.N. (1967): *Collective Oscillations in a Plasma* (M.I.T. Press: Cambridge, Mass.)

Arons, J. (1983): "Pair creation above pulsar polar caps: geometrical structure and energetics of slot gaps" *Astrophys. J.* **266**, 215–241

Arons, J. (1992): "Magnetospheric structure of rotation-powered neutron stars" in (eds) Hankins, T.H., Rankin, J.M., Gil, J.A., *The Magnetospheric Structure and Emission Mechanisms of Radio Pulsars* (Pedagogical University Press: Zielona Góra) pp. 56–77

Arons, J., Barnard, J.J. (1986): "Wave propagation in pulsar magnetospheres: dispersion relations and normal modes of plasmas in superstrong magnetic fields" *Astrophys. J.* **302**, 120–137

Asseo, E., Pelletier, G., Sol, H. (1990): "A non-linear radio pulsar emission mechanism" *Mon. Not. R. Astron. Soc.* **247**, 529–548

Axford, W.I. (1992): "Particle acceleration on galactic scales" in (eds) Zank, G.P., Gaisser, T.K., *Particle Acceleration in Cosmic Plasmas*, (AIP: New York) pp. 45–56

Barbosa, D.D. (1979): "Stochastic acceleration of solar flare protons" *Astrophys. J.* **233**, 383–394

Bekefi, G. (1966): *Radiation Processes in Plasmas* (John Wiley & Sons: New York)

Benson, R.F., Calvert, W., Klumpar, D.M. (1980): "Simultaneous wave and particle observations in the auroral kilometric radiation source region" *Geophys. Res. Lett.* **7**, 959–962

Benz, A.O. (1986): "Millisecond radio spikes" *Solar Phys.* **104**, 99–110

Benz, A.O. (1993): *Plasma Astrophysics* (Kluwer Academic Publishers: Dordrecht)

Beskin, V.S., Gurevich, A.V., Istomin, Ya.N. (1993): *Physics of the Pulsar Magnetosphere* (Cambridge University Press)

Blandford, R.D. (1975): "Amplification of radiation by relativistic particles in a strong magnetic field" *Mon. Not. R. Astron. Soc.* **170**, 551–557

Blandford, R.D., Netzer, H., Woltjer, L. (1990): *Active Galactic Nuclei* (Springer-Verlag: Berlin)

Boström, R., Gustafsson, G., Holback, B., Holgren, G., Koskinen, H., Kintner, P. (1988): "Characteristics of solitary waves and weak double layers in the magnetospheric plasma" *Phys. Rev. Lett.* **61**, 82–85

Cargill, P.J., Papadopoulos, K. (1988): "A mechanism for strong shock electron heating in supernova remnants" *Astrophys. J.* **329**, L29–L32

Carr, T.D., Gulkis, S. (1969): "The magnetosphere of Jupiter" *Ann. Rev. Astron. Astrophys.* **7**, 577–618

Cesarsky, C.J. (1980): "Cosmic-ray confinement in the galaxy" *Ann. Rev. Astron. Astrophys.* **18**, 289–319

Cheng, K.S., Ho, C. Ruderman, M.A. (1986a & b): "Energetic radiation from rapidly spinning pulsars. I & II" *Astrophys. J.* **300**, 500–521 & 522–539

Chupp, E.L. (1983): "High energy particle acceleration in solar flares – Observational evidence" *Solar Phys.* **86**, 383–393

Davis, L., Jr (1956): "Modified Fermi mechanism for the acceleration of cosmic rays" *Phys. Rev.* **101**, 351–358

Dessler, A.J. (ed.) (1983): *Physics of the Jovian Magnetosphere* (Cambridge University Press)

Dragt, A.J. (1961): "Effect of hydromagnetic waves of the lifetime of Van Allen radiation protons" *J. Geophys. Res.* **66**, 1641–1649

Dulk, G.A. (1967): "Apparent changes in the rotation rate of Jupiter" *Icarus* **7**, 173–182

Dulk, G.A. (1985): "Radio emission from the Sun and stars" *Ann. Rev. Astron. Astrophys.* **23**, 169–224

Dungey, J. W. (1961): "Interplanetary magnetic field and the auroral zones" *Phys. Rev. Lett.* **6**, 47–48

Dungey, J. W. (1963): "Loss of Van Allen electrons due to whistlers" *Planet. Space Sci.* **11**, 591–595

Fermi, E. (1949): "On the origin of cosmic radiation" *Phys. Rev.* **75**, 1169–1174

Field, G.B. (1986): "Theory of the interstellar medium" in (eds) Shapiro, S.T., Teukolsky, S.A., *Highlights of Modern Astrophysics* (John Wiley & Sons: New York) pp. 235–265

Gaisser, T.K. (1990): *Cosmic Rays and Particle Physics* (Cambridge University Press)

Ginzburg, V.L., Syrovatskii, S.I. (1965): "Cosmic magnetobremsstrahlung (synchrotron radiation)" *Ann. Rev. Astron. Astrophys.* **3**, 297–350

Ginzburg, V.L., Zheleznyakov, V.V. (1975): "On the pulsar emission mechanisms" *Ann. Rev. Astron. Astrophys.* **13**, 511–535

Goldman, M.V. (1983): "Progress and problems in the theory of type III solar radio emission" *Solar Phys.* **89**, 403–442

Goldman, M.V. (1984): "Strong turbulence of plasma waves" *Rev. Mod. Phys.* **56**, 709–735

Goldman, M.V., Smith, D.F. (1986): "Solar radio emission" in (ed.) Sturrock, P.A., *Physics of the Sun, Volume II*
(D. Reidel: Dordrecht) pp. 325–376

Goldreich, P., Julian, W.H. (1969): "Pulsar electrodynamics" *Astrophys. J.* **157**, 869–880

Goldreich, P. Lynden-Bell, D. (1969): "Io, a Jovian unipolar inductor" *Astrophys. J.* **156**, 59–78

Goldstein, H. (1959): *Classical Mechanics* (Addison-Wesley: Reading, Mass.)

Goldstein, M.L., Goertz, C.K. (1983): "Theories of radio emissions and plasma waves" in (ed.) Dessler, A.J., *Physics of the Jovian Magnetosphere* (Cambridge University Press) pp. 317–352

Grognard, R.J.M. (1985): "Propagation of electron streams" in (eds) McLean, D.J., Labrum, N.R., *Solar Radiophysics* (Cambridge University Press) pp. 253–286

Gurnett, D.A. (1974): "The Earth as a radio source. Terrestrial kilometric radiation" *J. Geophys. Res.* **79**, 4277–4238

Hall, D.E., Sturrock, P.A. (1967): "Diffusion, scattering, and acceleration of particles by stochastic electromagnetic fields" *Phys. Fluids* **10**, 2620–2628

Hanasz, J. (1966): "Chains of type I solar radio bursts" *Aust. J. Phys.* **19**, 635–647

Hasselmann, K., Wibberenz, G. (1968): "Scattering of charged particles by random electromagnetic fields" *Z. Geophys.* **34**, 353–388

Helliwell, R.A. (1967): "A theory of discrete VLF emissions from the magnetosphere" *J. Geophys. Res.* **72**, 4773–4790

Hewitt, R.G., Melrose, D.B., Rönnmark, K.G. (1982): "The loss-cone driven electron-cyclotron maser" *Aust. J. Phys.* **35**, 447–471

Hillas, A.M. (1984): "The origin of ultra-high-energy cosmic rays" *Ann. Rev. Astron. Astrophys.* **22**, 425–444

Holloway, N.J. (1973): "p-n junctions in pulsar magnetospheres?" *Nature Phys. Sci.* **246**, 6–9

Holman, G.D., Eichler, D., Kundu, M.R. (1980): "An interpretation of solar flare microwave spikes as gyrosynchrotron masering" in (eds.) Kundu, M.R., Holman, G.D., *Unstable Current Systems and Plasma Instabilities in Astrophysics* (D. Reidel: Dordrecht) pp. 457–459

de Jager, C., Kuijpers, J., Correia, E., Kaufmann, P. (1987): "A high-energy solar flare burst complex and the physical properties of its source region" *Solar Phys.* **110**, 317–326

Jokipii, J.R. (1966): "Cosmic-ray propagation. I. Charged particles in a random magnetic field" *Astrophys. J.* **146**, 480–487

Jokipii, J.R. (1971): "Propagation of cosmic rays in the solar wind" *Rev. Geophys. Space Phys.* **9**, 27–87

Jokipii, J.R. (1982): "Particle drift, diffusion, and acceleration at shocks" *Astrophys. J.* **255**, 716–720

Jokipii, J.R. (1987): "Rate of energy gain and maximum energy in diffusive shock acceleration" *Astrophys. J.* **313**, 842–846

Kaplan, S.A., Tsytovich, V.N. (1973): *Plasma Astrophysics* (Pergamon Press: Oxford)

Kardashev, N.S. (1962): "Nonstationariness of spectra of young sources of nonthermal radio emission" *Soviet. Astron. AJ* **6**, 317–327

Kazbegi, A.Z., Machabeli, G.Z., Melikidze, G.I. (1991): "On the circular polarization of pulsar emission" *Mon. Not. R. Astron. Soc.* **253**, 377–387

Kellerman, K.I., Pauliny-Toth, I.I.K. (1969): "The spectra of opaque radio sources" *Astrophys. J.* **155**, L71–L78

Kennel, C.F., Coroniti, F.V. (1984a): "Confinement of the Crab pulsar's wind by its supernova remnant" *Astrophys. J.* **283**, 694–709

Kennel, C.F., Coroniti, F.V. (1984b): "Magnetohydrodynamic model of Crab nebula radiation" *Astrophys. J.* **283**, 710–730

Kennel, C.F., Petschek, H.E. (1966): "Limit on stably trapped particle fluxes" *J. Geophys. Res.* **71**, 1–28

Kuijpers, J. (1985): "Radio observable processes in stars" in (eds) Hjellming, R.M., Gibson, D.M., *Radio Stars* (D. Reidel: Dordrecht) pp. 3–31

Kulsrud, R.M., Ferrari, A. (1971): "The relativistic quasilinear theory of particle acceleration by hydromagnetic turbulence" *Astrophys. Space Sci.* **12**, 302–318

Kulsrud, R.M., Pearce, W.P. (1969): "The effect of wave-particle interactions on the propagation of cosmic rays" *Astrophys. J.* **156**, 445–469

Lacombe, C. (1977): "Acceleration of particles and plasma heating by turbulent Alfvén waves in a radiogalaxy" *Astron. Astrophys.* **54**, 1–16

Lee, M.A. (1983): "Coupled hydromagnetic wave excitation and ion acceleration at interplanetary traveling shocks" *J. Geophys. Res.* **88**, 6109–6119

Lee, M.A. (1992): "Particle acceleration in the heliosphere" in (eds) Zank, G.P., Gaisser, T.K., *Particle Acceleration in Cosmic Plasmas* (AIP: New York) pp. 27–44

Leroy, M.M., Mangeney, A. (1984): "A theory of energization of solar wind electrons by the Earth's bow shock" *Annales Geophys.* **2**, 449–455

Lifshitz, E.M., Pitaevskii, L.P. (1981): *Physical Kinetics* (Pergamon Press: Oxford)

Lin, R.P., Potter, D.W., Gurnett, D.A., Scarf, F.L. (1981): "Energetic electrons and plasma waves associated with a solar type III radio burst" *Astrophys. J.* **251**, 364–373

Longair, M.S. (1992): *High Energy Astrophysics Volume 1 Particles, Photons and their Detection* (Cambridge University Press)

Louarn, P., Roux, A., de Féraudy, H., Le Quéau, D., André, M., Matson, L. (1990): "Trapped electrons as a free energy source for the auroral kilometric radiation" *J. Geophys. Res.* **95**, 5983–5995

Luo, Q., Melrose, D.B. (1992): "Coherent curvature emission and radio pulsars" *Mon. Not. R. Astron. Soc.* **258**, 616–620

Machabeli, G.Z. (1991): "On the impossibility of wave generation due to plasma motion in the weakly inhomogeneous magnetic field of pulsars" *Plasma Physics and Controlled Fusion* **33**, 1227–1234

Matsumoto, H. (1979): "Nonlinear whistler-mode interactions and triggered emissions in the magnetosphere: A review" in (ed.) Palmadesso, P.J., Papadopoulos, K., *Wave Instabilities in Space Plasmas* (D. Reidel: Dordrecht) pp. 163–190

McKean, M.E., Winglee, R.M., Dulk, G.A. (1989): "Propagation and absorption of electron-cyclotron maser radiation during solar flares" *Solar Phys.* **122**, 53–89

McLean, D.J., Labrum, N.R. (1985): *Solar Radiophysics* (Cambridge University Press)

Melrose, D.B. (1968): "The emission and absorption of waves by charged particles in magnetized plasmas" *Astrophys. Space Sci.* **2**, 171–235

Melrose, D.B. (1976): "An interpretation of Jupiter's decametric radiation and the terrestrial kilometric radiation as direct amplified gyro-emission" *Astrophys. J.* **207**, 651–662

Melrose, D.B. (1978): "Amplified linear acceleration emission applied to pulsars" *Astrophys. J.* **225**, 557–573

Melrose, D.B. (1980): *Plasma Astrophysics Volume I Emission, Absorption and Transfer of Waves in Plasmas* (Gordon & Breach: New York)

Melrose, D.B. (1980): *Plasma Astrophysics Volume II Astrophysical Applications* (Gordon & Breach: New York)

Melrose, D.B. (1981): "Evolution and radiation in pulsar polar cap models" in (eds) Sieber, W., Wielebinski, R., *Pulsars, 13 years of Research on Neutron Stars* (D. Reidel: Dordrecht) pp. 133–140

Melrose, D.B. (1983): "Prompt acceleration of $\geq$ 30 MeV per nucleon ions in solar flares" *Solar Phys.* **89**, 149–162

Melrose, D.B. (1986): *Instabilities in Space and Laboratory Plasmas* (Cambridge University Press)

Melrose, D.B. (1991): "Emission at cyclotron harmonics due to coalescence of z-mode waves" *Astrophys. J.* **380**, 256–267

Melrose, D.B., Cramer, N.F. (1989): "Quasilinear relaxation of electrons interacting with an inhomogeneous distribution of Langmuir waves" *Solar Phys.* **123**, 343–365

Melrose, D.B., Dulk, G.A. (1982): "Electron-cyclotron masers as the source of certain solar and stellar radio burst" *Astrophys. J.* **259**, 884–858

Melrose, D.B., Dulk, G.A. (1988): "Implications of the Liouville's theorem on the apparent directivities and brightness temperatures of solar radio bursts" *Solar Phys.* **116**, 141–156

Melrose, D.B., Dulk, G.A. (1993): "Electron cyclotron maser emission at oblique angles" *Planet. Space Sci.* **41**, 333-339

Melrose, D.B., Goldman, M.V. (1987) "Microstructures in type III events in the solar wind" *Solar Phys.* **107**, 329–350

Melrose, D.B., McPhedran, R.C. (1991): *Electromagnetic Processes in Dispersive Media* (Cambridge University Press)

Melrose, D.B., Rönnmark, K.G., Hewitt, R.G. (1982): "Terrestrial kilometric radiation: the cyclotron theory" *J. Geophys. Res.* **87**, 5140–5150

Mestel, L. (1993): "Pulsar magnetospheres" in (eds) Blandford, R.D., Hewish, A., Lyne, A.G., Mestel, L., *Pulsars as Physics Laboratories* (Oxford University Press) pp. 93–104

Mestel, L., Robertson, J.A., Wang, Y.-M., Westfold, K.C. (1985): "The axisymmetric pulsar magnetosphere" *Mon. Not. R. Astron. Soc.* **217**, 443–484

Michel, F.C. (1982): "Theory of pulsar magnetospheres" *Rev. Mod. Phys.* **54**, 1–66

Michel, F.C. (1991): *Theory of Neutron Star Magnetospheres* (University of Chicago Press)

Mozer, F.S., Cattell, C.A., Hudson, M.K., Lysak, R.L., Temerin, M., Torbert, R.B. (1980): "Satellite measurements and theories of low altitude auroral particle acceleration" *Space Sci. Rev.* **27**, 155–213

Nambu, M. (1989): "Stability of a nonstationary and inhomogeneous plasma" *Plasma Physics and Controlled Fusion* **31**, 143–145

Neuhauser, D., Koonin, S.E., Langanke, K. (1987) "Hartree-Fock calculations of atoms and molecular chains in strong magnetic fields" *Phys. Rev.* **A36**, 4163–4175

Nicholson, D.R. (1983): *Introduction to Plasma Theory* (John Wiley & Sons: New York)

Omidi, N., Gurnett, D.A. (1982): "Growth rate calculations of auroral kilometric radiation using the relativistic resonance condition" *J. Geophys. Res.* **87**, 2377–2383

Parker, E.N. (1957): "Acceleration of cosmic rays in solar flares" *Phys. Rev.* **107**, 830–836

Reames, D.V. (1992): "Particle acceleration in the heliosphere" in (eds) Zank, G.P., Gaisser, T.K., *Particle Acceleration in Cosmic Plasmas*, (AIP: New York) pp. 213–222

Rickett, B.J. (1990): "Radio propagation through the turbulent interstellar medium" *Ann. Rev. Astron. Astrophys.* **28**, 561–605

Robinson, P.A. (1989): "Escape of fundamental electron cyclotron maser emission from the Sun and stars" *Astrophys. J.* **341**, L99–L102

Robinson, P.A. (1993): "Stochastic-growth theory of Langmuir growth-rate fluctuations in type III solar radio sources" *Solar Phys.* **146**, 357–363

Robinson, P.A., Newman, D.L. (1990): "Two-component model of strong Langmuir turbulence: scalings, spectra and statistics of Langmuir waves" *Phys. Fluids B* **2**, 2999–3016

Roederer, J.G. (1970): *Dynamics of Geomagnetically Trapped Radiation* (Springer-Verlag: Berlin)

Roux, A., Hilgers, A., de Feraudy, H., Le Quéau, D., Louarn, P., Perraut, S., Bahnsen, A., Jespersen, M., Ungstrup, E., André, M. (1993): "Auroral kilometric radiation sources: in situ and remote observations from Viking" *J. Geophys. Res.* **98**, 11,657–11,670

Rowe, E.T. (1992a & b): "Particle motion in longitudinal waves. I. & II." *Aust. J. Phys.* **45**, 1–20 & 21–37

Ruderman, M.A. (1981): "Evolution and radiation in pulsar polar cap models" in (eds) Sieber, W., Wielebinski, R., *Pulsars, 13 years of Research on Neutron Stars* (D. Reidel: Dordrecht) pp. 87–98

Ruderman, M.A., Sutherland, P.G. (1975): "Theory of pulsars: polar gaps, sparks and coherent microwave emission" *Astrophys. J.* **196**, 57–72

Rybicki, G.B., Lightman, A.P. (1979): *Radiative Processes in Astrophysics* (John Wiley & Sons: New York)

Schott, G.A. (1912): *Electromagnetic Radiation* (Cambridge University Press)

Schulz, M., Lanzerotti, L.J. (1974): *Particle Diffusion in the Radiation Belts* (Springer-Verlag: Berlin)

Shabad, A.Ye. (1992) *Polarization of the Vacuum and a Quantum Relativistic Gas in an External Field* (Nova Science Publishers: New York)

Shabad, A.E., Usov, V.V. (1984): "Propagation of γ-radiation in strong magnetic fields of pulsars" *Astrophys. Space Sci.* **102**, 327–358

Shu, F.H. (1991): *The Physics of Astrophysics Volume I: Radiation* (University Science Books: Mill Valley, California)

Skilling, J.A. (1975a): "Cosmic ray streaming–I Effect of Alfvén waves on particles" *Mon. Not. R. Astron. Soc.* **172**, 557–566

Skilling, J.A. (1975b): "Cosmic ray streaming–II Effect of particles on Alfvén waves" *Mon. Not. R. Astron. Soc.* **173**, 245–254

Skilling, J.A. (1975c): "Cosmic ray streaming–III Self-consistent solutions" *Mon. Not. R. Astron. Soc.* **173**, 255–269

Spitzer, L., Jr (1990): "Theories of the hot interstellar gas" *Ann. Rev. Astron. Astrophys.* **28**, 71–101

Stix, T.H. (1962): *The Theory of Plasma Waves* (McGraw-Hill: New York)

Sturrock, P.A. (1971): "A model of pulsars" *Astrophys. J.* **164**, 529–556

Sturrock, P.A. (ed.) (1980): *Solar Flares* (Colorado Associated Press: Boulder)

Sturrock, P.A., Kaufmann, P., Moore, R.L., Smith, D.F. (1984): "Energy release in solar flares" *Solar Phys.* **94**, 341–357

Tsytovich, V.N. (1966): "Statistical acceleration of particles in a turbulent plasma" *Soviet Phys. Usp.* **9**, 370–404

Ursov, V.N., Usov, V.V. (1988): "Plasma flow nonstationarity in pulsar magnetospheres and two-stream instability" *Astrophys. Space Sci.* **140**, 325–336

Vlahos L. (1987): "Electron cyclotron maser emission from solar flares" *Solar Phys.* **111**, 155–166

Warwick, J.W. (1967): "Radiophysics of Jupiter" *Space Sci. Rev.* **6**, 841–891

Wdowczyk, J., Wolfendale, A.W. (1989): "Highest energy cosmic rays" *Ann. Rev. Nucl. Part. Sci.* **39**, 43–71

Wentzel, D.G. (1961): "Hydromagnetic waves and the trapped radiation. Part 1. Breakdown of the adiabatic invariance" *J. Geophys. Res.* **66**, 359–369

Wentzel, D.G. (1974): "Cosmic-ray propagation in the galaxy: collective effects" *Ann. Rev. Astron. Astrophys.* **12**, 71–96

Wild, J.P., Smerd, S.F., Weiss, A.A. (1963): "Solar bursts" *Ann. Rev. Astron. Astrophys.* **1**, 291–366

Wild, J.P., Smerd, S.F. (1971): "Radio bursts from the solar corona" *Ann. Rev. Astron. Astrophys.* **10**, 159–196

Winglee, R.M., Dulk, G.A., Pritchett, P.L. (1988): "Fine structure of microwave spike bursts and associated cross-field energy transport" *Astrophys. J.* **328**, 809–823

Wu, C.S. (1986): "Kinetic cyclotron and synchrotron maser instabilities. Radio emission processes by direct amplification of radiation" *Space Sci. Rev.* **41**, 215–298

Wu, C.S., Lee, L.C. (1979): "A theory of the terrestrial kilometric radiation" *Astrophys. J.* **230**, 621–626

Zheleznyakov, V.V., Shaposhnikov, V.E. (1979): "Absorption of curvature radiation" *Aust. J. Phys.* **32**, 49–59

Zhou, Y., Mattheus, W.H. (1990): "Transport and turbulence modeling of solar wind fluctuations" *J. Geophys. Res.* **95**, 10,291–10,311

Zlobec, P. (1975): "Intermediate polarization of type I bursts" *Solar Phys.* **43**, 453–461

Particle Acceleration

John G. Kirk

Max-Planck-Institut für Kernphysik
Postfach 10 39 80
69029 Heidelberg
Germany

1 Introduction

In this series of lectures I do not intend to provide a comprehensive overview of the particle acceleration mechanisms of interest in astrophysical plasmas. There are many of these, but, unfortunately, it is often very difficult to be specific about which of them really lies at the root of a particular observed phenomenon. Instead, I have chosen to concentrate on a few examples in which a theory has been developed all the way from the basics up to a testable model. Although I hope to provide a reasonably complete account of each example, it is nevertheless necessary to choose a starting point which is somewhat more advanced than just elementary electrodynamics. The kinetic theory of plasma astrophysics underlies the mechanisms to be discussed, and in this introductory lecture I shall sketch the physics of the two fundamental transport equations on which the rest of the course is based – those describing scatter-free propagation and propagation under the influence of pitch-angle scattering. (For a thorough treatment of these equations and a discussion of their ranges of validity, the reader should refer to the companion series of lectures by D.B. Melrose.) These two transport equations recur as twin themes throughout the course, and the first lecture ends with a short, but hopefully instructive discussion of how they conspire to accelerate particles in perhaps the oldest mechanism of all – magnetic pumping.

1.1 Nonthermal Particles

Despite the second law of thermodynamics, it quite obvious that the universe is not in a state of thermodynamic equilibrium, and there are few more dramatic examples of this than the energetic particles which are the subject of this course. The evidence for these reaches us in several ways, ranging from direct detection of cosmic rays incident on the atmosphere to observation of synchrotron emission from distant radio galaxies. Energetic particles are also detected directly by satellite experiments close to the site of their origin e.g., at the Earth's bow shock. Somehow, Nature contrives a way of avoiding equipartition of energy for these particles. This aspect we can understand at least in a qualitative sense.

Energy is shared out effectively between particles if many channels of interaction are open, or if collisions are frequent. Astrophysical plasmas are, however, so thin, that collisions are extremely rare. A cosmic ray particle, for example, which spends a million years or so wandering around in the disk of our galaxy, has only about a one-in-eight chance of colliding with another nucleus during that time. Such collisions are interesting and important for the synthesis of elements such as Lithium or Boron, and they may even be a significant danger to living cells at altitude in the Swiss Alps, but they are not effective in bringing cosmic rays into thermodynamic equilibrium with the interstellar medium. Of course, energetic charged particles interact with and are confined by the interstellar medium, but this happens only via the average electromagnetic field they feel, and it is to this force to which we first must turn our attention.

1.2 Lorentz Force

A particle of charge e and mass m in an electric field $\mathbf{E}$ and magnetic field $\mathbf{B}$ feels the 'Lorentz force' and has an equation of motion

$$\frac{\mathrm{d}\mathbf{p}}{\mathrm{d}t} = e\left(\mathbf{E} + \frac{1}{c}\mathbf{v}\wedge\mathbf{B}\right) \ , \tag{1}$$

where $\mathbf{v}$ is the particle velocity and $\mathbf{p} = \gamma m\mathbf{v}$ is the momentum, with $\gamma = 1/\sqrt{1 - v^2/c^2}$ the particle's Lorentz factor. From this expression it is clear, amongst other things, that I will be using Gaussian units. However, there is another important reason for examining this well-known equation here. If the electric field vanishes, $\mathbf{E} = 0$, then the scalar product of (1) with $\mathbf{p}$ reveals that the magnitude of the momentum p is constant. The particle energy $E = \sqrt{m^2c^4 + p^2c^2}$ $(= \gamma mc^2)$ is thus a conserved quantity. In nonrelativistic language, we can say that the force $e\mathbf{v}\wedge\mathbf{B}/c$ is normal to the velocity, so that no work is done on the particle. One can also make an analogous covariant argument. This is the basic property which enables energetic particles to avoid sharing out their energy: electric fields are rare in the type of highly conducting, fully ionised plasma we encounter in astrophysics. However, although this makes it easy for a particle to keep its energy, it makes it correspondingly difficult for it to acquire it in the first place. In fact, any model of particle acceleration must ultimately rely on an electric field to energise the particles. In many cases, however, the electric field does not appear explicitly in the theory, and it is a highly instructive exercise to locate it.

1.3 Liouville Equation

The most useful quantity we can calculate from an acceleration theory is the single particle distribution function $f(\mathbf{p}, \mathbf{x}, t)$ giving the number density of particles in the six-dimensional phase space $(\mathbf{p}, \mathbf{x})$ at time t. This is sufficient to enable us to compute, for example, the synchrotron radiation from accelerated electrons or the gamma-rays produced by cosmic rays passing through an interstellar cloud.

One case is particularly simple – that in which the particles do not interact amongst themselves. Then the distribution function obeys the *Liouville equation*, which is a simple consequence of the conservation of particle number combined with Hamiltonian mechanics. Consider the rate of change of the number of particles in an element of phase space $(\partial f/\partial t)\mathrm{d}^3x\mathrm{d}^3p$. This is given by the difference between the rate at which particles enter and leave opposite sides of the six-dimensional cube $\mathrm{d}^3x\mathrm{d}^3p$:

$$\frac{\partial f}{\partial t}+\frac{\partial}{\partial \mathbf{x}}[\dot{\mathbf{x}}\cdot f(\mathbf{p},\mathbf{x},t)]+\frac{\partial}{\partial \mathbf{p}}[\dot{\mathbf{p}}\cdot f(\mathbf{p},\mathbf{x},t)]=0\ . \tag{2}$$

The symbols $\dot{\mathbf{x}}$ and $\dot{\mathbf{p}}$ denote the time derivatives of position and momentum along a particle trajectory. The speed changes from place to place, and according to the Lorentz force (1), the acceleration varies with particle momentum, so that the derivatives of these quantities are important in (2). However, things simplify if $\mathbf{x}$ and $\mathbf{p}$ are canonically conjugate variables i.e., if the equations of motion can be written in Hamiltonian form:

$$\dot{\mathbf{x}}=\frac{\partial H}{\partial \mathbf{p}}\qquad \dot{\mathbf{p}}=-\frac{\partial H}{\partial \mathbf{x}}\ , \tag{3}$$

where $H(\mathbf{p},\mathbf{x})$ is the Hamiltonian. Substituting these relations into (2) leads immediately to the Liouville equation:

$$\frac{\partial f}{\partial t}+\dot{\mathbf{x}}\cdot\frac{\partial}{\partial \mathbf{x}}f(\mathbf{p},\mathbf{x},t)+\dot{\mathbf{p}}\cdot\frac{\partial}{\partial \mathbf{p}}f(\mathbf{p},\mathbf{x},t)=0\ . \tag{4}$$

The left-hand side of this equation is simply the derivative of f along a trajectory, so that (4) implies Liouville's theorem, one statement of which is: *the distribution function is constant along particle trajectories.*

This theorem is very useful in the analysis of shock-drift acceleration, for example. However, although the final statement is correct, the derivation of (4) has ignored the fact that for particles moving under the Lorentz force, $\mathbf{x}$ and $\mathbf{p}$ are not, in fact, canonically conjugate – the variable conjugate to $\mathbf{x}$ is $\mathbf{p}+e\mathbf{A}/c$, where $\mathbf{A}$ is the vector potential of the magnetic field ($\mathbf{B}=\nabla\wedge\mathbf{A}$). This subtlety need not concern us here, since direct calculation using the equation of motion (1) and the conservation equation (2) verifies the correctness of (4). In this special case, $\mathbf{A}$ is independent of $\mathbf{p}$, and so f remains constant along the trajectories in $(\mathbf{p},\mathbf{x})$ space. [An elegant treatment of particle transport can be based on the use of non-canonically conjugate variables and the associated Lie algebra – see, for example the monograph by Balescu (1988).]

1.4 Scattering

The situation is, of course, much more difficult when the interaction between particles must be taken into account. The interaction is mediated by the electromagnetic field, but because of the long-range character of the Coulomb interaction, the field at the position of a particular particle depends on the positions and velocities of all other particles within a Debye radius, of which there are a very large number. It is then a hopeless task to solve the Liouville equation. Fortunately, though, one can approach the problem using the Vlasov equation, which looks just like (4), except that the equation of motion of the trajectories refers not to the exact fields, but to 'averaged' or 'self-consistent' ones which are to be calculated from the single particle distribution function. Even with this simplification, the problem is not solved, because the equations of motion are still highly complicated and nonlinear. To make analytic progress, we are driven to attempt a linearisation by dividing the (averaged) fields into specified external ones plus a small fluctuating part arising from the collective effects of the particles. It is not my intention to go into this procedure in detail, since this would take us too deeply into kinetic plasma physics. I will merely quote the resulting transport equation and try to make the terms it contains physically plausible.

The first point to note is that we are interested only in accelerated or non-thermal particles and these have in general only a very small contribution to the total number density of charged particles. The energy density of cosmic rays in the interstellar medium (ISM), for example, is roughly $1\,\mathrm{eV\,cm^{-3}}$, and their number is dominated by particles of $1 - 10\,\mathrm{GeV}$. Thus, their number density is some nine to ten orders of magnitude lower than the average density of thermal particles in the ISM ($\sim 1\,\mathrm{cm^{-3}}$). The thermal plasma is responsible for the character of the normal modes of oscillation of the system and can treated for our purposes simply as a background fluid. Nevertheless, energetic particles interact with the background plasma through fluctuations in the fields, i.e., through these same normal modes. Here, an important simplification can be made. Because the gyroradius of an energetic particle is larger than the important microscopic length scales associated with the background plasma, it is the low frequency, long wavelength oscillations which are most important. In addition to magnetosonic disturbances of long wavelength, relativistic particles can undergo a resonant interaction with Alfvén waves of wavelength comparable to their gyroradius. In fact, the relativistic particles themselves can give rise to the Alfvén wave turbulence with which they resonate (see Melrose, Lecture 4, this volume). One can picture the interaction of a particle with a long wavelength MHD wave as the scattering of an energetic particle off a large heavy clump of background plasma, which is slowly distorting the local magnetic field lines through its inertia. The clump of plasma is undisturbed by the scattering particle, so that in its rest frame the particle changes direction but not energy. Since the Alfvén speed is generally small compared to flow speeds typical of the background, the distortion of the field lines moves with almost the same speed as the background plasma, and we can treat the collision as approximately elastic in the local rest frame of the fluid. This picture is the one used by Fermi (1949, 1954) in his seminal papers on particle acceleration in the interstellar medium.

The gyroradius of an energetic particle, whilst large compared to other microscopic lengths, is nevertheless usually small in macroscopic terms. To a first approximation we shall neglect the effects of macroscopic gradients on the particle distribution function, which is tantamount to neglecting those drifts of a particle trajectory which arise from inhomogeneities in the fields. In a frame in which the electric field vanishes, the distribution function is then a function of only two variables in momentum space, instead of three – the missing one being the phase of gyration about a field line. Essentially, our assumptions imply that the particles are 'gyrotropic' and that the magnetic moment of a particle is, in the absence of scattering, invariant. For the remaining momentum coordinates one may choose, for example, spherical polars (p, μ), where p is the magnitude of the momentum and μ the cosine of the pitch angle (the angle between the velocity vector and the magnetic field). Then, because in quasi-linear theory the scattering centres – the blobs of plasma or Alfvén waves – each cause only a small deflection of the particle, we obtain a Fokker-Planck type collision term representing *pitch-angle scattering*:

$$\frac{\mathrm{d}f}{\mathrm{d}t} = \frac{\partial}{\partial \mu}\left(D_{\mu\mu}\frac{\partial f}{\partial \mu}\right) \quad . \tag{5}$$

The derivative on the left-hand side of this equation must be taken along a particle trajectory as in (4). However, since we consider only distributions which are gyrotropic in the plasma rest frame, and assume there is no electric field in this frame, the third term on the left-hand side of (4), which contains the Lorentz force: $e(\mathbf{E} + \mathbf{v} \wedge \mathbf{B}/c)\partial f/\partial \mathbf{p}$, vanishes, provided the plasma rest frame is an inertial reference frame. In this case we can write the derivative as

$$\frac{\mathrm{d}f}{\mathrm{d}t} = \frac{\partial f}{\partial t} + \mathbf{v} \cdot \nabla f \ . \tag{6}$$

It is important to remember that this simple form is valid only for distributions which are independent of the gyrophase, which will turn out to be sufficient for the acceleration mechanisms discussed here. Another generalisation is necessary if the plasma rest frame (in which μ is measured) is *accelerating*, but this is postponed until Chapter 3.

Of course, the pitch-angle scattering term in (5) does not tell the whole story. For example, if we were to leave the energetic particles for a long time in a homogeneous background, (5) would ensure they became isotropic, but then nothing further would happen. In such a case, we must include also the slower processes which arise because the energy change in scattering, though small, is non-zero. The quasi-linear theory then provides us with a Fokker-Planck type equation for diffusion in p–space. For the case of an isotropic, homogeneous distribution in the absence of an external electric field this equation reads:

$$\frac{\partial f}{\partial t} = \frac{1}{p^2}\frac{\partial}{\partial p}\left(D_{pp}\frac{\partial f}{\partial p}\right) \quad . \tag{7}$$

Here, D_{pp} is the momentum diffusion coefficient. Like $D_{\mu\mu}$ it is given in the quasi-linear theory in terms of the fluctuations in the electromagnetic fields – i.e., the spectrum of the waves present in the plasma. This is why the equation is only of Fokker Planck *type*: the wave intensity itself depends on the particle distribution, so that (5) and (7) are strictly speaking not linear, but only 'quasi-linear'.

Expressions for these diffusion coefficients have been worked out by many authors (Hall & Sturrock 1967, Melrose 1969, Luhmann 1976, Achatz et al. 1991). There are several situations where acceleration is thought to occur by diffusion in momentum space according to (7) e.g., the 'impulsive' acceleration of ions in solar flares (Miller et al 1990). However, I propose to concentrate on situations in which momentum diffusion is unimportant. Whenever, the velocity of the scattering centres through the background plasma is small compared to the velocity differences within this plasma over the length scale given by the particle's mean free path, we can expect this approximation to hold. A supernova explosion, for example, ejects matter at a speed of several thousands of kilometers per second, much larger that the speed of scattering centres (Alfvén waves) in the interstellar medium, which is typically less than $100\,\mathrm{km\,s^{-1}}$.

Having said this, it will be clear that the acceleration processes I will discuss form only a relatively small part of a large subject. Most emphasis will be placed on what happens in the neighbourhood of a shock front, because the energetic astrophysical phenomena we wish to understand can almost invariably be identified with the propagation of a shock front. For energetic particles, there are then two effects which are important, and they are of fundamentally different character. The first of these is usually called shock-drift acceleration. It is a deterministic process in the sense that one considers the orbits of particles in a prescribed electromagnetic field. Interactions between the particles are assumed to be unimportant, and the only reason particles emerge with a range of energies, rather than being monochromatic, is that they are allowed to enter with different initial conditions (phases). Although this introduces a kind of randomness into the situation, such mechanisms are essentially nonstochastic. On the other hand, the second important process – often called 'diffusive acceleration' – hinges upon the pitch-angle scattering process. In this sense, it is a stochastic mechanism. Whereas the Liouville equation is useful in the analysis of shock-drift acceleration, it is the pitch-angle diffusion equation which is employed in the stochastic case.

1.5 Magnetic Pumping

Before turning to shock fronts, however, I would like to start off with a short example which does not involve them. One of the oldest mechanisms of particle acceleration, and one which appears under many guises is that of magnetic pumping, or the betatron effect. Early discussions are given by Swann (1933), Schlüter (1957) and Parker (1958). This mechanism is interesting because it provides an instructive example of how the interplay of a reversible effect and stochasticity can lead to acceleration.

Consider a particle in a homogeneous plasma containing a uniform magnetic field $\mathbf{B}$ directed along the z-axis, and allow this field to vary slowly in magnitude as a function of time. In concrete terms, we could imagine a particle in a very long wavelength low frequency MHD wave (a magnetosonic wave rather than an Alfvén wave, since the latter implies a change in magnetic field direction not magnitude). From elementary electrodynamics, we know that the gyromotion of the particle about the field conserves the first adiabatic invariant or 'magnetic moment' under such circumstances:

$$\frac{{p_\perp}^2}{B} = \text{constant} \ , \tag{8}$$

where $p_\perp$ is the component of the momentum perpendicular to $\mathbf{B}$. In addition, since the system is translationally invariant the z component of momentum remains constant:

$$p\mu = \text{constant} \ . \tag{9}$$

The equation of motion of the trajectory is therefore

$$\frac{\dot{p_\perp}}{p_\perp} = \frac{\dot{B}}{2B} \qquad \dot{p}_z = 0 \ , \tag{10}$$

or, in spherical polars

$$\frac{\dot{p}}{p} = \frac{(1-\mu^2)}{2}\frac{\dot{B}}{B} \qquad \frac{\dot{\mu}}{\mu} = -\frac{(1-\mu^2)}{2}\frac{\dot{B}}{B} \ , \tag{11}$$

and the equation governing a homogeneous distribution of such particles is just the Liouville equation:

$$\frac{\partial f}{\partial t} + \frac{\dot{B}}{2B} p_\perp \frac{\partial f}{\partial p_\perp} = 0 \ . \tag{12}$$

Clearly, if B increases, the energy of each particle in the system increases too. This is the first example of acceleration without the explicit mention of an electric field. (It is present, of course, and arises from the time variation of B through the Maxwell equation $\nabla \wedge \mathbf{E} = -\dot{\mathbf{B}}/c$). As an acceleration mechanism, however, this process is not so effective, since sooner or later (especially in an MHD wave) B can be expected to return to its initial value. The process described so far is reversible, so that no net energy gain would then result. But what happens if we allow for a small level of fluctuations to produce pitch-angle scattering as described above? The transport equation becomes (in spherical coordinates)

$$\frac{\partial f}{\partial t} + \frac{\dot{B}}{2B}(1-\mu^2)\left(p\frac{\partial f}{\partial p} - \mu\frac{\partial f}{\partial \mu}\right) = \left.\frac{\partial f}{\partial t}\right|_{\text{scatt}} \ , \tag{13}$$

where the right-hand side describes the change of the distribution due to scattering.

In the limit that the pitch-angle scattering is very strong, the distribution is completely isotropic. The term $\partial f/\partial\mu$ then vanishes, and we can integrate over pitch angles to remove the right-hand side (since $\partial[\int \mathrm{d}\mu f]/\partial t|_{\text{scatt}} = 0$), leaving a reversible equation for the isotropic distribution: the system behaves just like an ideal gas under adiabatic compression.

Taking the picture of an ideal gas further, let us assume the particles under consideration are relativistic – synchrotron emitting electrons, perhaps. Then we can write down expressions for the pressure. Actually, since the problem at hand is anisotropic, we must distinguish between the z direction on the one hand, and the x and y directions on the other (These two are equivalent because of gyrotropy). The zz and xx components of the stress tensor (i.e., the components of the anisotropic pressure) are defined as

$$\begin{aligned} P_{zz} &= \int \mathrm{d}^3p\, p\mu^2 f \\ &= 2\pi \int \mathrm{d}p\,\mathrm{d}\mu\, p^3\mu^2 f(p,\mu) \\ P_{xx} &= \int \mathrm{d}^3p\, p(1-\mu^2)\cos^2\phi f \\ &= \frac{1}{2}\left[2\pi \int \mathrm{d}p\,\mathrm{d}\mu\, p^3(1-\mu^2) f(p,\mu)\right] \end{aligned} \tag{14}$$

and the energy density is just $E = 2P_{xx} + P_{zz}$. For an isotropic distribution, we have $P_{xx} = P_{zz} = E/3$. We can now substantiate our claim about adiabatic compression by multiplying (13) by p and integrating over d^3p whilst assuming isotropy ($\partial f/\partial\mu = 0$). We easily derive the relation

$$P_{xx}B^{-4/3} = \text{constant} \ , \tag{15}$$

which reflects the fact that the particles are relativistic (the ratio of specific heats is 4/3) and that their number density is proportional to B (since they are tied to the field lines).

To find the dependence of pressure on B in the absence of pitch-angle scattering, on the other hand, we must revert to Liouville's theorem. Suppose we start off with an isotropic distribution $f(p)$ and field strength B_A. Sometime later, the field strength is B, and we define the parameter b, proportional to the specific volume, as $b = B_A/B$. A particle which started at $b = 1$ with momentum components $p_\perp, p_z$ follows the trajectory given by (10) and has new components $p_\perp' = p_\perp/b^{1/2}$ and $p_z' = p_z$. The distribution $f'(p_\perp', p_z')$ is anisotropic. The pressure (xx-component) is therefore

$$P_{xx} = \frac{1}{2}\left[2\pi \int \mathrm{d}p_\perp'\,\mathrm{d}p_z' \frac{p_\perp'^3 f'(p_\perp', p_z')}{\sqrt{p_\perp'^2 + p_z'^2}}\right] \ . \tag{16}$$

But using Liouville's theorem we can write $f'(p_\perp', p_z') = f(p)$, because the distribution function is constant along trajectories in the absence of scattering. Transforming the remaining terms in the integrand into the variables $p_\perp$ and p_z and then expressing these in spherical polar coordinates, we can rewrite (16) as

$$P_{xx} = \frac{1}{2}\left[2\pi \int \mathrm{d}p\, p^3 f(p)\right]\left[b^{-3/2}\int \mathrm{d}\mu \frac{1-\mu^2}{\sqrt{1-(1-b)\mu^2}}\right]$$

$$= P_A \left[\frac{3}{4b^{3/2}} \int d\mu \frac{1-\mu^2}{\sqrt{1-(1-b)\mu^2}} \right] , \tag{17}$$

where P_A is the (isotropic) pressure at $b = 1$.

Consider now the cycle depicted in Fig. (1). The particles begin at the point A with an isotropic distribution, brought about by a small amount of pitch-angle scattering present in the system. They then suffer a compression which, although sufficiently slow to ensure the adiabatic invariant (8) is conserved, is, nevertheless, too fast to enable pitch-angle scattering to be effective. To a good approximation, the distribution evolves along the line A–B without suffering any scattering, and the pressure P_{xx} grows as b decreases according to (17). Having reached its maximum, the magnetic field is assumed to stay there long enough for pitch-angle scattering to return the plasma to isotropy, moving along the line B–C. The energy density remains constant during this phase, since b does not change, and pitch-angle scattering conserves the particle energy. To complete the cycle, the plasma then slowly decompresses and returns the magnetic field to its initial value at D. During this final phase, pitch-angle scattering is sufficiently rapid to keep the distribution isotropic, so that the final pressure is $P_D = P_C b^{4/3}$. A simple calculation leads to the relation connecting the initial and final pressures:

$$\begin{aligned} \frac{P_D}{P_A} &= \frac{1}{2b^{1/6}} \left(\frac{\arcsin(\sqrt{1-b})}{\sqrt{1-b}} + \sqrt{b} \right) \\ &\geq 1 \; . \end{aligned} \tag{18}$$

This process is therefore a bona-fide acceleration mechanism. It is especially noteworthy that acceleration results independent of the amplitude of the change of field between A and B. Thus, if pitch-angle scattering were not negligible during a compression, we could divide up the process into a sequence of very small compressions, each of which would result in an energy increase which was slightly greater than that achieved by adiabatic compression alone.

If we now perform the same cycle in the reverse direction i.e., scatter-free expansion, followed by isotropisation and slow compression, we find after complete cycle:

$$\begin{aligned} \frac{P_D}{P_A} &= \frac{1}{2b^{1/6}} \left(\frac{\operatorname{arcsinh}(\sqrt{b-1})}{\sqrt{b-1}} + \sqrt{b} \right) \\ &\geq 1 \; . \end{aligned} \tag{19}$$

Thus, repeating the arguments presented above, scattering produces an irreversible energy gain compared to adiabatic change not just on compression, but also on expansion. We therefore expect a long wavelength oscillation to produce a net increase in the energy density of nonthermal particles each period. This is the mechanism of 'magnetic pumping', first presented consistently by Schlüter (1957) for distributions which remain close to isotropy. Physically, one expects this to be the case, at least for very low frequency waves, since any

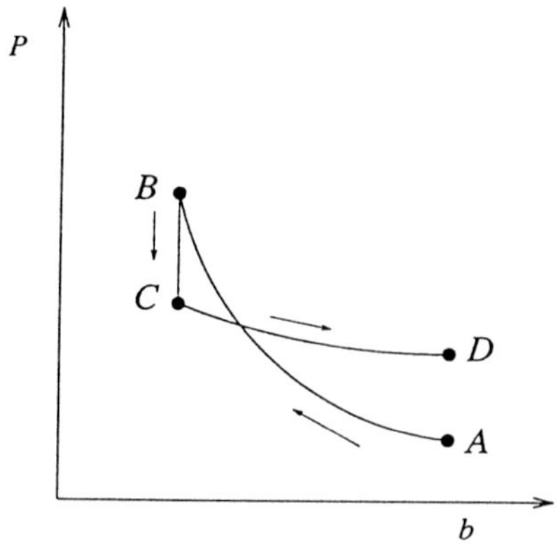

Fig. 1. The pressure vs $b = B_A/B$ diagram for magnetic pumping. (Note the abscissa is proportional to the inverse of the particle density.) Starting at A, the gas of energetic particles undergoes scatter-free compression to B whilst the particles conserve their magnetic moments. At B, scattering takes effect and the xx component of the pressure decreases as the gas isotropises at constant energy and particle density, moving to C. There follows a slow adiabatic expansion phase from C to D during which pitch-angle scattering keeps the distribution isotropic. After one cycle, the isotropic distribution of particles at D has a higher pressure than the isotropic distribution at A.

anisotropy in the particles will generate waves which in turn are responsible for pitch-angle scattering. Using this approach, one can derive a momentum diffusion coefficient and reformulate the problem to obey (7) (see Melrose 1980 and Sect. 5.4 of the lectures in this volume).

Of course, we have treated a very simple spatially homogeneous system, ignoring the spatial gradients which are inherent in waves. We have also chosen not to discuss the spectrum of accelerated particles, or the maximum energy to which they can be accelerated. In later lectures we will look at similar problems in connection with acceleration at shocks, where it turns out that such questions often have surprisingly clear-cut answers.

2 Shock-Drift Acceleration I

2.1 Shock Fronts as Discontinuities

Shock fronts in astrophysical plasmas can usually be assumed to be collisionless. The mean free path for Coulomb interaction is generally larger than the macroscopic scales of interest, even for those particles which make up the bulk of the thermal plasma. Nevertheless, sudden transitions or shock fronts are observed in situ around the Earth's magnetosphere, in interplanetary space, and around other solar system bodies. The structure and properties of these 'collisionless shocks' is an active field of research, which is intimately connected with some questions relevant to theories of particle acceleration. On the other hand, many aspects of particle acceleration are only indirectly concerned with the detailed structure of collisionless shocks. An energetic particle, whose gyroradius is large compared to the shock thickness, cannot interact resonantly with the fluctuations responsible for thermalising particles in the shock front. Even the steady potential differences set up in some shocks (see, for example, Leroy and Mangeney 1984) are of the order of the kinetic energy of a thermal particle, and so unimportant for energetic particles. To a good approximation, such particles notice merely a discontinuous change in the velocity and density of the background flow and in the electromagnetic fields on encountering a shock. On each side of the shock front the particle distribution function is then determined by one of the transport equations discussed in the previous Chapter (4) or (5).

Of course, there is a major shortcoming with this approach: there can be no attempt to describe the mechanism by which thermal particles are *injected* into the acceleration process. If we choose to neglect the complicated interactions inside collisionless shocks, we cannot hope to describe particles of energy less than several times the thermal energy. Nevertheless, it is a long way from 'several times thermal energy' to Lorentz factors of hundreds to millions observed from synchrotron radiating electrons, or to energies of over 10^{13} eV achieved by particles accelerated at a supernova remnant shock. Most of the energy input into these nonthermal populations occurs within the domain accessible to our simple picture. In other words, the injection mechanism provides the fuel, the acceleration mechanism burns it.

2.2 The Kinematics of Shock Fronts

Before discussing the effect they have on energetic particles, we must first look carefully at the properties of shock fronts in hydrodynamics and magnetohydrodynamics, assuming they can be considered as infinitely thin transition layers between regions in which the usual fluid equations are valid. In Fig. 2 a plane shock front is depicted. The first question which arises is the reference frame in which the shock is to be considered. In the case of a supernova explosion, for example, we usually prefer to think of the shock front as moving rapidly outwards from the site of the explosion. The interstellar medium into which it propagates can be taken to be at rest. For obvious reasons, we will call this frame of reference the 'upstream rest frame'. Geometrically, the shock front is a surface, so that its velocity $u_{\rm sh}$ is necessarily directed along the normal $\hat{\mathbf{n}}$. As we know from MHD, the electric field must vanish in the upstream rest frame, simply because the plasma is a very good electrical conductor. The magnetic field, however, is not constrained. In Fig. 2 a uniform magnetic field $\mathbf{B}$ is shown, making an angle $\Phi_{\rm up}$ with the normal: $\mathbf{B}\cdot\hat{\mathbf{n}} = B\cos\Phi_{\rm up}$. In this reference frame, the region downstream of the shock front has a complicated appearance, with both electric and magnetic fields. This is not the only disadvantage of the reference frame from a theoretical point of view. More serious is that the shock front is not at rest. The acceleration processes to be considered take place at the shock front, so that it is convenient to transform to a reference frame in which it is at rest. The only problem is to decide which one; the 'shock rest frame' is not unique, because any Lorentz boost along the shock surface leaves it at rest in the new reference frame.

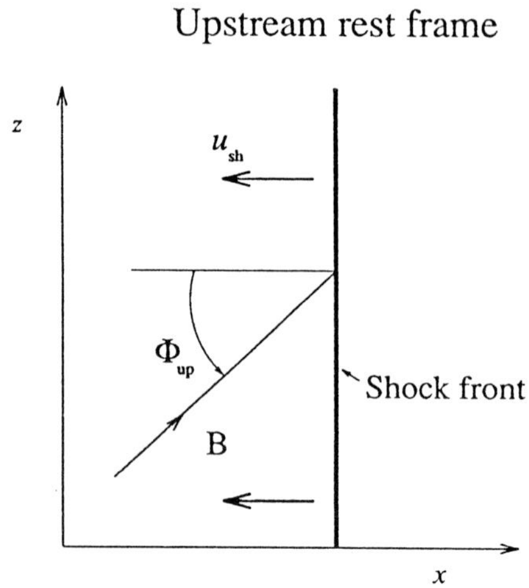

Fig. 2. The shock front as seen in the upstream rest frame

Fortunately, there exists a special frame for each shock in which it is not only at rest, but also has a particularly simple appearance. To see this, we must first review how electromagnetic fields transform under Lorentz boosts. We need keep only three properties in mind:

1. The quantity $\mathbf{E} \cdot \mathbf{B}$ is a Lorentz scalar
2. So too is $|\mathbf{E}|^2 - |\mathbf{B}|^2$
3. In any Lorentz boost the components of $\mathbf{E}$ and $\mathbf{B}$ parallel to the boost direction remain unchanged.

Return now to the upstream rest frame. Any boost along the magnetic field leaves it unchanged, according to 3. Therefore, according to 2, the electric field remains zero. Clearly, then, a boost along the field with speed such that the shock appears stationary is rather special. In geometrical terms, this is achieved by boosting with the velocity (speed *and* direction) of the point of intersection between the shock front and a magnetic field line. Of course, this is not always possible, since the point of intersection moves at superluminal speed if $u_{\rm sh}/\cos\Phi_{\rm up} > 1$. Shocks which do permit the transformation are called 'subluminal' and those which do not are called 'superluminal'. In the case of subluminal shocks, it was first noticed by de Hoffmann and Teller (1950) that this transformation leads to a very special reference frame. To see this, we need to add to the properties of electromagnetic fields under Lorentz boosts the simple properties they have at shock fronts. From the Maxwell equations $\nabla \cdot \mathbf{B} = 0$ and $\nabla \wedge \mathbf{E} = -\dot{\mathbf{B}}/c$ we know that in the steady state:

4. the normal component of **B** immediately upstream of a shock is the same as it is immediately downstream and
5. the tangential components of **E** are also unchanged across a shock front.

In the upstream region, plasma flows along the magnetic field lines and the electric field vanishes. The components of **E** tangential to shock must, according to 5, vanish also on the downstream side of the shock. There remains the possibility of a downstream electric field component along the shock normal. But plasma must cross the shock front, so the component of **B** normal to the shock does not vanish, and, therefore, the existence of a downstream electric field would imply $\mathbf{E} \cdot \mathbf{B} \neq 0$. This is impossible in a plasma of infinite conductivity, since then 1 would imply a non-vanishing electric field even in the rest frame of the downstream plasma. We conclude that in this special reference frame (called the de Hoffmann/Teller frame, see Fig. 3) the shock front is stationary, the electric field vanishes both upstream and downstream, and the plasma flows everywhere along the magnetic field lines. The de Hoffmann/Teller frame is unique (ignoring rotations).

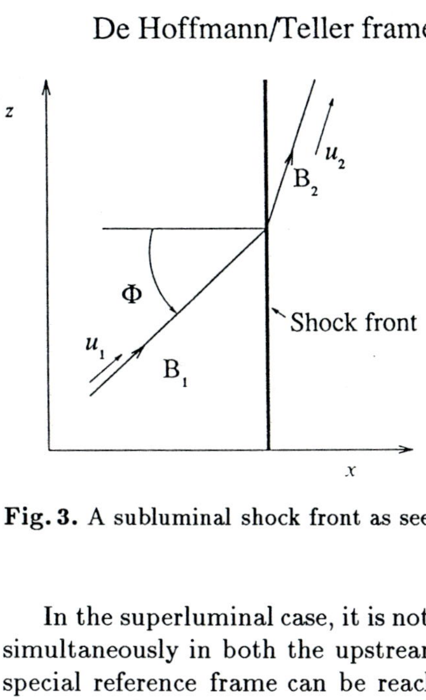

Fig. 3. A subluminal shock front as seen in the de Hoffmann/Teller frame of reference

In the superluminal case, it is not possible to transform away the electric field simultaneously in both the upstream and downstream regions. Nevertheless, a special reference frame can be reached by a Lorentz transformation along the magnetic field, starting in the upstream rest frame. Denoting the speed of the boost by v_{t}, the velocity addition formula gives the speed v_{int} of the intersection point in the new frame of reference as:

$$v_{\mathrm{int}} = \frac{v_{\mathrm{t}} - u_{\mathrm{sh}} / \cos \Phi_{\mathrm{up}}}{1 - v_{\mathrm{t}} u_{\mathrm{sh}} / \cos \Phi_{\mathrm{up}}} \tag{20}$$

(expressing all speeds in units of the speed of light). Thus, as $u_{sh} \rightarrow \cos \Phi_{up}/v_t$ (< 1) the speed of intersection $v_{int} \rightarrow \infty$ and the field lines turn to be in the plane of the shock front. The upstream electric field is, of course, still zero, but, unfortunately, the shock front is not stationary, since we did not transform with the speed of the point of intersection. In this frame, the upstream plasma flows along the field lines, which themselves drift in towards the shock front, remaining parallel to it, suggesting the name 'upstream drift frame' (Begelman and Kirk 1990). The upstream drift frame is unique.

In order to bring the shock to rest, we can, for example, perform another Lorentz boost, this time along the shock normal (i.e., perpendicular to the field lines). According to 3, the magnetic field remains in the plane of the shock, but an electric field is generated, and the plasma no longer flows along the field lines. This situation is depicted in Fig. 4. According to 4, the downstream magnetic field is also perpendicular to the shock normal, so that an appropriate name for this frame is the 'perpendicular shock frame'. The perpendicular shock frame is *not* uniquely defined by requiring the magnetic field to lie in the plane of the shock. However, it is unique if one also requires that the three vectors: shock normal $\mathbf{n}$, magnetic field $\mathbf{B}$ and plasma velocity $\mathbf{v}$ be coplanar, implying that the electric field is also perpendicular to the shock normal (and, of course, to both $\mathbf{v}$ and $\mathbf{B}$). It is then easy to see that the situation downstream is essentially the same – a boost along the shock normal leads to the 'downstream drift frame' in which $\mathbf{E} = 0$, and the plasma flows along the magnetic field.

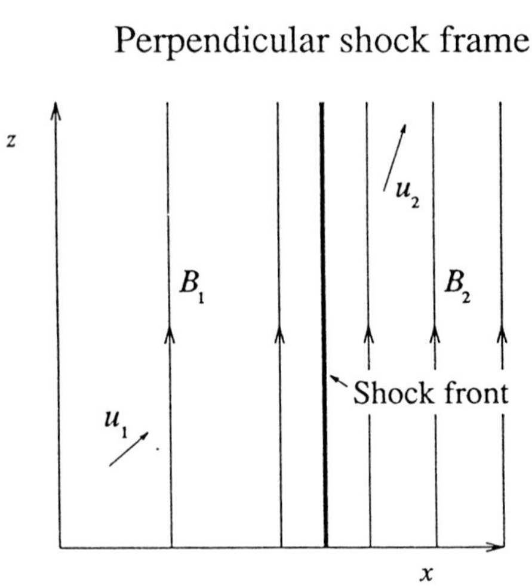

Fig. 4. A superluminal shock front as seen in the perpendicular shock frame

To summarise, subluminal shocks possess a frame (de Hoffmann/Teller) in which $\mathbf{E} = 0$, everywhere, whereas superluminal shocks have a frame (perpendicular shock) in which both $\mathbf{B}$ and $\mathbf{E}$ lie normal to each other in the shock plane. In each case the shock is stationary.

2.3 Particle Trajectories

Superluminal Shocks In circumstances in which plasma flows at highly relativistic speeds, such as in the MHD wind of the Crab Nebula (Kennel & Coroniti 1984), superluminal shock fronts are the rule rather than the exception. A shock front which is stationary in a relativistic flow must, unless it falls almost along a streamline of the fluid, have a speed u_{sh} in the upstream rest frame which is relativistic $\Gamma_{sh} = 1/(1 - u_{sh}^2)^{1/2} \gg 1$. The speed of the point of intersection of a field line and the shock moves with $u_{sh}/\cos\Phi_{up}$ which exceeds that of light unless **B** is almost along the shock normal, or $\Phi_{up} \lesssim 1/\Gamma_{sh}$. Except for those with very accurately aligned magnetic fields, all shock fronts will be superluminal. Of course, superluminal shocks occur also in plasmas of more modest speed, provided the magnetic field is suitably aligned and the shock surface is free of irregularities. However, in reality, only an exceedingly small part of a shock front in, say, the solar wind, could at any given instant be superluminal, since alignment to within about a tenth of a degree is required of both field and shock front. Nevertheless, several early theoretical investigations looked at non-relativistic, exactly perpendicular shocks, which are simply superluminal shocks seen in the perpendicular shock frame (e.g., Schatzman 1963).

At such a shock front, it is particularly easy to visualise the process of shock-drift acceleration. Consider the trajectory shown in Fig. 5, where the plasma is assumed to flow perpendicular to the magnetic field, which is directed perpendicular to the plane of the page. The electric field is therefore in the z direction, and causes the upstream trajectory to have the typical cycloidal shape arising from the superposition of gyration about the magnetic field and an $\mathbf{E} \wedge \mathbf{B}$ drift across the field. The rate at which the 'guiding centre' of the orbit drifts because of the electric field is $u_1 = \mathbf{E} \wedge \mathbf{B}/B^2$, which is, of course, equal to the speed with which the plasma flows across the field lines in this frame. Let us assume that this speed is small compared to the speed v of the particle, i.e., that the particle's orbit moves only a fraction of a gyroradius closer to the shock on each loop. In Fig. 5, this ratio was chosen to be 1:10. On hitting the shock front, there is no change (within our approximations) of the magnitude or direction of the particle's momentum. But, if the shock is compressive, the higher magnetic strength in the downstream region causes a tightening of the loop performed by the particle. As shown in the figure, a positively charged particle is steadily pushed in the direction of the electric field as its orbit slowly crosses over the shock front, whereas a negative particle would be pushed against the field. Thus, in each case, the electric field performs work on the particle, leading to a net gain in energy. The motion of the guiding centre is very similar to the guiding centre motion in an inhomogeneous magnetic field, which gives rise to a 'gradient drift'. Because of this, the acceleration process is termed *shock-drift acceleration*. According to the classification used in Chapter 1, shock-drift acceleration is non-stochastic; it depends solely on the particle motion in a specified electromagnetic field.

It is obvious from the orbit shown in Fig. 5 that even when cutting the shock front, there is only a slight change in the shape of the orbit from one loop to the next, leading one to suspect the presence of an adiabatic invariant.

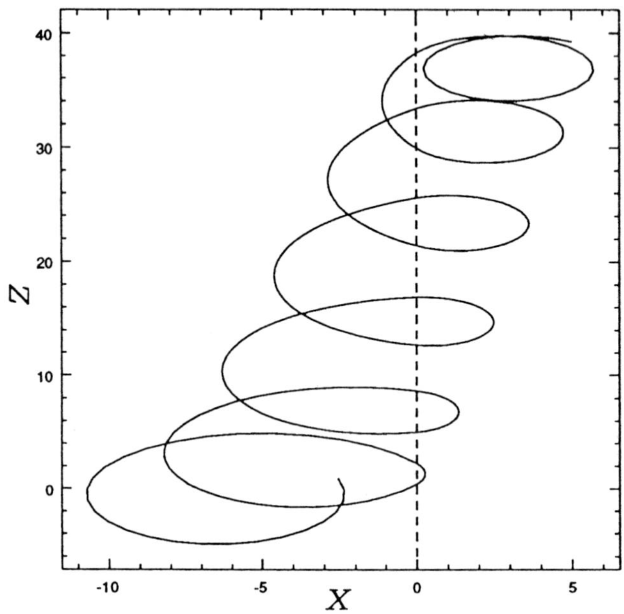

Fig. 5. The trajectory of a particle crossing a nonrelativistic shock front. The magnetic field is in the y direction and the electric field in z direction. Plasma flows across the shock front in the positive x direction. (Note that different scales are used for the x and y directions.) The trajectory shown is that of a relativistic particle, at a shock front of compression ratio 3, with upstream speed $u_1 = c/10$.

However, the theory of adiabatic invariance and particle drifts as developed by Kruskal (1962) breaks down in the face of a discontinuous change in magnetic field strength and the adiabatic invariance of the magnetic moment is then no longer guaranteed. Thus, it was somewhat puzzling that calculations by Parker (1958) and Schatzman (1963) showed explicitly that the value of the quantity ${p_\perp}^2/B$ is the same on the upstream part of a trajectory as on the downstream part, to lowest order in the small parameter u_1/v (see also Toptyghin 1980). The resolution of the puzzle was found by Whipple et al (1986), who formulated an adiabatic invariant which is conserved during interaction with the shock front provided $u_1/v \ll 1$, and which is approximately equal to the magnetic moment on those parts of the trajectory which are well away from the discontinuity. Thus, in spite of the sudden jump in magnetic field, it turns out that, to a good approximation, the magnetic moment is conserved, at least when the particle speed is much faster than the flow speed. This assumption is the key to an analytic treatment of shock-drift acceleration (Webb et al 1983). An immediate consequence for superluminal shocks is that the maximum energy a particle can gain is rather small. In Fig. 5 there are no forces acting on the particle in the y direction, so that p_y is unaffected. Denoting the upstream (downstream) quantities by the suffix 1 (2) we have

$$\frac{p_{\perp 1}^2}{B_1} = \frac{p_{\perp 2}^2}{B_2} \; , \tag{21}$$

leading to a maximum energy gain for relativistic particles of

$$\frac{\gamma_1}{\gamma_2} = b^{-1/2} \; , \tag{22}$$

where, to conform with the notation of Chapter 1, we have defined $b = B_1/B_2$. If the background plasma can be treated as an ideal gas whose ratio of specific heats is 5/3, and if the magnetic field is too weak to exert a significant pressure, the maximum compression gives $b = 1/4$, leading to a mere doubling of the energy of a relativistic particle (increasing to a factor 4 for the kinetic energy of a nonrelativistic particle). These results apply in the limit of low flow speeds, but it is possible to push the adiabatic assumption further, provided some care is taken in defining the magnetic moment. The problem is that in the perpendicular shock frame, the quantity $p_\perp$ oscillates about its mean value, so that the conserved magnetic moment is imprecisely defined by (21). The solution is to define the magnetic moment in the upstream (or downstream) drift frame, where the electric field vanishes. To first order in $u_{1,2}/v$ the definitions are equivalent.

Subluminal Shocks The situation for subluminal shocks is somewhat different. In fact, because of the existence of the de Hoffmann/Teller (dHT) reference frame, in which the energy of each particle is conserved, it is not readily apparent that there is any particle acceleration mechanism at all! Nevertheless, the particle trajectories are similar in that when the orbits intersect the discontinuity, the guiding centre drifts along the shock front. The important difference is that a particle incident from upstream can either emerge into the downstream plasma (as in the superluminal case) or be reflected. It is easy to understand this process as similar to reflection by the compressed magnetic field at the ends of a magnetic bottle. (Reflection of particles incident on the shock from the downstream side does not occur.) Physically, the conditions required for there to be an adiabatic invariant are that the orbit intersects the shock front many times i.e., that the motion of its gyrocentre across the discontinuity is slow compared to its gyration period. In the dHT frame (Fig. 3), simple geometry gives an estimate of this condition:

$$\tan\alpha \tan\Phi \gg 1 \; , \tag{23}$$

where α is the particle's pitch angle. Clearly, no matter what the fluid speed, this condition is violated for particles of low pitch angle, or for shocks which have a small angle Φ between the magnetic field and the shock normal. Nevertheless, extensive numerical work has shown that the assumption of conservation of magnetic moment is not as bad as might be expected. A comprehensive review of this topic is given by Decker (1988). Figure 6 is taken from this paper and shows the fractional number of particles undergoing a given energy change on interaction. It is assumed that particles start with a random distribution of pitch angles and phases. For the kind of quantity shown in Fig. 6, which is averaged over angles, the adiabatic test particle theory (i.e., conservation of magnetic moment) is remarkably good.

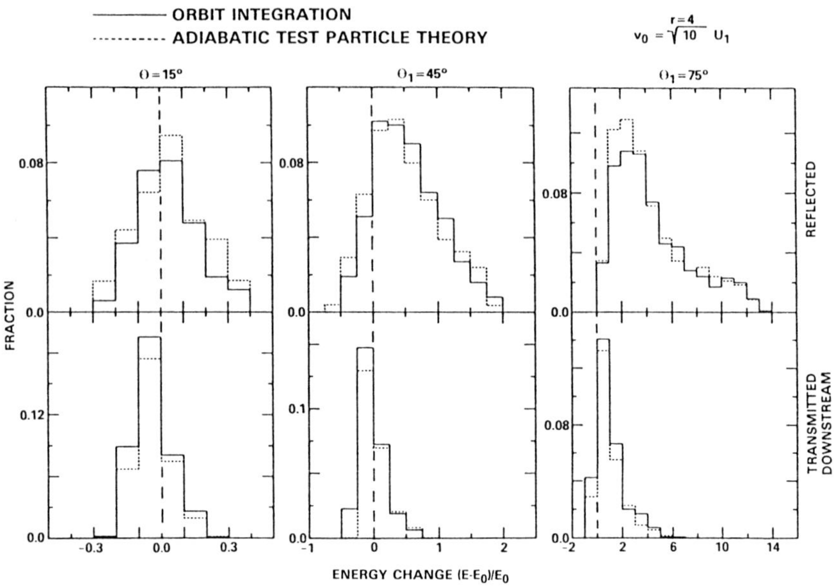

Fig. 6. Calculations by Decker (1980) showing the fraction of particles incident on a shock front which are transmitted (lower panels) or reflected (upper panels) as a function of the change of energy undergone. The compression ratio of the shock is 4, and the incoming particle speed is $\sqrt{10}$ times the upstream fluid speed. Three different angles between the shock normal and the magnetic field are used. The numerically calculated result is shown (solid line) together with the analytic result found assuming conservation of the magnetic moment (dotted line). Note that some particles – especially transmitted ones – lose energy. Reflected ones tend to gain, particularly at high obliquity. In general there is little deviation of the analytic approximation from the numerical result.

Encouraged by these results, consider the case of a subluminal shock front seen in a conventional frame of reference, which we can call the 'laboratory' frame (Fig. 7). This frame is one of those in which the shock front is stationary, and it is distinguished by the fact that the upstream plasma approaches the shock front along the direction of the normal **n**. Such a picture pertains, for example, when from the Earth's rest frame one looks at that region of the bow shock where it meets the the Solar Wind 'head-on'. Let us calculate the energy gain of a particle which is reflected from the shock front. The first point to note is that the particle kinetic energy is not constant in the lab. frame, but oscillates.

If the upstream flow is nonrelativistic (such as the solar wind) the amplitude of the oscillation is small, corresponding to the potential difference across the particle's gyroradius. In terms of the particle Lorentz factor $\gamma' = 1/\sqrt{1-v'^2}$, (we adopt here the convention that primed quantities are measured in the lab. frame) the amplitude of the oscillation is

$$\Delta\gamma' = \gamma' v' u' \sin\Phi' \ , \tag{24}$$

where u' is the speed of the fluid in the lab. frame, which equals the speed of the shock in the upstream rest frame $u_{\rm sh}$. In order to transform into the dHT frame we must move with the intersection point of a magnetic field and the shock front. In so doing, the magnetic field lines are considered to be frozen in to the plasma i.e., they move with a velocity given by the component of the fluid speed normal to $\mathbf{B}$. (This is, in fact an extension of the transformation rule discussed in section 2.2 – it is relatively straightforward to prove that the dHT frame can be reached from *any* other by boosting with the velocity of the intersection point.) Thus the required transformation is one with speed $\beta' = -u' \tan\Phi'$ in the z direction. Once in the dHT frame (Fig. 3), we see that the reverse transformation is one of speed $\beta = -\beta' = u \sin\Phi$ in the z direction. The Lorentz factor of a particle in the lab. frame is related to that in the dHT frame by

$$\gamma' = \Gamma(\gamma - \beta\gamma v_z) \ , \tag{25}$$

where $\Gamma = 1/\sqrt{1-\beta^2}$. After reflection, the component of the particle's velocity along the field is reversed in the dHT frame, whilst the perpendicular components remain constant. In terms of v_z, this is rather difficult to express. However, since the phase change of the perpendicular velocity components is irrelevant for our considerations, we can choose it to be π, which means a reflection reverses every component of the particle's velocity (including v_z). It is then easy to write the Lorentz factor in the lab. frame after reflection:

$$\bar{\gamma}' = \Gamma(\gamma + \beta\gamma v_z) \ . \tag{26}$$

(Quantities after reflection are denoted by a bar: $\bar{\gamma}'$ etc.) The ratio of kinetic energy after reflection to kinetic energy before reflection is therefore (in the lab. frame)

$$\frac{\bar{\gamma}'}{\gamma'} = \frac{1+\beta v_z}{1-\beta v_z} \ . \tag{27}$$

We have already noted that the kinetic energy in the lab. frame is not constant along the trajectory, but oscillates. Equation (27) also makes this clear: v_z oscillates in the dHT frame, so that the ratio $\bar{\gamma}'/\gamma'$ must oscillate too. If we wish to compute an average value of this ratio, we can do it by performing the average over the phase of v_z. A particle which has speed $v_\parallel$ along the field and $v_\perp$ perpendicular to it in the dHT frame has a z component of velocity given by:

$$v_z = v_\parallel \sin\Phi - v_\perp \cos\Phi \cos\phi \ , \tag{28}$$

where ϕ is the phase associated with gyration about the magnetic field. The cosine of the pitch angle μ is then

$$\begin{aligned}\mu &= \frac{v_{\parallel}}{v} \\ &= \frac{v_z}{v \sin \Phi} + \frac{v_{\perp}}{v} \cot \Phi \cos \phi \ .\end{aligned} \tag{29}$$

Averaging over phase, we can write

$$\mu = \frac{\langle v_z \rangle}{v \sin \Phi} \ , \tag{30}$$

leading to

$$\left\langle \frac{\bar{\gamma}'}{\gamma'} \right\rangle = \frac{1 + \beta \mu v \sin \Phi}{1 - \beta \mu v \sin \Phi} \ . \tag{31}$$

For $\beta \sin \Phi \rightarrow 1$, this quantity is potentially very large. However, we have so far not considered the conditions under which reflection (as opposed to transmission) will occur. This is easily accomplished. Because the magnitude of the momentum is conserved in the dHT frame, (21) can be written:

$$\frac{1 - \mu_1^2}{B_1} = \frac{1 - \mu_2^2}{B_2} \ . \tag{32}$$

There exists no real solution for μ_2 (i.e., reflection occurs) for

$$|\mu_1| < \sqrt{1 - b} \ . \tag{33}$$

Furthermore, in order to have the particle approach the shock initially, we require $\mu_1 > 0$. Thus the energy gain is limited to

$$\left\langle \frac{\bar{\gamma}'}{\gamma'} \right\rangle < \frac{1 + v\beta \sin \Phi \sqrt{1 - b}}{1 - v\beta \sin \Phi \sqrt{1 - b}} \ . \tag{34}$$

For relativistic particles ($v = 1$) the average energy gain reaches a maximum for an orientation of the shock front such that $\beta \sin \Phi \rightarrow 1$:

$$\left\langle \frac{\bar{\gamma}'}{\gamma'} \right\rangle_{\max} = \frac{1 + \sqrt{1 - b}}{1 - \sqrt{1 - b}} \ . \tag{35}$$

The maximum field compression is equal to the compression ratio of the shock front, which, for an ideal gas of specific ratio 5/3 is 4. This gives the result $\langle \bar{\gamma}'/\gamma' \rangle_{\max} = 13.93$, which can be compared with the result of 7 obtained using a nonrelativistic approximation scheme (Toptyghin 1980).[1] Nevertheless, shock-drift acceleration by reflection results in only a modest increase in the particle

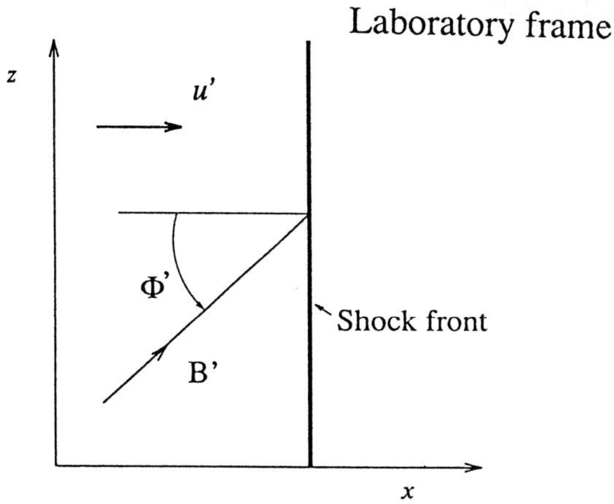

Fig. 7. The shock front as seen in the laboratory frame of reference

energy. In fact, the situation is even worse for transmitted particles – whose maximum energy is only half that of reflected ones.

Shock-drift acceleration – meaning essentially reflection at a subluminal shock front – is more frequently applied to electrons than ions. The reasons are twofold. Firstly, thermal ions upstream of a shock front generally have their pitch angle inside the loss-cone ($|\mu| > \sqrt{1-b}$) for orientations of the field sufficient to give appreciable acceleration, whereas the faster moving electrons can in principle be picked up directly from the thermal population. Secondly, the gyroradius of an electron may in fact be much smaller than the shock thickness, in which case its motion in the increasing magnetic field is adiabatic to a very good approxi-

[1] For the above argument one needs to prove $\beta \sin \Phi \to 1$ is possible for at least some u', and show $b \to 1/\rho_c$ (with ρ_c the compression ratio) for an overlapping range of u'. From Kirk and Heavens (1989)

$$\begin{aligned}\beta &= u' \tan \Phi_{\rm up}/\sqrt{1-u'^2} \qquad (36)\\ &= u' \tan \Phi' \ . \qquad (37)\end{aligned}$$

Therefore,

$$\beta \sin \Phi = \beta/\sqrt{1+(1-\beta^2)/\tan^2 \Phi'} \ , \qquad (38)$$

which clearly goes to unity for $\beta \to 1$. The expression for b from KH is

$$\begin{aligned}b &= \left[\rho_c^2 - \frac{(\rho_c^2-1)}{(1-u'^2)}\left(\cos^2 \Phi_{\rm up} - u'^2\right)\right]^{-1/2} \qquad (39)\\ &= \left[\rho_c^2 - \frac{(\rho_c^2-1)}{(1-u'^2)}\left(\frac{1}{1+(1-u'^2)\beta^2/u'^2} - u'^2\right)\right]^{-1/2} \ , \qquad (40)\end{aligned}$$

and the term in parentheses goes to zero for all u' as $\beta \to 1$.

mation. Thermal ions, on the other hand, have a gyroradius of the same order as the shock thickness, and may resonate with the waves present there. Only if an ion has a much larger gyro radius is it plausible that its kinematics can be approximated using the adiabatic invariance of the magnetic moment.

The most important applications of the mechanism are to produce

1. the energetic electron beams observed upstream of the bow shock, where it is called 'fast Fermi acceleration' (Wu 1984) and
2. the energetic electrons responsible for type II solar radio bursts (Holman & Pesses 1983, Melrose & Dulk 1987, Benz & Thejappa 1988).

Because of the strict limitation of the maximum energy, variations have been invented which rely on trapping the particle in the vicinity of the shock front (in the case of '$\mathbf{v} \wedge \mathbf{B}$' acceleration), or on the enhanced magnetic field inside the shock front due to magnetic overshoot (see Decker 1988 for a review). More recently, the importance of geometrical effects which might provide multiple encounters with the same shock front have been stressed (Decker 1990, Kirk & Wassmann 1992).

3 Shock-Drift Acceleration II

Shock-drift acceleration in its simplest form is capable of increasing the energy of a particle by only a factor of the order of 10. This makes it quite inadequate as a mechanism for accelerating, for example, cosmic rays. It is also unable to explain the origin of the relativistic electrons which we observe in many parts of the cosmos via their synchrotron radiation. Nevertheless, there are circumstances in which we may be reasonably sure that shock-drift acceleration is at work. In astrophysical settings, however, we can detect only very energetic particles, and the only chance of identifying the shock-drift mechanism in action is if we have a means of comparing rather sensitively the energy of these particles before and after the acceleration event. For this purpose, the properties of synchrotron radiation are important. In this lecture, we first look at the effects of adiabatic expansion and compression on synchrotron emission. This discussion provides an introduction to the particle transport equation which will be used later on in the course. It also sets the scene for the discussion of enhanced radio emission to be expected downstream of both nonrelativistic and relativistic shocks, at which the shock-drift mechanism is at work.

3.1 Adiabatic Expansion: the Synchrotron Bubble

Consider a volume containing thermal plasma, a (for simplicity homogeneous) magnetic field and a population of relativistic electrons. Assume the latter are maintained isotropic by pitch-angle scattering, and have a power-law spectrum such that the number of particles of Lorentz factor γ moving with pitch angle θ and at phase ϕ per unit volume is

$$\mathrm{d}N = C\gamma^{2-s}\mathrm{d}\gamma\mathrm{d}(\cos\theta)\mathrm{d}\phi \ . \tag{41}$$

The distribution function is related to C by $f(\mathbf{r},p) = C\gamma^{-s}/(mc)^3$. The synchrotron emissivity $\epsilon(\omega,\theta)$, which is the energy radiated per second per unit volume per unit solid angle in a direction making an angle θ with the magnetic field, and per unit frequency interval at frequency ω, is

$$\epsilon(\omega,\theta) = C\frac{\sqrt{3}e^2\Omega_0\sin\theta}{4\pi c}\left(\frac{2\omega}{3\Omega_0\sin\theta}\right)^{(3-s)/2} \left(\frac{3s+1}{6}\right)\Gamma\left(\frac{3s+1}{12}\right)\Gamma\left(\frac{3s-7}{12}\right) , \tag{42}$$

where $\Gamma(x)$ is the gamma-function, e is the electronic charge and $\Omega_0 = eB/mc$, with B the magnitude of the magnetic field and and m the particle (electron) mass. Equation (42) assumes the frequency of emission is unaffected by departures from a pure power-law electron spectrum at low or high energies, and is summed over the two polarisation states of emission (Ginzburg and Syrovatskii 1965). For our purposes, the important points to note are that the spectrum is a power-law in frequency $\epsilon \propto \nu^{-\alpha}$ with index

$$\alpha = (s-3)/2 \tag{43}$$

and the volume emissivity at a given frequency is proportional to $B^{(s-1)/2}$ and to C. The dependence of the synchrotron emission on the volume of the system arises, therefore, from changes in both the magnetic field B and the distribution function f.

Flux-freezing Consider first B. Assuming the magnetic field to be frozen into the background plasma, we can write

$$\frac{\partial\mathbf{B}}{\partial t} - \nabla\wedge(\mathbf{u}\wedge\mathbf{B}) = 0 , \tag{44}$$

which can be rearranged into the form

$$\left(\frac{\partial}{\partial t} + \mathbf{u}\cdot\nabla\right)\mathbf{B} = (\mathbf{B}\cdot\nabla)\mathbf{u} - \mathbf{B}(\nabla\cdot\mathbf{u}) \tag{45}$$

(noting that $\nabla\cdot\mathbf{B} = 0$). The left-hand side of this equation is convenient for Lagrangian coordinates, since it is the derivative along the trajectory of a fluid element. To be specific, define the coordinate $\mathbf{R}$ which labels a fluid element by the position it had at the instant $t = 0$:

$$\mathbf{r} = \mathbf{R} + \int_0^t \mathrm{d}t'\,\mathbf{u}(\mathbf{R},t') . \tag{46}$$

We then have

$$\frac{\mathrm{d}\mathbf{B}}{\mathrm{d}t} \equiv \left(\frac{\partial\mathbf{B}}{\partial t}\right)_{\mathbf{R}} = (\mathbf{B}\cdot\nabla)\mathbf{u} - \mathbf{B}(\nabla\cdot\mathbf{u}) . \tag{47}$$

The equation of continuity of the plasma also takes on a simple form in these 'comoving' coordinates:

$$\frac{\mathrm{d}\rho}{\mathrm{d}t} \equiv \frac{\partial \rho}{\partial t} + (\mathbf{u} \cdot \nabla)\rho = -\rho \nabla \cdot \mathbf{u} \ . \tag{48}$$

We can now specialise to two simple cases. For one-dimensional compression or expansion with $\mathbf{u}$ a function only of the coordinate perpendicular to the direction of $\mathbf{B}$ (and with this direction, $\mathbf{B}/|\mathbf{B}|$, constant), the first term on the right-hand side of (47) vanishes, and we can compare with (48) to find

$$B \propto \rho \ . \tag{49}$$

The second case, that of homologous expansion or contraction ($\mathbf{u} \propto r\hat{\mathbf{r}}$) is most easily treated by first assuming a purely radial field. This gives a term $-2Bu/r$ on the right-hand side of (47), from which it follows that $B \propto r^{-2}$, or $B \propto \rho^{2/3}$. However, a homologous expansion or contraction is invariant to translation of the coordinate origin, so that the scaling $B \propto \rho^{2/3}$ must also apply to nonradial field components.

One-dimensional compression is similar to that experienced at a perpendicular shock front, except that the present treatment assumes all quantities undergo only adiabatic variations. Homologous expansion is more appropriate for sources such as supernova remnants.

The Particle Transport Equation Now consider the effect on the relativistic electrons. In the presence of pitch-angle scattering, their distribution is described by (5), where the time derivative must be interpreted as a comoving derivative along a particle trajectory.

$$\frac{\mathrm{d}f}{\mathrm{d}t} \equiv \frac{\partial f}{\partial t} + \mathbf{v} \cdot \nabla f \ . \tag{50}$$

However, because we are now considering a background plasma which is in motion, a subtlety arises. The pitch-angle scattering term attempts to impose isotropy on the particles *in the frame of the scattering centres.* If we assume scattering is predominantly by Alfvén waves, then the rest frame of the scattering centres is essentially that of the plasma. It is here that the scattering term has the simple form written in (5). In general, though, it is more convenient to use a lab. frame, in which the plasma (and therefore the scattering centres) are in motion. There is only one way in which we can do this and keep the simple form of the pitch-angle diffusion operator, and that is by using a mixed system of phase-space coordinates to describe our system. This means that we take momentum space variables $\mathbf{p}$, as measured in the local rest frame of the background plasma, together with position variables $\mathbf{x}$ measured in the lab. system, in which the plasma flows with a (not necessarily constant) velocity $\mathbf{u}$. Contenting ourselves with a Galilean rather than a Lorentz transformation, we can write the relation between the momentum $\mathbf{p}'$ in the lab. frame and $\mathbf{p}$ in the plasma rest frame as

$$\mathbf{p} = \mathbf{p}' - m\mathbf{u} \ , \tag{51}$$

where m is the electron mass. Now we can transform the right-hand side of (50) to take account of the fact that p is measured in an accelerating frame:

$$\left(\frac{\partial}{\partial x_i'}\right)_{\mathbf{p}'} = \left(\frac{\partial}{\partial x_i'}\right)_{\mathbf{p}} + \left(\frac{\partial p_j}{\partial x_i'}\right)_{\mathbf{p}'} \left(\frac{\partial}{\partial p_j}\right) , \tag{52}$$

where the indices indicate cartesian components and we have used the summation rule for repeated indices. The quantity $(\partial/\partial x_i')_{\mathbf{p}'}$ indicates the derivative at constant $\mathbf{p}'$. Using a similar transformation for the time derivative, we find

$$\begin{aligned}\frac{\mathrm{d}f}{\mathrm{d}t} &= \frac{\partial f}{\partial t} - m\frac{\partial u_i}{\partial t}\frac{\partial f}{\partial p_i} + \left(\frac{p_i}{m} + u_i\right)\left[\frac{\partial f}{\partial x_i'} - m\frac{\partial u_j}{\partial x_i'}\frac{\partial f}{\partial p_j}\right] \\ &= \frac{\partial f}{\partial t} + u_i\frac{\partial f}{\partial x_i'} - \frac{\partial u_j}{\partial x_i'}p_i\frac{\partial f}{\partial p_j} \\ &\quad -m\frac{\partial u_i}{\partial t}\frac{\partial f}{\partial p_i} - mu_i\frac{\partial u_j}{\partial x_i'}\frac{\partial f}{\partial p_j} + \frac{p_i}{m}\frac{\partial f}{\partial x_i'} .\end{aligned} \tag{53}$$

This expression, combined with the pitch-angle scattering equation (5) gives the full angular dependent transport equation in the nonrelativistic limit.[2] For the moment, we will assume pitch-angle scattering to be very strong, so that the distribution function is isotropic in the local plasma rest frame. This leads to a major simplification, since only the first three terms on the right-hand side of (53) survive and one arrives at the transport equation

$$\frac{\partial f}{\partial t} + \mathbf{u}\cdot\nabla f - \frac{1}{3}(\nabla\cdot\mathbf{u})p\frac{\partial f}{\partial p} = 0 , \tag{54}$$

where we have now dropped the primes on the spatial derivatives, since they are invariably taken in the lab. frame.

The transport equation (54) states that the derivative of the function f is zero along 'trajectories' defined by $\dot{\mathbf{r}} = \mathbf{u}$ (i.e., comoving with the fluid) and $\dot{p} = -p\nabla\cdot\mathbf{u}/3$. From the continuity equation (48) we see that along this trajectory, we have $\dot{\rho} = \nabla\cdot\mathbf{u}$, so that

$$p\rho^{-1/3} = \text{constant} . \tag{55}$$

For the power-law distribution of electrons assumed in (41), this relation implies

$$\begin{aligned} C &\propto fp^s \\ &\propto \rho^{s/3} .\end{aligned} \tag{56}$$

We are now in a position to put the factors (49) and (56) together to find the dependence of the differential luminosity $\mathrm{d}L(\omega)/\mathrm{d}\omega\mathrm{d}\Omega$ emitted by our synchrotron bubble at frequency ω into the direction of the unit vector Ω. Assuming

[2] This simplified derivation of the transport equation applies only to nonrelativistic particles. In fact, (53) is valid for relativistic particles too, provided the fluid flow is nonrelativistic. (It is accurate to order $(u/c)^2$).

homologous expansion or contraction one finds:

$$\frac{\mathrm{d}L(\omega)}{\mathrm{d}\omega\mathrm{d}\Omega} \propto \rho^{(2s-4)/3} \\ \propto r_s^{4-2s} \quad , \tag{57}$$

whereas a one-dimensional contraction/expansion leads to a stronger dependence:

$$\frac{\mathrm{d}L(\omega)}{\mathrm{d}\omega\mathrm{d}\Omega} \propto \rho^{(5s-9)/6} \quad , \tag{58}$$

provided the plasma is compressed normal to the magnetic field direction. Most synchrotron sources have power-law spectra $I_\omega \propto \omega^{-\alpha}$, with index α between 0.5 and 1, which, since $\alpha = (s-3)/2$ corresponds to $4 < s < 5$. Equation (57) shows us that the differential luminosity decreases quite rapidly as the radius of the bubble increases, typically as r_s^{-4} or faster.

This model (for homologously expanding sources) is known as the 'Synchrotron Bubble' and was first investigated by Shklovskii (1960) and van der Laan (1966). Radio astronomers have applied the bubble model to supernova remnants to predict a relationship between the surface-brightness Σ and linear diameter D of a remnant (the 'Σ–D' relation). Dividing (57) by the projected surface area of the source and using (43) one finds

$$\Sigma \propto D^{-4(1+\alpha)} \quad . \tag{59}$$

However, direct comparison of this prediction with observation is hampered by selection effects (e.g., Green 1991).

The discussion given here assumes the source to be optically thin at the wavelengths of interest. Finite optical depth effects have been included more recently by Hjellming & Johnston (1988), and the model has been discussed in connection with transient radio sources by Han & Hjellming (1992) and Ball et al (1994).

3.2 Nonrelativistic, Perpendicular Shocks

As we have seen, synchrotron radiation is very sensitive to compression or expansion of the source. Although this is important for sources where the total compression or decompression may be several orders of magnitude, the mere passage of plasma over a shock front which has a compression ratio ρ_c of 4 (appropriate for a strong shock in a nonrelativistic ideal gas with just translational degrees of freedom) brings with it, according to (58) an increase in the differential luminosity of a plasma blob of only a factor of $\rho_c^{11/6} \approx 13$ (for $\alpha = 0.5$). If, however, we are lucky enough to have a line of sight which lies in the shock front and can resolve the emitting regions up and downstream, then the observed surface brightness increases by an extra factor of ρ_c, because each element of plasma covers an area of the image which is smaller downstream than upstream. The maximum attainable is an amplification of about 51 (again assuming $\alpha = 0.5$), and this is achieved only if the magnetic field lies exactly in the plane of the shock.

Shock-drift acceleration changes this picture somewhat. Consider a superluminal shock front, into which the fluid drifts at a nonrelativistic velocity. As discussed in Chapter 2, the magnetic moment of each nonthermal particle is conserved in the absence of scattering. Far upstream we can assume a small amount of pitch-angle scattering is present, which enables us to take the distribution to be isotropic. (There is a conceptual subtlety here, which we shall return to in Sect. 4.1.) Denoting upstream quantities by the suffix 1 and downstream by 2, we can relate the upstream and downstream momenta of a particle using the conservation of magnetic moment (32) and the constancy of the parallel (to $\mathbf{B}$) momentum. If the distribution function of particles on the upstream side is $f_1(p_1)$, Liouville's theorem tells us that its value on the downstream side $f_2(p_2, \mu_2)$ is unchanged, provided the arguments lie on the same trajectory:

$$\begin{aligned} p_1^2(1-\mu_1^2) &= b\,p_2^2(1-\mu_2^2) \\ p_1 &= \mu_2 p_2/\mu_1 \\ &= p_2\sqrt{b+(1-b)\mu_2^2}\ , \end{aligned} \tag{60}$$

so that

$$\begin{aligned} f_2(p_2,\mu_2) &= f_1(p_1) \\ &= f_1\left(p_2\sqrt{b+(1-b)\mu_2^2}\right)\ . \end{aligned} \tag{61}$$

Of course, the downstream distribution is anisotropic on leaving the vicinity of the shock front. However, we can once again appeal to a small amount (in the sense of being unimportant during interaction with the shock) of pitch-angle scattering to return the distribution to isotropy in the downstream plasma. Because pitch-angle scattering is elastic, the final form of the distribution is obtained by simply averaging over μ_2.

$$\langle f_2\rangle(p_2) = \frac{1}{2}\int_{-1}^{+1} d\mu_2\, f_1\left(p_2\sqrt{b+(1-b)\mu_2^2}\right)\ . \tag{62}$$

Adopting the power-law distribution (41), we have

$$f_1 = C_1(p_1/mc)^{-s}/(mc)^3$$

i.e.,

$$\frac{C_2}{C_1} = \int_0^{+1} d\mu_2\, \left(b+(1-b)\mu_2^2\right)^{-s/2}\ . \tag{63}$$

For $s = 4$ ($\alpha = 0.5$) the integral in (63) is expressible in terms of elementary functions. Putting $b = 1/\rho_c$, we arrive at the result given by van der Laan (1962):

$$\frac{C_2}{C_1} = \frac{\rho_c}{2}\left(1+\frac{\rho_c}{\sqrt{\rho_c-1}}\arcsin\sqrt{\frac{\rho_c-1}{\rho_c}}\right)\ . \tag{64}$$

Combining this expression with that for the increase of the magnetic field we find for the amplification factor η of the surface brightness of a superluminal shock observed edge-on (equivalent to the increase in volume emissivity)

$$\begin{aligned}\eta &= \frac{C_2}{C_1}\rho_c^{(s-1)/2} \\ &\approx 55 \ ,\end{aligned} \tag{65}$$

only slightly greater than that of 51, obtained in the case of pure adiabatic compression.[3]

From an observational point of view, shock-drift acceleration at nonrelativistic superluminal shocks is essentially indistinguishable from simple adiabatic compression. This conclusion holds also for subluminal shocks, provided one can neglect the reflected particles. Either of these mechanisms is adequate to explain the enhanced synchrotron radiation seen from the shock fronts associated with old supernova remnants, such as IC443 (Duin & van der Laan 1975) or the Cygnus loop (Green 1984), provided the upstream population of relativistic electrons is assumed to consist of ambient cosmic rays (van der Laan 1962).

If the magnetic field is not perpendicular to the shock normal, the value of b rises, making the increase in volume emissivity smaller. An orientation effect also enters, since the angle between **B** and the line of sight changes (Laing 1981). A more fundamental effect is introduced by particles reflected by the shock front. These are probably unimportant if they remain in the relatively weak magnetic field upstream of the shock front. However, they can affect things significantly if they are able to return to the shock front – a topic which forms the subject of Chapter 4.

3.3 Relativistic, Perpendicular Shocks

If the upstream plasma drifts into the shock front in the perpendicular shock frame at a relativistic speed, the situation changes dramatically. Because the velocity of the energetic electrons is now not much larger than the drift velocity, individual loops of the trajectory are stretched out in the direction of the drift (normal to the shock in our case). As a result, a particle crosses the shock

---

[3] An expression for the amplification factor η of surface brightness, or, more precisely, of the synchrotron volume emissivity for general s is given by Begelman and Kirk (1990):

$$\eta = \frac{\rho_c}{2\sqrt{\rho_c - 1}} B_{\bar\omega}\left(\frac{1}{2}, \frac{s}{2} + \frac{1}{2}\right) \ , \tag{66}$$

where $\bar\omega = (\rho_c - 1)/\rho_c$ and $B_{\bar\omega}$ is the incomplete Beta function (Abramowitz & Stegun 1972, page 944). A useful approximation to this formula is

$$\eta \approx \rho_c^{s+1} \frac{\Gamma(3/2)\Gamma(s/2 + 1/2)}{\Gamma(s/2 + 1)} \ . \tag{67}$$

front only a few times before drifting off downstream. The individual loops at the shock bear little resemblance to each other, and the adiabatic invariance is clearly lost. The energy gained by a given particle is now a sensitive function of the phase at which it first encounters the shock front. In fact, a few particles actually lose energy.[4] Some examples are shown in Fig. 8. In some ways, the situation is analogous to that of shock-drift acceleration at subluminal shocks. There adiabatic invariance is gradually lost as the magnetic field direction turns towards the shock normal and some particles can pass across the shock front in a few gyrations.

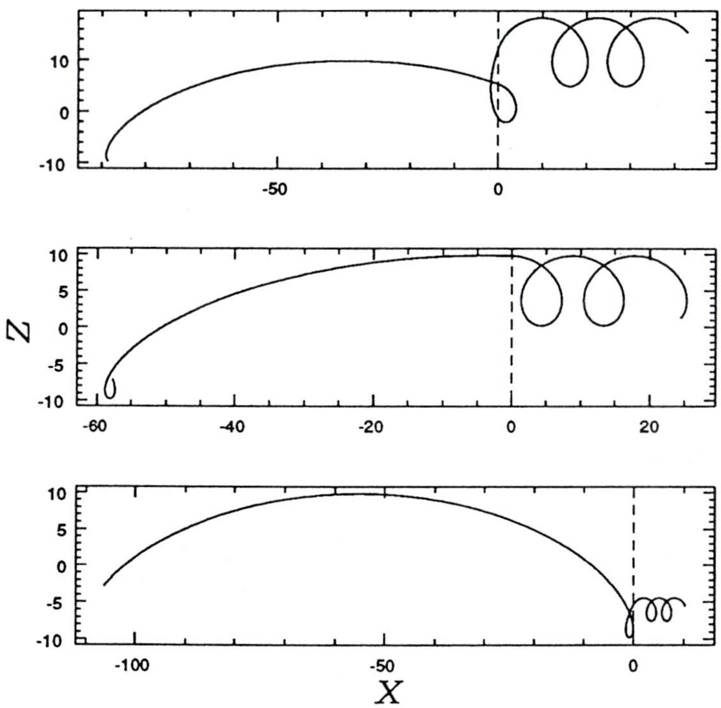

Fig. 8. Particle trajectories crossing a relativistic shock

The only recourse in such a situation is to numerical tracing of the orbits, which, however, is almost trivially simple in the absence of scattering. The basic approach is to look at trajectories in the upstream and downstream drift frames. There the electric field vanishes, and the particle follows a helical path. If we start in the upstream drift frame with values γ, μ, and ϕ for the Lorentz factor,

[4] However, although the existence of decelerated particles at superluminal shocks is a direct consequence of the loss of adiabatic invariance, this is not the case for subluminal shocks. At these shocks, Webb et al (1983) show that the curvature drift associated with the change of magnetic field direction can lead to a reduction in particle energy in the shock frame.

cosine of the pitch angle and phase respectively, the first thing to calculate is the point of intersection with the shock front, which is a plane advancing at uniform speed normal to the field lines. On hitting the shock, a Lorentz transformation of the instantaneous momentum into the downstream drift frame is made. There the situation is similar, except that the shock front is now a plane normal to the field lines which *recedes* from the particle's helical orbit at uniform speed. If another intersection occurs, another section of orbit in the upstream drift frame is needed. If not, one obtains values of $\bar{\gamma}/\gamma$, $\bar{\mu}$ and $\bar{\phi}$ for the transmitted particle. Liouville's theorem then provides a simple way of relating the upstream distribution to the downstream distribution.

To find the increase in synchrotron volume emissivity, one assumes an isotropic incoming distribution, and averages over the angular dependence of the outgoing one. Numerically, this amounts to a two-dimensional integral (over $\bar{\mu}$ and $\bar{\phi}$). The integrand (i.e., the downstream distribution at a particular $\bar{\gamma}$) is found by tracing a trajectory *backwards* from the chosen $\bar{\mu}$, $\bar{\phi}$ to the corresponding initial (upstream) values. For a power-law distribution of incoming particles one can define a compression ratio for the electron distribution:

$$R_e = f(p)/\bar{f}(p) \ , \tag{68}$$

where each distribution is measured in the respective plasma rest frame. (In the nonrelativistic limit, this quantity is equivalent to $(C_2 \Delta V_1)/(C_1 \Delta V_2)$ of (64).) The integrand appropriate for a calculation of R_e is simply $(\gamma/\bar{\gamma})^{-s}$.

It is instructive to compare the results obtained by numerical trajectory tracing with those found using the adiabatic approximation. This is done in Fig. 9 for a superluminal shock front in which the upstream plasma flows perpendicular to the magnetic field in the perpendicular shock frame (see Begelman & Kirk 1990). The compression ratio of the shock, defined as the ratio of downstream to upstream density measured in the shock frame, is chosen to be 4. The value of R_e obtained assuming conservation of the magnetic moment is shown as a solid line, plotted against the spatial component of the plasma 4–velocity $\gamma_1 u_1$ (measured in the 'perpendicular shock frame'). Although the compression ratio is constant, the ratio of the proper densities downstream and upstream increases with increasing fluid speed. For this reason, the value of ρ_c obtained using the adiabatic invariance of the magnetic moment also increases. The dotted line shows the result of a numerical evaluation. At low speed, invariance of the magnetic moment is a good approximation, as expected. At higher speeds the 'exact' result starts to develop a series of oscillations of increasing amplitude. The lower plot in Fig. 9 shows the reason. Here the average number of crossings experienced by an incoming particle before transmission is plotted. This is of course, a large number at low speeds. But at high speeds the number decreases, and it is possible to identify each peak in the oscillations with an odd average number of crossings. The last maximum, at around $\gamma_1 u_1 = .34$ corresponds roughly to 3 crossings. The average number of crossings must obviously remain larger than unity, so no further maximum is attained.

Although interesting, the numerical treatment of semirelativistic shocks shown in Fig. 9 is not dramatic. Figure 10, on the other hand shows that for relativistic

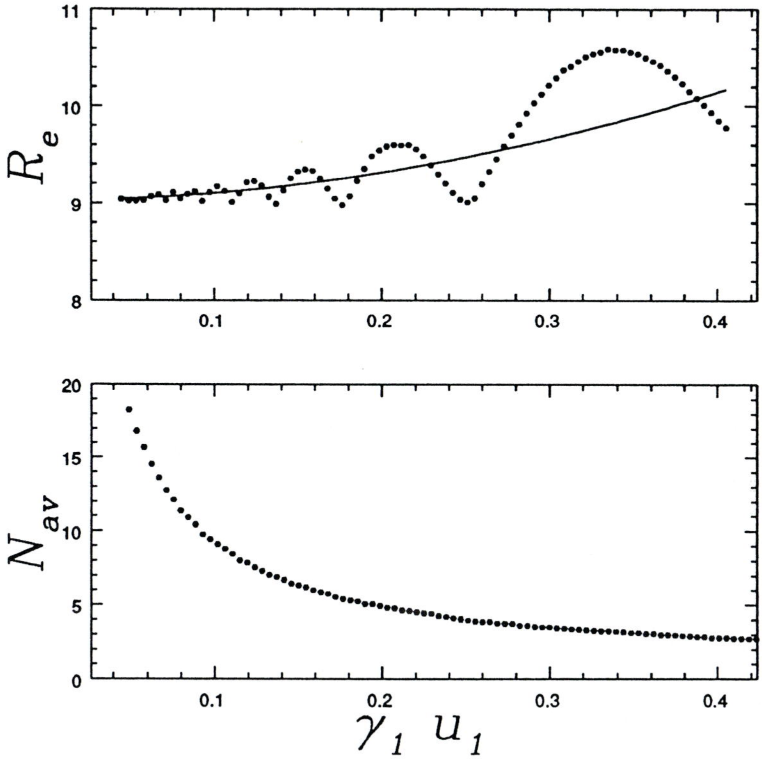

Fig. 9. The amplification factor R_e at a semi-relativistic perpendicular shock (compression ratio $\rho_c = 4$) of a power-law electron distribution with $s = 4.5$, plotted against the spatial component of the upstream four-velocity $\gamma_1 u_1$. The solid line shows the adiabatic approximation, the dots are a numerical evaluation. The lower plot shows the average number of shock crossings N_{av}.

shocks, the quantity R_e can be substantially larger than for the case of simple adiabatic compression. The compression ratio in this figure is 3 in the shock frame, which is more appropriate to a relativistic flow. The ratio of the proper densities is therefore roughly $3\gamma_1$ (because the downstream flow is always sub-relativistic). According to (56) adiabatic compression yields $R_e \approx (3\gamma_P)^{s/3}$, so that for $\gamma_1 = 10$ and $s = 4.5$, $R_e \approx 160$. In fact, it can be seen from Fig. 10 that shock-drift acceleration gives about $R_e \approx 600$. Clearly, relativistic shocks can be effective particle accelerators. For synchrotron sources which contain relativistic shocks, this result is very important, but also rather difficult to interpret directly. The volume emissivity, as can easily be verified by adding in the compression of the magnetic field, rises in the case considered by roughly 2×10^5, a factor which greatly exceeds the dynamic range of most radio maps. The difficulty arises, however, in evaluating the effects of Doppler boosting on the observed flux. Of course, this depends on the obliquity of the shock front, and whether or not the deflection of the flow happens to bring it into or move it out of motion along the line of sight. Nevertheless, it appears that this mechanism is capable of explaining the bright compact hot spots observed in extragalactic radio jets (Begelman & Kirk 1990).

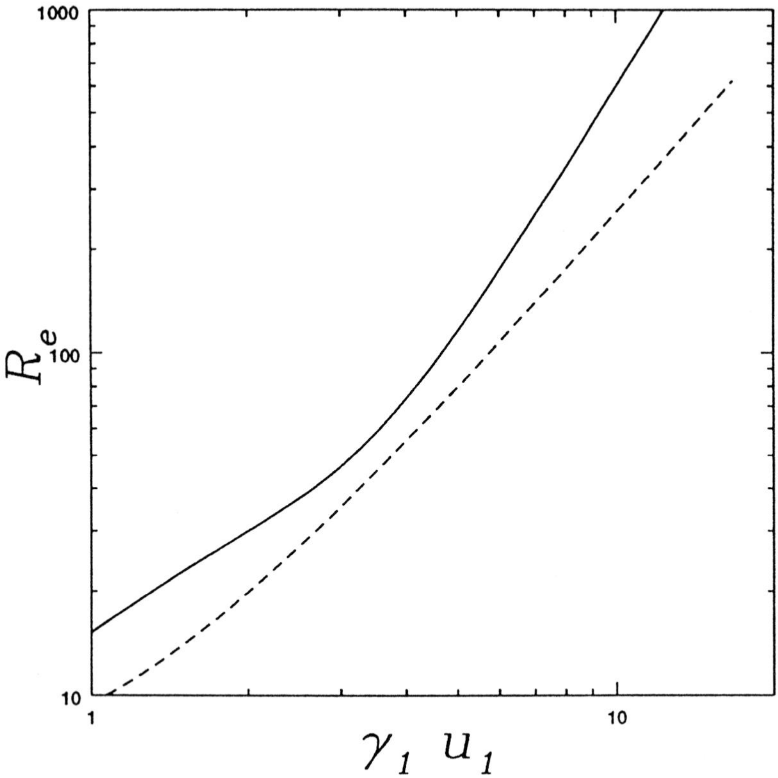

Fig. 10. The amplification factor R_e at a relativistic perpendicular shock (compression ratio $\rho_c = 3$) of a power-law electron distribution with $s = 4.5$, plotted against the spatial component of the upstream four-velocity $\gamma_1 u_1$. The dashed line shows the adiabatic approximation, the solid line a numerical evaluation

4 The First-Order Fermi Process at Shocks I

A potential problem with the shock-drift mechanism is that motion in an electromagnetic field is time-reversible. We might expect, therefore, that shock-drift deceleration is just as frequent as shock-drift acceleration, i.e., that the reduction of magnetic field strength in a slow-mode shock or rarefaction wave would offset the acceleration by the magnetic compression associated with a fast-mode shock. This problem does not arise in our approach, because we have been careful to permit pitch-angle scattering to occur far upstream or downstream of the shock front. The incoming distribution is specified to be isotropic, and it is assumed that the outgoing one will also ultimately be isotropised. This, of course, introduces irreversibility into the system. If we were to allow the downstream distribution to reexpand slowly to the upstream density, we would recover the magnetic pumping mechanism – at least insofar as shock-drift acceleration can be assumed to conserve a particle's magnetic moment.

The treatment of shock-drift acceleration in the presence of scattering far up and downstream in Chapter 3 is, however, rather restricted, since reflected particles are ignored. This means that every particle is assumed to drift through the shock front once and only once. The analogy with the mechanism of magnetic pumping is then apposite. For superluminal shocks, there are indeed no reflected particles, but in most practical applications, (e.g., the acceleration of solar wind particles at the Earth's bow shock) an encounter with a superluminal section of a shock front is very rare. Only when highly relativistic flow speeds are present do superluminal shocks become the rule rather than the exception.

The question naturally arises as to what happens when the shock front is subluminal. This case was discussed in Chapter 2 assuming pitch-angle scattering in the far upstream and downstream regions to be unimportant. We now turn to the implications of scattering for acceleration at such shocks.

4.1 Isotropy and Pitch-Angle Scattering

Let us look in a little more detail at the assumption that the incoming distribution function is isotropic. Figure 11 depicts a shock front which is bordered on its upstream and downstream sides by a zone in which pitch-angle scattering is absent. Further upstream and further downstream are zones in which scattering may be important. Of course, we do not really expect pitch-angle scattering to be any weaker in the vicinity of a shock front than elsewhere – quite the reverse. Nevertheless, the mean free path for scattering should be much longer than the length scale over which the background plasma randomizes its kinetic energy i.e., much longer than the thickness of the shock front. Figure 11 is just a convenient way of representing this ordering of length scales: the length scale of the figure is the thickness of the scatter-free zone, which is much larger than the shock thickness.

Imagine that we are able to manipulate the scattering in the upstream zone so as to prepare any chosen particle distribution for injection across the border into the scatter-free zone. We can, for example, ensure that the particles incident on this boundary from upstream form part (slightly more than a hemisphere because of the plasma motion) of an isotropic distribution. However, we are not at liberty to specify the distribution of those particles which return to the boundary from the scatter-free zone, since these may be influenced by interactions with the shock front (which we do not manipulate). Consequently, we cannot specify the entire distribution function on the boundary a priori. A similar argument applies to the border on the downstream side. Here we are at liberty to decide how particles should return to the scatter-free zone from further downstream, but cannot specify how they arrive there from the direction of the shock front.

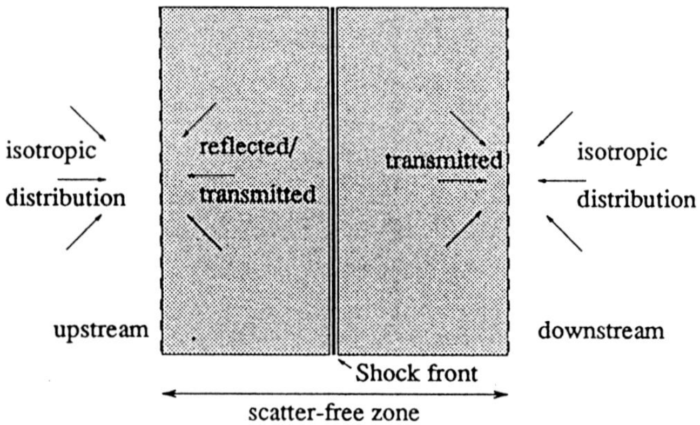

Fig. 11. The scatter-free zone around a shock front

There is one exception to this rule, and it is this which enabled us to assume a fully isotropic incoming distribution in treating superluminal shocks: if the magnetic field is perpendicular to the normal to the shock and if the boundaries of the scatterfree zone are located more than two gyroradii away from the shock front, then all particle trajectories incident on boundary A from the right have experienced only the upstream magnetic and electric fields since entering the scatter-free zone. None of them has had an opportunity to encounter the shock front. However, we know that in the rest frame of the upstream plasma there is a uniform magnetic field, so that in this frame each particle returns to the boundary after performing a partial gyration about a field line and so has the component of its velocity along the shock normal reversed. Thus, a distribution which is isotropic for particles in the hemisphere of velocities pointing to the right is returned as isotropic within the hemisphere of velocities pointing to the left (i.e., of particles arriving from the scatter-free zone).

Clearly, this is a special case. If the magnetic field is not strictly perpendicular then the particles returning to A may have been reflected at the shock front, or even have travelled along the magnetic field all the way from the boundary B. Alternatively, if the scattering is so strong that the borders A and B must lie within a gyroradius of the shock, the trajectory will intersect the shock before return. In each of these cases, the particle distribution at A and B will not be strictly isotropic. We must then face the problem of computing the distribution function using the fully angle-dependent transport equation, which, for pitch-angle scattering, is given by (5).

4.2 The Diffusion Approximation

The first attempt at a solution of this problem involves assuming that the departure from isotropy is small. This leads to a description of the particle distribution which corresponds to spatial diffusion. The transport equation we start from is

(5), in which the derivative along a trajectory is written using the mixed coordinate system (53):

$$\frac{\partial f}{\partial t}+u_i\frac{\partial f}{\partial x_i'}-\frac{\partial u_j}{\partial x_i'}p_i\frac{\partial f}{\partial p_j}$$
$$-m\frac{\partial u_i}{\partial t}\frac{\partial f}{\partial p_i}-mu_i\frac{\partial u_j}{\partial x_i'}\frac{\partial f}{\partial p_j}+\frac{p_i}{m}\frac{\partial f}{\partial x_i'}=\frac{\partial}{\partial\mu}\left(D_{\mu\mu}\frac{\partial f}{\partial\mu}\right)\ . \tag{69}$$

It is important to recall that the distribution function f in this equation is assumed to be independent of gyrophase, i.e., f is 'gyrotropic'. Let us examine a particularly simple case, in which the fluid speed is constant and directed along the magnetic field. Then, in a coordinate system with x along the magnetic field, we have

$$\frac{\partial f}{\partial t}+(u+\mu v)\frac{\partial f}{\partial x}=\frac{\partial}{\partial\mu}D_{\mu\mu}\frac{\partial f}{\partial\mu}\ . \tag{70}$$

Now consider solutions which are stationary in the lab. frame and are almost isotropic in the fluid frame (recalling the mixed coordinate system in use):

$$f(\mathbf{x},\mu,p)=f^{(0)}(\mathbf{x},p)+f^{(1)}(\mathbf{x},\mu,p)\ , \tag{71}$$

where

$$\int_{-1}^{+1}\mathrm{d}\mu f^{(1)}(\mathbf{x},\mu,p)=0\ . \tag{72}$$

Substituting into (70) gives

$$(u+\mu v)\left(\frac{\partial f^{(0)}}{\partial x}+\frac{\partial f^{(1)}}{\partial x}\right)=\frac{\partial}{\partial\mu}D_{\mu\mu}\frac{\partial f^{(1)}}{\partial\mu}\ . \tag{73}$$

This equation can be integrated from $\mu=-1$ to μ to give

$$\Bigg[\underbrace{\frac{\mu^2-1}{2}v}_{(1)}+\underbrace{(\mu+1)u}_{(2)}\Bigg]\frac{\partial f^{(0)}}{\partial x}+\underbrace{v\int_{-1}^{\mu}\mathrm{d}\mu'\,\mu'\frac{\partial f^{(1)}}{\partial x}}_{(3)}+\underbrace{u\int_{-1}^{\mu}\mathrm{d}\mu'\,\frac{\partial f^{(1)}}{\partial x}}_{(4)}$$
$$=D_{\mu\mu}\frac{\partial f^{(1)}}{\partial\mu}\ . \tag{74}$$

Because pitch-angle diffusion should not give rise to a flux of particles over the boundaries $\mu=\pm1$, we require $D_{\mu\mu}$ to vanish there. In (74), the constant of integration has been chosen so that the left-hand side of this equation vanishes at $\mu=-1$. However, for it to vanish at $\mu=1$, terms (2) and (3) must cancel exactly at that point. Provided $u/v\ll1$, this suggests the ordering $f^{(1)}\sim f^{(0)}u/v$, in which case terms (1), (2), (3) and (4) are of the order 1, u/v, u/v, and $(u/v)^2$

respectively. To lowest order, only (1) is retained, in which case (74) may be integrated once more to obtain an explicit expression for the anisotropy:

$$f^{(1)} = C - \frac{v}{2}\frac{\partial f^{(0)}}{\partial x}\int_{-1}^{\mu} \mathrm{d}\mu' \, \frac{1-\mu'^2}{D_{\mu'\mu'}} \ , \tag{75}$$

where C is a constant, determined by the condition (72). The flux $\mathcal{F}(p,x)\mathrm{d}p$ of particles in the momentum interval $\mathrm{d}p$ through unit area perpendicular to the magnetic field is

$$\begin{aligned}\mathcal{F}(p,x) \equiv 2\pi p^2 v \int_{-1}^{+1} \mathrm{d}\mu \, \mu f^{(1)} &= -\pi p^2 v^2 \frac{\partial f^{(0)}}{\partial x}\int_{-1}^{+1} \mathrm{d}\mu \, \mu \int_{-1}^{\mu} \mathrm{d}\mu' \, \frac{1-\mu'^2}{D_{\mu'\mu'}} \\ &= -\pi p^2 \frac{v^2}{2}\frac{\partial f^{(0)}}{\partial x}\int_{-1}^{+1} \mathrm{d}\mu \, \frac{(1-\mu^2)^2}{D_{\mu\mu}} \\ &= -\left[\frac{v^2}{8}\int_{-1}^{+1} \mathrm{d}\mu \, \frac{(1-\mu^2)^2}{D_{\mu\mu}}\right]\frac{\partial \mathcal{N}}{\partial x} \ ,\end{aligned} \tag{76}$$

where the density $\mathcal{N}(p,z)\mathrm{d}p$ of particles in the momentum interval $\mathrm{d}p$ is defined as

$$\mathcal{N} = 4\pi p^2 f^{(0)} \ . \tag{77}$$

A relation of the type (76), in which a flux is linearly proportional to the gradient of a density is fundamental to diffusion processes. It is known as Fick's Law. In order to arrive at it, we had to assume the particle velocity to be large compared to the fluid velocity, a condition which is certainly fulfilled for cosmic rays at supernova blast waves, for example, but violated if the fluid speed becomes relativistic. From Fick's Law we can identify the spatial diffusion coefficient:

$$\kappa = \frac{v^2}{8}\int_{-1}^{+1} \mathrm{d}\mu \, \frac{(1-\mu^2)^2}{D_{\mu\mu}} \ . \tag{78}$$

We can now proceed to the transport equation for the isotropic part of the distribution $f^{(0)}$ by integrating (73) from $\mu = -1$ to $\mu = +1$

$$\frac{\partial}{\partial x}\left(\kappa \frac{\partial f^{(0)}}{\partial x}\right) - u\frac{\partial f^{(0)}}{\partial x} = 0 \ . \tag{79}$$

This equation describes stationary transport in the *diffusion approximation* for constant fluid speed u. It expresses a balance between diffusion (the first term) and advection (second term) which must hold for stationary distributions. [5]

[5] The generalisation of (79) to include space and time dependence of u, and time dependent $f^{(0)}$, as well as the inclusion of cross-field diffusion (i.e., including a scattering in phase in (5)) is straightforward, providing the assumptions $f^{(1)} \ll f^{(0)}$ and $\partial f^{(1)}/\partial t \ll \partial f^{(0)}/\partial t$ can be made. The result is the well-known cosmic ray transport equation (Parker 1965, Dolginov & Toptyghin 1966, Gleeson & Axford 1967,

4.3 Test-Particle Acceleration at a Parallel Shock Front

Consider a shock front in which not only the magnetic field lies along the shock normal (a parallel shock) but also the plasma velocity. The frame in which the shock front is stationary is then automatically the de Hoffmann/Teller frame. We look for a stationary solution to the distribution function of energetic particles, which is described by (79) both upstream and downstream, provided u and κ are replaced in each region by the appropriate quantities: upstream u_1 , κ_1 and downstream u_2, κ_2. The general solution of the transport equation is

$$f^{(0)}_{1,2} = A_{1,2}(p) + C_{1,2}(p)\exp\left(\int_0^x \mathrm{d}x' u_{1,2}/\kappa_{1,2}\right) \ . \tag{81}$$

Since $u_{1,2}$ and $\kappa_{1,2}$ are positive, this solution decays exponentially upstream (towards negative x) and increases exponentially downstream (towards positive x). This is simply because diffusion can only balance advection when it is directed against the fluid flow. Therefore, in the absence of any natural boundaries, a physically acceptable solution can only be found for $C_2 = 0$.

In the special case of a parallel shock, an unscattered particle undergoes no acceleration on encountering the front – it also cannot be reflected. Returning to the picture in Fig. 11 of a shock surrounded by a scatter-free zone, we see that in this special case the size of the zone is irrelevant and the particle enters the downstream boundary B with the same pitch angle and momentum with which it departed the upstream boundary A. However, in applying this condition to match our solution across the shock front, we must remember that a mixed coordinate system is in use in which p and μ are measured in the rest frame of the local plasma. Because the plasma speed changes across the shock front, p and μ change too, despite the fact that the particle itself is unaffected. We can express the relationship between momentum space coordinates up and downstream for a nonrelativistic particle as follows:

$$p_{z1} = p_{z2} \qquad p_{y1} = p_{y2} \qquad p_{x1} = p_{x2} - m(u_1 - u_2) \quad , \tag{82}$$

leading to

$$p_2 \approx p_1\left(1 + \frac{\Delta u}{v_1}\mu_1\right) \ , \tag{83}$$

where $\Delta u = u_1 - u_2$. It is easily checked that this formula applies also to relativistic particles to first order in u/v. Across the scatter-free zone (which can be arbitrarily thin for a parallel shock) Liouville's theorem applies, so that

$$f_1(p_1, \mu_1) = f_2(p_2, \mu_2) \ . \tag{84}$$

Skilling 1975):

$$\frac{\partial f^{(0)}}{\partial t} + u_i\frac{\partial f^{(0)}}{\partial x_i} - \frac{1}{3}\frac{\partial u_i}{\partial x_i}p\frac{\partial f^{(0)}}{\partial p} - \frac{\partial}{\partial x_i}\left(\kappa_{ij}\frac{\partial f^{(0)}}{\partial x_j}\right) = 0 \ , \tag{80}$$

where κ_{ij} is the diffusion tensor.

In the diffusion approximation, we already know the form of the distributions upstream and downstream from (75). In particular, the downstream distribution is isotropic, since it must be independent of x. For $\Delta u/v_1 \ll 1$, we can write

$$f_1(p_1,\mu_1) = f_2^{(0)}(p_2) \approx f_2^{(0)}(p_1) + \mu_1 \frac{\Delta u}{v_1} p_1 \left.\frac{\partial f_2^{(0)}}{\partial p}\right|_{p_1} , \tag{85}$$

so that integrating over μ_1, we find

$$\mathcal{N}_1(p, x=0) = \mathcal{N}_2(p, x=0) \tag{86}$$

and hence

$$A_1 + C_1 = A_2 \ . \tag{87}$$

Thus, the relative velocity of the downstream plasma with respect to upstream does not affect the density, but gives rise to an apparent anisotropy in the distribution, which, for speeds slow compared to the particle velocity, is simply proportional to μ_1. Of course, we cannot expect the anisotropy to match $f^{(1)}$ exactly, since this quantity depends on the form of the pitch-angle diffusion coefficient. Nevertheless, we can find an approximate match by demanding that, in addition to the density, the other quantity of importance in the diffusion approximation – the flux – be given exactly. At the shock front, Fick's Law gives:

$$-\kappa_1 \frac{\partial f_1^{(0)}}{\partial x} = \frac{v_1}{2} \int_{-1}^{+1} \mathrm{d}\mu_1 \mu_1 f_1$$

so that, according to (85) and (81)

$$-u_1 C_1 \approx \frac{\Delta u}{3} p_1 \left.\frac{\partial f_2^{(0)}}{\partial p}\right|_{p_1, x=0} . \tag{88}$$

Downstream, the flux vanishes. Thus, whereas the density is unaffected to lowest order by the relative motion of the two frames, the flux gains a term proportional to the momentum derivative of the density. This is a simple consequence of the mixed coordinate system we are using and could also be derived by a formal integration of (80). Using the matching conditions (87) and (88), the problem reduces to the ordinary differential equation

$$\frac{\Delta u}{3u_1} p \frac{\mathrm{d}f_2^{(0)}}{\mathrm{d}p} + f_2^{(0)} = A_1 , \tag{89}$$

the solution of which is

$$f_2^{(0)}(p) = ap^{-s} + s \int_0^p \frac{\mathrm{d}p'}{p} \left(\frac{p'}{p}\right)^s A_1(p') , \tag{90}$$

where a is a constant and the power-law index is given by

$$s = \frac{3u_1}{\Delta u} \quad . \tag{91}$$

Clearly, if we exclude the possibility of particles entering our system from far upstream, then $A_1 = 0$. To obtain a nontrivial solution, we must set an inhomogeneous boundary condition at some value of the momentum, say $f(p_0, x = 0) = f_0$. If we choose p_0 to be as low as possible consistent with our approximations about the transport process (diffusive and not interacting with the shock front), then we can imagine that some unknown process (to be called 'injection') succeeds in boosting a small number of thermal particles to a momentum p_0. At higher momenta, our acceleration mechanism takes over. In fact, this procedure is equivalent to modifying the matching conditions at the shock front by adding an extra flux of particles $-\Delta u p_0 f_0/3$ to the right-hand side of the matching condition (88).

The index s is characteristic of the shock front, and depends only on the compression ratio $\rho_c = u_1/u_2$. If particles are present far upstream with a power law spectrum $A_1 = A_0 p^{-q}$, then the final result at high momentum is a power law of index given by the smaller of s and q. Thus, if a flatter spectrum is advected into the shock, it remains unchanged in shape, but is amplified:

$$f_2^{(0)} = \frac{s}{s-q} A_0 p^{-q} \quad . \tag{92}$$

There are several interesting points concerning this mechanism:

- The spectrum of accelerated particles is a power law of index s which is determined solely by the compression ratio of the shock and is independent of the diffusion coefficient κ. This arises, of course, from our neglect of any boundaries to the system. In general, we might expect the mean free path of an energetic particle to increase with energy, so that a stationary solution would feel the boundaries at some momentum (Eichler 1984, a more detailed treatment of this topic can be found in the review article by Blandford & Eichler 1987).
- The increase in energy on crossing the shock front is of first order in the small quantity $\Delta u/v$ (see (83)), implying that particles which have undergone an amplification p/p_0 must have crossed the shock front roughly $pv/(\Delta u\, p_0)$ times. In the simple picture there is no treatment of the upper limit to the particle energy.
- Although our derivation has been based on a parallel shock, we could just as well have started from the more general equation (80), which (for constant up and downstream speeds) differs only in that the diffusion coefficient contains a combination of diffusion along and across the magnetic field. The matching conditions applied at the boundary of the scatter-free zones can equally well be applied at the shock front itself, provided the scattering is effective enough to ensure the distribution is nearly isotropic everywhere. With this proviso, the spectrum is independent of the orientation of the

magnetic field. However, the question of the requirements this places on the scattering process and, more generally, on the nature of cross-field diffusion itself, is a topic which is still undergoing development (e.g., Jokipii, 1987, Achterberg & Ball 1994)

- The assumption of isotropy cannot hold for particles just above thermal energy, or for any particles at a relativistic shock. We shall look at the generalisation of the mechanism to this case in the next chapter.

5 The First-Order Fermi Process at Shocks II

In addition to the treatment based on solving the particle transport equation, diffusive shock acceleration can also be considered in an equivalent microscopic picture, given originally by Bell (1978). In fact, the physical concepts lying at the root of the process emerge more clearly in this treatment. This is the subject of the first part of the present lecture. In the remainder, two embellishments of the theory are discussed – the application to relativistic flows and to oblique shocks. In both cases it is necessary to abandon the assumption of isotropy of the distribution function. Nevertheless, the microscopic picture of particles being repeatedly scattered across the shock front still holds, it is just the technicalities involved with a computation of the escape probability and the average amplification which become complex.

5.1 Microscopic Treatment

First of all, consider the energy gained by a particle which crosses and then recrosses a parallel shock front. Going from upstream to downstream we have $1 > \mu_1 > -u_1/v_1$ and, from (83)

$$p_2 = p_1(1 + \mu_1 \Delta u/v_1) \ , \tag{93}$$

to first order in $\Delta u/v_1$. Scattering in the downstream medium results only in changes in the pitch angle, so that if the same particle returns to the upstream medium it has momentum $\bar{p}_1$ given by:

$$\bar{p}_1 = p_2(1 - \mu_2 \Delta u/v_2) \ , \tag{94}$$

with $-1 < \mu_2 < -u_2/v_2$. The average momentum gain on performing the cycle is found by assuming the particle distribution to be isotropic, in which case the probability of a particle crossing the shock front is proportional to the relative velocity between it and the front, i.e., $|\mu v + u|$:

$$\left\langle \frac{\Delta p}{p_1} \right\rangle = \frac{\int_{-u_1/v_1}^{1} \mathrm{d}\mu_1 |\mu_1 v_1 + u_1| \int_{-1}^{-u_2/v_2} \mathrm{d}\mu_2 |\mu_2 v_2 + u_2| (\bar{p}_1 - p_1)/p_1}{\int_{-u_1/v_1}^{1} \mathrm{d}\mu_1 |\mu_1 v_1 + u_1| \int_{-1}^{-u_2/v_2} \mathrm{d}\mu_2 |\mu_2 v_2 + u_2|} \ . \tag{95}$$

To first order in $\Delta u/v_1$, we find

$$\left\langle \frac{\Delta p}{p} \right\rangle = \frac{\Delta u \int_0^1 \mathrm{d}\mu_1 \int_{-1}^0 \mathrm{d}\mu_2 |\mu_1\mu_2|(\mu_1 - \mu_2)}{v \int_0^1 \mathrm{d}\mu_1 \int_{-1}^0 \mathrm{d}\mu_2 |\mu_1\mu_2|}$$

$$= \frac{4\Delta u}{3v} , \tag{96}$$

where to this order we can neglect the difference between v_1 and v_2. The gain in momentum is thus a stochastic quantity, whose average value is of first order in $\Delta u/v$. This is the reason for the name 'first-order Fermi process'. In contrast, randomly moving scattering centres lead to an average energy gain per interaction which is *second* order in the ratio of the speed of the scatters to that of the particle. Nevertheless, to gain energy appreciably, a particle must perform many cycles of crossing and recrossing. On each cycle it wanders through the downstream medium. However, the further away from the front it goes, the less likely it is to return. In fact, we know from the macroscopic solution that the downstream distribution is constant, so that even at very large distances from the shock there are still particles being advected further away by the background fluid. This is not the case upstream, where the particle density falls off exponentially with distance from the shock.

To calculate the fraction of particles crossing the shock which subsequently escape without returning, we can compare the number crossing from upstream to downstream per second, with the number crossing an imaginary border far downstream per second. The flux of particles across the imaginary boundary is

$$\dot{n}_{\mathrm{esc}} = 2\pi p_2^2 \int_{-1}^{+1} \mathrm{d}\mu_2 (\mu_2 v_2 + u_2) f_2(p_2)$$

$$= \mathcal{N}_2 u_2 , \tag{97}$$

whereas the number crossing the shock front from upstream to downstream per second is

$$\dot{n}_{\mathrm{cross}} = 2\pi p_2^2 \int_{-u_2/v_2}^{+1} \mathrm{d}\mu_2 |\mu_2 v_2 + u_2| f_2(p_2)$$

$$\approx \frac{v}{4} \mathcal{N}_2 . \tag{98}$$

Consequently, the escape probability per cycle is

$$P_{\mathrm{esc}} = \frac{\dot{n}_{\mathrm{esc}}}{\dot{n}_{\mathrm{cross}}}$$

$$= 4u_2/v . \tag{99}$$

We can now easily determine the steady state spectrum, since the total number of particles crossing into the downstream region per second with momentum larger than $p+\langle\Delta p\rangle$ is equal to the total number crossing with momentum larger than p, minus those which have escaped during a single cycle:

$$u_2 \int_{p+\langle\Delta p\rangle}^{\infty} \mathrm{d}p'\, \dot{n}_{\mathrm{cross}}(p') = (1 - P_{\mathrm{esc}})\, u_2 \int_p^{\infty} \mathrm{d}p'\, \dot{n}_{\mathrm{cross}}(p') \ , \tag{100}$$

so that, using (98),

$$\frac{\langle\Delta p\rangle}{p} p\mathcal{N}_2(p) = P_{\mathrm{esc}} \int_p^{\infty} \mathrm{d}p' \mathcal{N}_2(p') \ , \tag{101}$$

which has the solution

$$\begin{aligned} \mathcal{N}_2(p) &\propto p^{-1-P_{\mathrm{esc}}/(\langle\Delta p\rangle/p)} \\ &= p^{-1-(3u_2/\Delta u)} \ , \end{aligned} \tag{102}$$

in agreement with (91).

5.2 Relativistic Shocks

We can see from the microscopic approach, that the faster a shock becomes, the harder it will be for a particle wandering about in the downstream medium to catch up with the front and return to the upstream medium. This will be reflected in the formalism by the violation of the assumption of isotropy of the distribution. Even if pitch-angle scattering succeeds in making the distribution function isotropic in the local fluid frame, it will appear highly distorted to a shock front approaching with a speed comparable to the particle speed. In fact, one might naively expect that the first order Fermi process should become less 'efficient' when $u \approx v$. Less efficient in the case of a stationary solution means that particles should escape more frequently, i.e., that the spectrum should be steeper. The situation is complicated, however. It is not only the escape probability which enters into the calculation of the spectrum in (101) but also the average amplification per cycle $\langle\Delta p/p\rangle$. The escape probability rises for faster shocks, but the amplification rises too, because of the larger relative velocity between the upstream and downstream fluids.

There are two physical situations in which the particle velocity may not be large compared to the shock speed, so that the anisotropy of the distribution is important:

1. Relativistic shocks, which are thought to be present in astrophysical objects such as extragalactic radio jets or pulsar driven supernova remnants, are possible sites for the acceleration of the highly relativistic electrons ($\gamma = 10^2 \ldots 10^6$) responsible for the synchrotron radiation from these objects. Extremely relativistic shocks are, however, almost certainly superluminal, so that the first order Fermi process does not work at all unless there is very effective transport across the magnetic field.

2. Mildly suprathermal particles are observed directly at the Earth's bow shock. They are presumably present at all shocks which accelerate particles and have speeds slightly larger than that of the shock front. Their acceleration is essentially part of the injection mechanism.

In each case, there is some doubt as to whether the simple picture of pitch-angle diffusion around a discontinuous background fluid flow can apply. One necessary condition for this is that the energetic particles evade interaction with the collective processes which mediate the shock front. Ultrarelativistic particles may achieve this by virtue of their large gyroradius, which controls the range of turbulent fluctuations with which they can interact resonantly. The same may be true for heavy ions of speed somewhat greater than the shock speed. However, it is difficult to determine whether there is a range of speeds for protons such that the transport can be described by pitch-angle scattering, even though the distribution is anisotropic. If more complicated transport is suspected, the only recourse is to computer simulation. This is a sizable field of research, an overview of which can be found in recent conference proceedings, such as IAU Colloquium #142 (1994). In this lecture we will restrict ourselves to relativistic shocks and investigate the consequences of the assumption that pitch-angle scattering is indeed the dominant process for energetic particle transport.

To deal with relativistic shocks, we must abandon the diffusion approximation and go back to the full transport equation complete with angular dependence. The nonrelativistic version of this equation is (53); it is a straightforward but lengthy calculation to repeat the derivation using Lorentz transformations where necessary (Kirk et al 1988). In the case of a plane parallel shock, when the distribution is a function of only the distance from the shock front, one finds that instead of (70), the transport equation reads:

$$\Gamma(1+uv\mu)\frac{\partial f}{\partial t}+\Gamma(u+v\mu)\frac{\partial f}{\partial x}=\frac{\partial}{\partial\mu}D_{\mu\mu}\frac{\partial f}{\partial\mu} , \tag{103}$$

where $\Gamma=\left(1-u^2\right)^{-1/2}$. To look for a stationary solution we can now no longer make the *Ansatz* of small anisotropy (71) but must consider a more general approach. Separating the variables μ and x leads directly to a form of the general solution in terms of an eigenfunction expansion:

$$f=\sum_{i=-\infty}^{\infty} g_i(p)Q_i^u(\mu)\exp(\Lambda_i x/\Gamma v) , \tag{104}$$

where the $g_i(p)$ are arbitrary functions of momentum. The eigenvalues Λ_i and eigenfunctions $Q_i^u(\mu)$ are solutions of the equation

$$\frac{\partial}{\partial\mu}D_{\mu\mu}\frac{\partial}{\partial\mu}Q_i^u(\mu)=\Lambda_i(u+\mu)Q_i^u(\mu) . \tag{105}$$

The faster the shock front, the more eigenfunctions are needed in order to represent the solution accurately, and for each eigenfunction we must use the boundary and matching conditions to determine the associated function $g_i(p)$. This is, of course, significantly more difficult than the analogous step in the diffusion approximation, where we had only two unknown functions in (81). The general method is, however, the same (Kirk & Schneider 1987a, Heavens & Drury 1988): first of all the eigenfunctions which lead to a divergent downstream solution (i.e., those with $\Lambda_i > 0$) are rejected, then a relativistic version of the matching conditions across the shock front (82) is used to write an approximate form of the upstream distribution (analogous to (85)). Finally, this approximate solution is required to show physical behaviour upstream – i.e., to vanish when projected onto those upstream eigenfunctions which diverge at $x \to -\infty$. This step is analogous to the nonrelativistic requirement that the particle flux be given exactly (88) and leads not to a single first-order differential equation, but to a system of these, one for each eigenfunction. If we are interested only in a power law distribution, such as will establish itself well above the injection energy – provided, that is, that no other process such as losses or escape introduces a momentum scale – the solution of the problem reduces to solving a set of linear homogeneous algebraic equations whose coefficients are functions of the unknown power law index s. The solution of these equations determines both the spectral index and the angular dependence of the distribution function.

The details of the method are well documented in the papers quoted, and would be out of place here. However, the results are of some interest. In particular, one would like to know whether or not relativistic shocks are more effective or less effective particle accelerators than nonrelativistic shocks. Even given the validity of the assumptions about particle transport discussed above, this question must be made more precise before a meaningful answer can be given. To illustrate this, Fig. 12 shows the power law index produced at four different kinds of parallel shock front as a function of speed of the incoming fluid. Two of these fronts are intrinsically relativistic. A relativistic gas, in which the pressure is one third of the energy density (i.e., the plasma consists of essentially massless particles both up and downstream) has particularly simple jump conditions at a shock front, namely $u_1 u_2 = 1/3$. The compression ratio thus increases as the upstream speed u_1 increases, tending to 3 as $u_1 \to 1$. The lowest speed at which this shock can exist is for an upstream speed just above the speed of sound $1/\sqrt{3}$. In this case the shock front is weak, in the sense that the upstream pressure is not negligible (the other shocks in Fig. 12 are assumed strong). A relativistic gas equation of state is appropriate in extragalactic jets if the plasma of which they are composed consists exclusively of electron/positron pairs. The resulting power law index is indicated by the dashed-dotted line, and is significantly steeper (i.e., the index s is larger) than those stemming from other types of shock. The other intrinsically relativistic shock is depicted by the dashed line. Here it is assumed that the plasma becomes so hot on the downstream side that electron pairs are produced spontaneously. In the figure, the effect of creating 100 pairs per incident proton is shown. This makes the equation of state of the plasma very soft. As a result, the compression ratio is high, and the spectrum of accelerated particles extremely hard. The remaining two curves in Fig. 12 show a shock front in a gas with (almost) cosmic abundances i.e., Hydrogen plus 25% Helium by mass, both fully ionised. The only difference is in the way the kinetic energy is shared amongst the electrons and ions after thermalisation at the shock front. (Unfortunately, the physics of collisionless, parallel shock fronts does not allow us to answer this question a priori.) The solid line shows the effect of putting all this energy into the electrons, which consequently become relativistic for quite low u_1. This implies a softening of the equation of state and an increase in the compression ratio, which is mirrored by a harder spectral index. If, on the other hand, the energy goes entirely into the ions, the dotted line results. This is the case which departs least from the nonrelativistic result. The compression ratio remains close to 4 until u_1 is well over 0.5. Although the equation of state softens, relativistic kinematics force the compression ratio to tend to the limiting value of 3 as $u_1 \rightarrow 1$ (where all shocks have a relativistic gas downstream), so that the spectrum steepens.

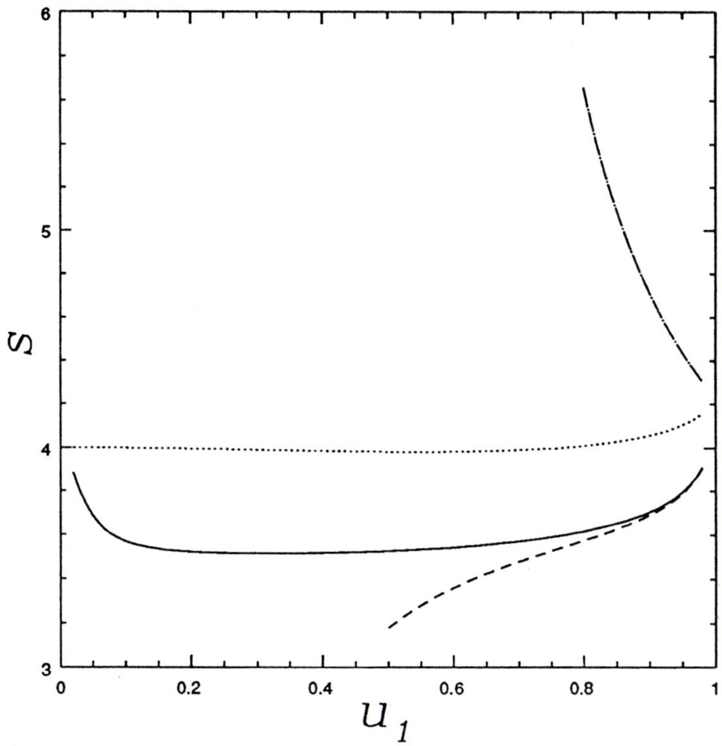

Fig. 12. The power law index s of particles accelerated at relativistic shocks. The jump conditions are calculated according to how the energy is distributed amongst the constituents of the downstream plasma: (i) hot electrons, cold ions (solid line), (ii) hot ions, cold electrons (dotted line), (iii) hot pairs (100 per proton), cold ions (dashed line). In addition the spectrum from a shock in a relativistic gas is shown (dashed-dotted line). Isotropic pitch-angle diffusion is used.

According to the theory of diffusive acceleration, a nonrelativistic shock produces a power law of index (see (91))

$$s = \frac{3u_1}{\Delta u} \tag{106}$$

or, in terms of the compression ratio ρ_c:

$$s = \frac{3\rho_c}{\rho_c - 1} \quad . \tag{107}$$

If, to compare the effectiveness of relativistic and nonrelativistic shocks, we want to extend one of these formulae into the relativistic regime, we have the freedom to choose either (106) and interpret Δu as a relative velocity, or to choose (107), keeping the definition of ρ_c as the ratio of the fluid density in the downstream to that in the upstream region, measured in the rest frame of the shock front. The former choice results in a substantial over-estimate of s. The latter choice is a reasonably good approximation for the particular form of pitch-angle diffusion used in calculating Fig. 12, but still overestimates s as $u_1 \rightarrow 1$. According to Fig. 12, relativistic shocks produce harder spectra than would be expected from the nonrelativistic formula. This conclusion, however, is by no means general, but depends on the type of pitch-angle scattering used.

The angular distribution of accelerated particles, as seen from the rest frame of the downstream plasma is shown in Fig. 13. This figure was computed using an isotropic pitch-angle diffusion coefficient,

$$D_{\mu\mu} \propto 1 - \mu^2 \quad , \tag{108}$$

as too was Fig. 12. (This kind of diffusion coefficient is called 'isotropic' because the resulting scattering operator is proportional to the μ–dependent part of the Laplacian ∇^2. However, the operator is not strictly invariant to rotation, because it does not contain any reference to scattering in phase.) An interesting property apparent in Fig. 13 is the lack of particles travelling in the direction $\mu = 1$ i.e., along the shock normal in the downstream direction. The reason for this is that a particle which crosses into the upstream plasma, undergoes relatively little deflection before it is advected back over the shock front by the relativistically flowing fluid. This property makes it clear that the solution found depends on the form of the pitch-angle scattering operator. Various possibilities have been investigated in the literature, such as pitch-angle diffusion in Kolmogorov turbulence (Heavens & Drury 1988) or even diffusion with the addition of a large-angle scattering term, in order to mimic the action of strong turbulence on particle orbits (Kirk & Schneider 1988). The conclusion to be drawn is that not only the angular distribution, but also the spectral index of the accelerated particles depends on the details of the scattering. Thus, one of the robust features of diffusive acceleration at shock fronts – that the predicted spectrum is independent of the details of the particle transport – is lost in the relativistic case. The probable range of spectral indices is not large, but it is big enough to vitiate a simple comparison with the nonrelativistic formula.

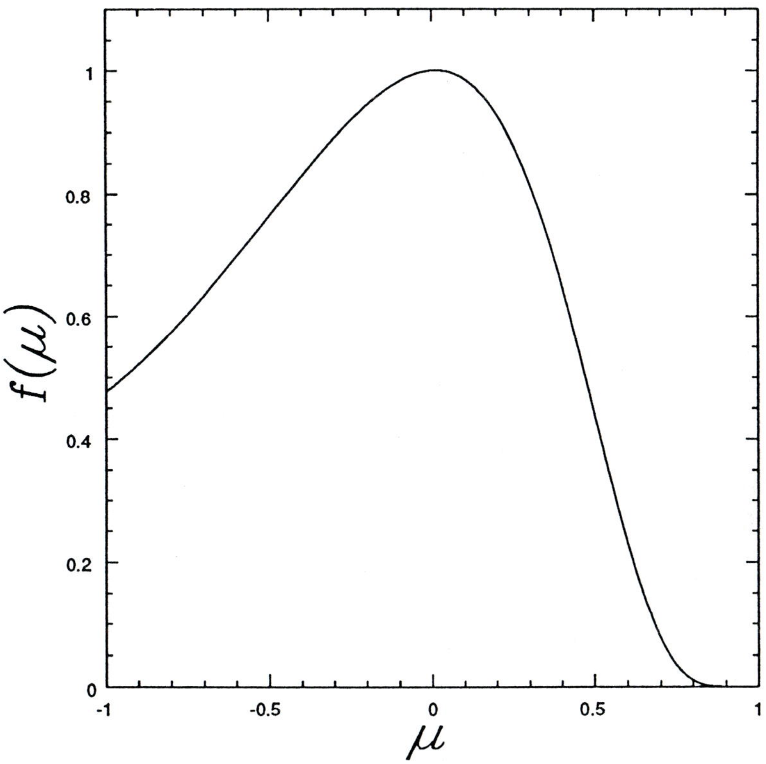

Fig. 13. The angular distribution of accelerated particles at a parallel relativistic shock front with $u_1 = 0.9$, and $u_2 = 0.37$, as seen in the rest frame of the downstream plasma. Isotropic pitch-angle diffusion is employed.

5.3 Oblique Shocks

The standard treatment of diffusive shock acceleration is based on the picture of a parallel shock front. However, although it is certainly easier to justify the treatment for a parallel shock, the theory applies also to an oblique shock, provided the transport in its vicinity can be described as diffusive. This means not only that the particle speed must be large compared to the fluid speed, but also that the anisotropy $f^{(1)}$ must be small compared to the isotropic part $f^{(0)}$. The first condition is easily satisfied at a nonrelativistic shock front, and, for a parallel shock, the anisotropy is dictated solely by the transformation of frames between the upstream and downstream media, and so is necessarily of the order of u/v. At an oblique shock, the shock-drift mechanism tends to make the distribution anisotropic. In the absence of scattering, the parameter b controls the degree of anisotropy, making the downstream distribution oblate in shape: the ratio of the distribution at $\mu = 0$ to that at $\mu = 1$ for a power law dependence on momentum can be found from (61) to be $b^{-s/2}$, a quantity much larger than unity under a wide range of conditions. Consequently, if diffusive acceleration is to apply at an oblique shock front, pitch-angle scattering must be sufficiently rapid to prevent shock-drift acceleration – the particle trajectory must be appreciably disturbed before the orbit has a chance to drift across the shock front. As noted in Chapter 4, the precise requirement on the scattering is a subject of current research interest. The question we turn to here, however, is of what happens when the scattering is insufficient to disturb the process of shock-drift acceleration.

As we have seen, the method of Section 5.2 is appropriate for dealing with anisotropic distribution functions. Suitably generalised, it can also be applied to oblique shocks (Kirk & Heavens 1989). The key assumption which makes the problem tractable is the neglect of cross-field diffusion. An additional but related assumption is that the distribution function is gyrotropic, and the magnetic moment conserved on crossing the shock. The treatment is based on computing the distribution in the de Hoffmann/Teller frame, and makes use of the concept of a scatter-free zone around the shock front, as depicted in Fig. 11. The basic steps are essentially the same as described in Section 5.2:

1. The downstream distribution is represented as a sum of those eigenfunctions which do not lead to divergent behaviour far downstream.
2. An approximation to the upstream distribution is then found using the matching conditions at the shock, which are given not by a Lorentz transformation, but by the conservation of the magnetic moment, combined with Liouville's theorem. However, only those parts of the upstream distribution lying in the 'loss cone' are found by this method.
3. Those parts of the upstream distribution lying outside the loss cone (i.e., pitch angles for which particles are reflected) cannot be found from the downstream distribution. Conservation of magnetic moment means this part of the distribution is an even function of the cosine μ of the pitch angle, when measured in the de Hoffmann/Teller frame, so that it can be represented as a sum of even Chebychev polynomials in μ.
4. Finally, the approximate upstream distribution is constrained to show physical behaviour far upstream by requiring the projection on a number of divergent eigenfunctions to vanish. As in the case of a parallel shock, this leads to a value of the power law index (assuming the accelerated particles to be essentially massless, so that there is no momentum scale in the problem) and also gives the angular dependence of the distribution function.

Figure 14 gives the results of applying this method to three shock fronts each of compression ratio 4, but with speeds 0.1, 0.3, and 0.5 times that of light, as seen in the upstream rest frame (Fig. 2). As the angle $\Phi_{\rm up}$ is varied from zero to $\arccos(u_{\rm sh})$, the speed of the upstream plasma in the de Hoffmann/Teller frame varies from $u_{\rm sh}$ to unity. This quantity is plotted as the abscissa in Fig. 14. The most remarkable property of this figure is that an oblique shock produces a *flatter* spectrum than the corresponding parallel shock. In fact, as the magnetic field orientation takes the shock close to being superluminal, the spectral index tends to 3. Equation (102) shows that this corresponds either to zero escape probability, or to infinite amplification, in the language of the microscopic approach of Bell. A careful consideration shows the latter to be the case. As the speed in the de Hoffmann/Teller frame tends to that of light, the amplification increases without limit.[6]

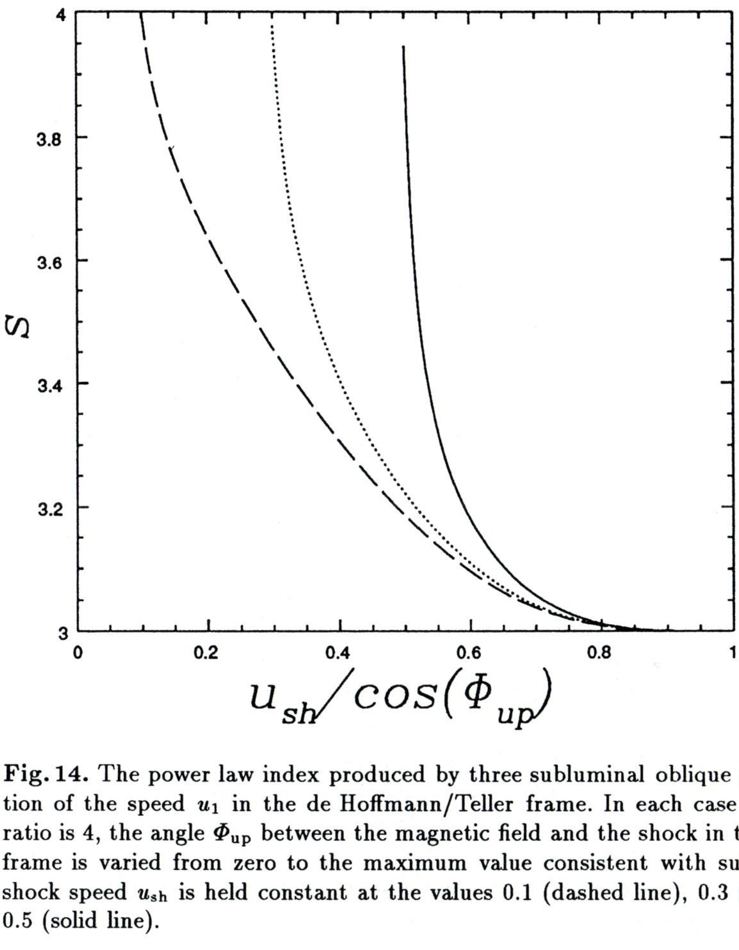

Fig. 14. The power law index produced by three subluminal oblique shocks as a function of the speed u_1 in the de Hoffmann/Teller frame. In each case the compression ratio is 4, the angle Φ_{up} between the magnetic field and the shock in the upstream rest frame is varied from zero to the maximum value consistent with subluminality. The shock speed u_{sh} is held constant at the values 0.1 (dashed line), 0.3 (dotted line) and 0.5 (solid line).

The angular distribution of particles is shown in Fig. 15. Both the particles immediately upstream of the shock and those immediately downstream are shown. Although the speed of the shock in this example is rather modest ($u_{\text{sh}} = 0.1$), the distribution is highly anisotropic. An interesting property of the upstream distribution is that it shows discontinuities (within the limits of the

[6] Note that arbitrarily large amplification does not conflict with the maximum amplification for the shock-drift mechanism given in (35). In the mechanism discussed here, amplification occurs in the de Hoffmann/Teller frame, in which the shock-drift mechanism (reflection) does not produce an energy change. It is instead the process of scattering in the fast moving upstream fluid which produces the large energy gain.

numerical method) at those pitch angles exactly on the loss cone ($\mu_s = \pm 0.85$ in this example). This is due to the assumption of adiabatic invariance of the magnetic moment, and would be smoothed out in a realistic treatment. However, a more basic property is that there are very few particles which return from the downstream fluid into the upstream region, because of the rapid speed of recession of the downstream fluid in the de Hoffmann/Teller frame. The acceleration mechanism in this example is thus based almost exclusively on the alternation of reflections at the shock front and scatterings in the upstream medium.

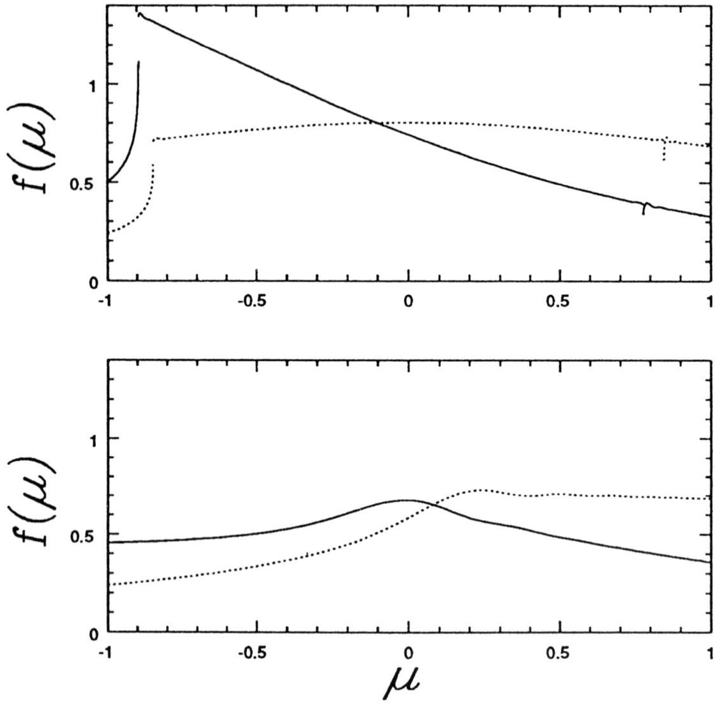

Fig. 15. The angular distribution of accelerated particles at a subluminal oblique shock of compression ratio 4, speed $u_{sh} = 0.1$ and angle of magnetic field to shock normal $\Phi_{up} = 60°$. The upper plot shows the distribution on the upstream side of the shock as a function of μ measured in the rest frame (solid line) and the de Hoffmann/Teller frame (dotted line). The lower plot shows the distribution immediately downstream of the shock.

6 Cosmic Ray Acceleration in Supernova Remnants I

One of the most successful applications of the theory of diffusive acceleration at shock fronts concerns the origin of cosmic rays of energy below about 10^5 GeV/nucleon. Although supernova remnants have long been suspected as the source of these particles, it is the diffusive acceleration mechanism which provides the most convincing explanation of the way in which energy is extracted from the explosion and channelled into cosmic rays.

6.1 The Spectrum of Cosmic Rays

The cosmic rays incident on the Earth's upper atmosphere have a spectrum which is remarkably close to a power-law over an energy range of over five orders of magnitude, from about 10 to about 10^6 GeV per nucleon. Just as we are protected from the full flux of cosmic rays by the atmosphere, so too is the atmosphere protected from cosmic rays in the interstellar medium by the solar wind. The outward flow of this highly ionised plasma and the magnetic field frozen into it tends to evacuate cosmic rays from a region around the Sun (the heliosphere); a charged particle from the interstellar medium can enter only by diffusing against the flow. Consequently, a model of particle transport in the heliosphere is needed to find the energy density of cosmic rays in the interstellar medium from measurements at Earth. The spectrum of cosmic rays calculated in this way is thought to reach a maximum at a few GeV. Particles of about this energy form the dominant contribution to the energy density, which is estimated to be about $1\,\mathrm{eV\,cm^{-3}}$ – comparable with the energy density of the interstellar magnetic field. Cosmic rays arrive at Earth from all directions in space – they are isotropically distributed to an accuracy of about 1 part in 10^4. The index of the spectrum between 100 and 10^6 GeV/nucleon is such that in the local interstellar medium, the differential density, as defined in (77) is

$$\mathcal{N}(p) \propto p^{2-s} \tag{109}$$

with

$$s = 4.71 \pm 0.05 \tag{110}$$

(see, for example, Gaisser 1990). Above about 10^7 GeV (the 'knee') the spectrum steepens to about $s = 5$, only to flatten again at 10^{10} GeV (the 'ankle') before disappearing in a forest of rapidly growing error bars, which extend up to about 10^{12} GeV. The part of this spectrum of relevance here is that between 100 and 10^6 GeV. These particles probably originate in supernova explosions in our galaxy.

The form of the spectrum given in (110) applies only to the major constituents of cosmic rays, such as protons, alpha particles and the nuclei of heavier elements such as carbon and oxygen. Other nuclei, such as Boron, are present in cosmic rays, but are not manufactured by stars and have very low cosmic abundances. They are thought to be 'secondaries' produced by the 'primaries' on their passage through the interstellar medium. From the composition of cosmic rays, one infers that the average column density (the 'grammage') traversed before reaching the Earth by a primary in the GeV range is about $5 - 10\,\mathrm{g\,cm^{-2}}$. Comparing this figure with the interaction length for a proton ($55\,\mathrm{g\,cm^{-2}}$) shows that, on average, cosmic rays escape from the galaxy, before being absorbed in the interstellar medium (except for low energy heavy nuclei). Comparing it, on the other hand, with the column density through the galactic disk ($10^{-3}\,\mathrm{g\,cm^{-2}}$) shows that trajectories cannot be ballistic, but that the particle motion must be diffusive in character.

Secondary cosmic rays have a significantly steeper spectrum ($s \approx 5$). Since the energy of a secondary reflects the energy of the responsible primary, this means that the higher the energy of a cosmic ray, the less matter it has passed through before reaching us. In other words, higher energy cosmic rays escape from the galaxy more quickly than lower energy ones; the escape time being roughly $t_{\mathrm{esc}} \propto p^{-0.6}$ between about 10 GeV and 1 TeV ($= 10^{12}$ eV). The simplest 'leaky-box' model for cosmic ray confinement in the galaxy can be written as a differential equation for the density:

$$\frac{\partial \mathcal{N}}{\partial t} + \frac{\mathcal{N}}{t_{\mathrm{esc}}} = Q(p) \ , \tag{111}$$

where $Q(p)$ is the rate at which the accelerators inject cosmic rays into the interstellar medium. In order to have a density $\mathcal{N} \propto p^{-4.7}$, the momentum dependence of t_{esc} requires of the accelerators that they inject cosmic rays with a somewhat flatter spectrum i.e., a putative cosmic ray accelerator must produce

$$s \approx 4.1 \ . \tag{112}$$

From the energy density and the (energy dependent) grammage, one can find the rate at which the galaxy loses energy via cosmic rays:

$$L_{\mathrm{CR}} \approx 5 \times 10^{40} \mathrm{erg\,s^{-1}} \ . \tag{113}$$

The corresponding average residence time in the galactic disk is about 6×10^6 years. To maintain the status quo, then, we need an acceleration mechanism which replenishes this energy and is more or less steady on the time scale of a few million years.

6.2 Supernova Remnants

As soon as they were recognised as exploding stars, it was suggested that supernovae might be the sources of cosmic rays (Baade & Zwicky 1934). This opinion has hardened into a consensus amongst astronomers and cosmic ray physicists, at least partly because there seems to be no competing source which can satisfy the energy requirements. Estimating the average supernova rate in our galaxy from observations of similar galaxies, (and assuming a low value of the Hubble constant) one arrives at 2 per century (van den Berg & Tammann 1991), rather lower than the rate implied by historical observations of galactic supernovae. Assuming the energy of an average supernova is 10^{51} erg, a rate of 2 per century corresponds to a power input into the interstellar medium of

$$L_{\mathrm{SN}} = 6 \times 10^{41}\,\mathrm{erg\,s^{-1}} \ . \tag{114}$$

Comparing this with (113) shows that if supernovae are the source of cosmic rays, the efficiency of the acceleration mechanism must be of the order of 10%. This is very high, but perhaps not impossibly so.

In order to understand how the theory of diffusive acceleration can be applied to a supernova, we must first have a rough idea of how such an explosion develops (see Fig. 16). The starting point is conveniently chosen avoiding the rather uncertain details of the explosion mechanism itself. Assume, then, that a supernova sets a few solar masses of matter into radial motion away from the site of the progenitor at several thousand kilometers per second into the previously undisturbed interstellar medium. The leading edge of this matter pushes out the surrounding material and forces it into supersonic motion. A shock wave is set up which moves outwards heating and accelerating the plasma. Meanwhile, the ejecta cool rapidly by adiabatic expansion so that, to maintain the pressure driving the outward shock, a 'reverse shock' develops, which eats its way into the outer edge of the ejecta, heating and decelerating it. At first, the reverse shock makes slow headway into the ejecta, which, since these are expected to have a very steep density profile at their outer edge, act as a piston driving the forward shock. One can understand this stage of evolution by considering momentum balance across the region between the two shocks (Chevalier 1982). Until the outer shock has passed over a mass which is comparable with that of the ejecta, it continues to expand at essentially constant velocity. This phase of development is called the 'free-expansion' or 'sweep-up' phase. During it, the energy of explosion remains untapped in the kinetic energy of the ejecta, so that even if particle acceleration were to occur, there could not be a significant contribution to the cosmic ray population.

Once the mass overtaken by the outer shock (the 'swept-up' mass) becomes comparable to the mass of the ejecta, the initial outward momentum, being shared between these two components, corresponds to a substantially lower velocity. The reverse shock propagates rapidly through the greatly expanded, and therefore somewhat under-dense interior, so that the entire region inside the outer shock front is transformed into a hot expanding bubble. The pressure in this bubble is still much greater than that of the surroundings, so that it expands rapidly. This stage of the evolution is the start of the 'adiabatic' or 'Sedov' phase, so-called because, on the one hand, the supernova energy is conserved and remains inside the shock front, and, on the other, because the dynamics can be described by a self-similar flow pattern found by Taylor (1950) and Sedov (1959). The shock radius r_s as a function of time is given in this solution by

$$r_s = \eta_0 \left(\frac{E_{SN} t^2}{\rho_1} \right)^{1/5} , \tag{115}$$

where η_0 is a dimensionless constant (equal to 1.17 for a gas of adiabatic index 5/3), E_{SN} is the energy of the explosion, and ρ_1 is the density of the surroundings. The energy in the self-similar solution is always divided between thermal energy and kinetic energy in the same ratio, which, for a gas of adiabatic index 5/3 is 4:1. The beginning of this phase of evolution of the supernova remnant ($t = t_1$)

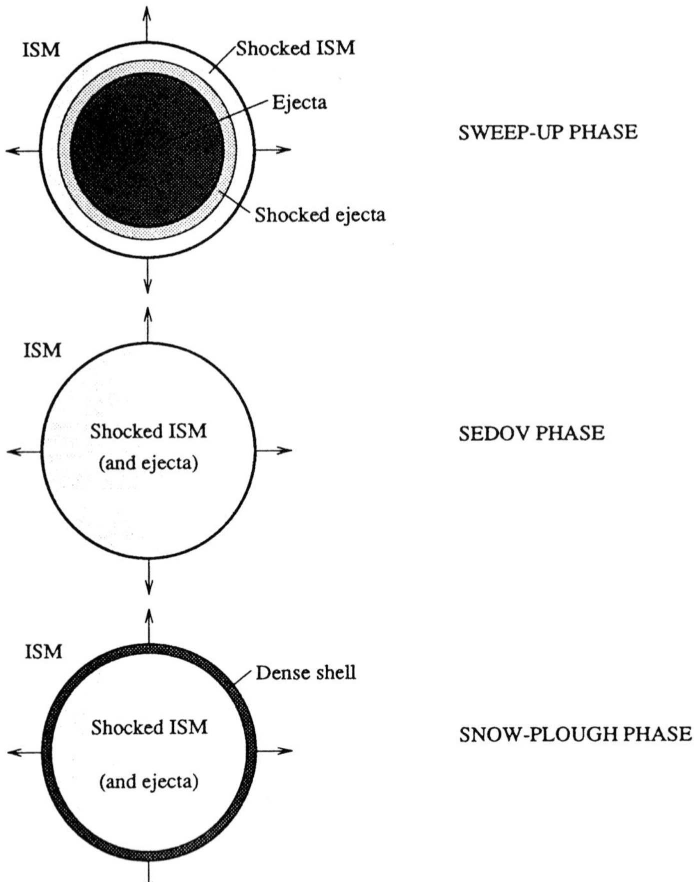

Fig. 16. The three phases of evolution of a supernova remnant. In the sweep-up phase, the shock expands at approximately constant velocity v_s, the radius is typically $r_s \lesssim 2\,\mathrm{pc}$. In the Sedov phase, $v_s \propto t^{-3/5}$ and $10\,\mathrm{pc} \lesssim r_s \lesssim 20\,\mathrm{pc}$, whereas in the snow-plough phase $v_s \propto t^{-5/7}$ and $20\,\mathrm{pc} \lesssim r_s \lesssim 60\,\mathrm{pc}$.

occurs very roughly when the shock has reached a radius such that the swept up mass equals the ejected mass:

$$r_s(t_1) = 2 \cdot (M_{ej}/M_\odot)^{1/3}(n/1\,\mathrm{cm}^{-3})^{-1/3}\,\mathrm{parsec} \ , \tag{116}$$

where M_{ej} is the mass of the ejecta and n is the number density of the surrounding medium. A typical value for the supernova age at this point is several hundred years. The end of the adiabatic phase occurs when the dynamical timescale, which increases as $r_s/v_s \propto t$ becomes comparable to the time taken for the plasma behind the shock front to cool, or, alternatively, when the internal pressure has decreased to roughly the pressure of the surroundings (this is relevant only for explosions in hot thin surroundings). In the former case, cooling then causes a dense shell of matter to form immediately behind the shock front (e.g., Cox & Reynolds 1987), and the remnant enters the so-called 'snow-plough' phase, in which the shock decelerates more rapidly ($r_s \propto t^{2/7}$). The remnant ends its life when the expansion speed decreases to the sound speed in the ambient medium. This occurs at a time

$$t_{\rm SNR} \approx 1.3 \times 10^6 (E_{\rm SN}/10^{51}\,{\rm erg})^{11/35} (n/1\,{\rm cm}^{-3})^{-13/35}\ {\rm years} \tag{117}$$

and radius

$$r_s \approx 64\,(E_{\rm SN}/10^{51}\,{\rm erg})^{11/35} (n/1\,{\rm cm}^{-3})^{-13/35}\ {\rm parsec} \tag{118}$$

(Dorfi 1993).

The important points of this sequence of events for particle acceleration are that it is during the adiabatic and snow-plough phases that the energy available at the shock front is greatest. In these phases the energy of the explosion is mostly contained in the internal energy of the shocked gas (80%), which performs work on the surrounding medium at the shock front. In principle, diffusive acceleration could channel some of this energy into cosmic rays. In the sweep-up phase, little energy is available to the shock, and even if energetic particles were present, the increase in radius of the remnant by a factor of 30 during the adiabatic and snow-plough phases would cause prohibitive energy losses (i.e., the energetic particles themselves would be forced to heat and accelerate the surroundings). At the end of the snow-plough phase, the expansion has essentially come to a standstill, so that adiabatic losses cease. However, it is clear that it is the competition between adiabatic losses and acceleration at a shock front which is continuously expanding and decelerating which finally determines the spectrum produced and the total energy put into cosmic rays. The problem is, therefore, intrinsically one with a strong time dependence.

6.3 Time Dependent Diffusive Acceleration

The time dependence of the diffusive acceleration process is easily understood in the microscopic picture (see Section 5.1). At a parallel shock front, the steady state distribution function decays exponentially upstream from the front (81), so that the differential number of particles in the upstream region is

$$\begin{aligned} n(p) \equiv \int_{-\infty}^{0} {\rm d}x \mathcal{N}_1 &= \mathcal{N}_2 \int_{-\infty}^{0} {\rm d}x \exp\left(-\int_{-\infty}^{x'} {\rm d}x'\, u_1 x'/\kappa_1\right) \\ &= \frac{\langle\kappa_1\rangle \mathcal{N}_2}{u_1}\ , \end{aligned} \tag{119}$$

where the last relation defines a spatially averaged diffusion coefficient $\langle\kappa_1\rangle$. However, according to (98) the number of particles crossing from upstream to downstream per second is $\dot{n}_{\rm cross} = v\mathcal{N}_2/4$. Therefore, the mean time spent by a particle between entering and leaving the upstream medium is

$$\langle \Delta t_{\rm up}\rangle = \frac{4\,\langle\kappa_1\rangle}{u_1 v} \tag{120}$$

(see, for example, Drury 1983). Such a trajectory is depicted in Fig. 17a. The average in (120) is taken over all possible trajectories which start at a point on the shock front (labelled A) . Since the density far upstream goes to zero, all of these trajectories also end on the shock front (e.g., at B). In order to find the time taken for a full crossing/recrossing cycle, one must also find the average time spent in the downstream region. Intuitively, a particle ought to spend longer downstream, since the scattering centres tend to drag it away from the shock front rather than push it back. But this argument is incorrect, because we are interested in the average time spent in the downstream region *by a particle which returns to the shock front.* Reversing the direction of time in Fig. 17a, we obtain, in Fig. 17b, a trajectory in a medium moving away from the shock front with speed u_1. The time taken for the particle to move from B to A is the same as the time taken in (a) to go from A to B. Consequently, if we average over all such trajectories which *end* at a point on the shock front (A), we arrive at the same answer as in (120). Thus, the average downstream residence time of trajectories which return to the shock front is just $\langle \Delta t_{\mathrm{down}} \rangle = 4 \langle \kappa_2 \rangle / u_2 v$. For a complete cycle we have

$$\begin{aligned} t_{\mathrm{c}} &= \langle \Delta t_{\mathrm{up}} \rangle + \langle \Delta t_{\mathrm{down}} \rangle \\ &= \frac{4}{v} \left(\frac{\langle \kappa_1 \rangle}{u_1} + \frac{\langle \kappa_2 \rangle}{u_2} \right) \quad . \end{aligned} \tag{121}$$

Of course, it is also possible to return to the transport equation (80) and derive this result more formally (Toptyghin 1980, Axford 1981). In fact, one can solve for the entire time dependent distribution function (Drury 1991), although only approximately when κ depends on x and/or p. The application to realistic situations such as supernova remnants, however, demands a simpler approach.

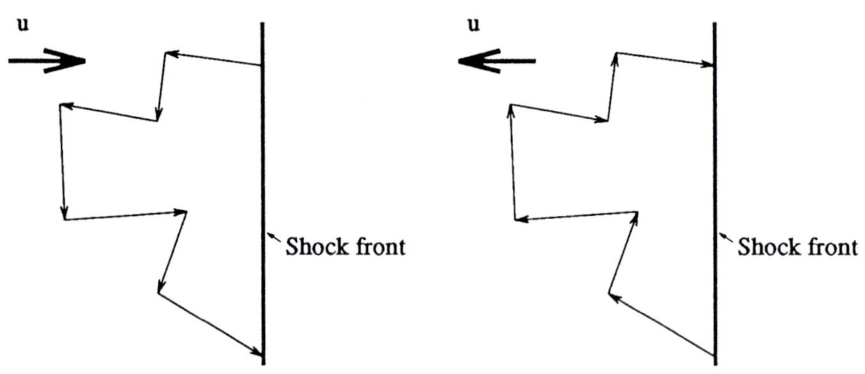

Fig. 17. A stochastic trajectory in the upstream region (a), and the same trajectory, time reversed, in the downstream medium (b)

One such, which has appeared in various guises (e.g., Axford 1981, Bogdan & Völk 1983, Moraal & Axford 1983, Lagage & Cesarsky 1983, Schlickeiser 1984, Völk & Biermann 1988, Ball & Kirk 1992) consists in treating the spatial variation of the distribution around the shock front in an approximate manner. The key to the method is to replace the escape of particles from the vicinity of the shock front by a 'catastrophic' loss-term in a Fokker-Planck type equation:

$$\frac{\partial n}{\partial t} = \underbrace{Q\delta\left(p-p_0\right)}_{\text{source term}} \underbrace{-\frac{P_{\text{esc}}}{t_{\text{c}}}n}_{\text{escape}} + \underbrace{\frac{\partial}{\partial p}\left(-\frac{\langle\Delta p\rangle}{t_{\text{c}}}n\right)}_{\text{acceleration}} \ . \tag{122}$$

By integrating over a small momentum range Δp, one can easily convince oneself that the escape term simply removes particles from the system at the rate suggested by the microscopic treatment (97). The equation is, nonetheless, a rather drastic simplification of the real situation. It clearly represents a spatially averaged picture, since in reality acceleration occurs only at the shock front, and escape only far downstream. We must, therefore, interpret n as an integral of the density $\mathcal{N}(p)$ over the volume of a region around the shock front which includes essentially all particles taking part in the acceleration process. Even so, in the real system the probability of any given particle undergoing acceleration is correlated with the probability of its escaping (both depend on position), so that (122) cannot be expected to give exact results.

It is a simple matter to write down the dynamic friction term in (122) from the microscopic picture (96):

$$\frac{\langle\Delta p\rangle}{t_{\text{c}}} = \frac{p\Delta u}{3(\langle\kappa_1\rangle/u_1 + \langle\kappa_2\rangle/u_2)} \tag{123}$$

$$\equiv \frac{p\mathcal{D}}{t_{\text{c}}} \tag{124}$$

(e.g., Schlickeiser 1984), where we have defined

$$\mathcal{D} = 4(u_1 - u_2)/3v \ . \tag{125}$$

In many applications, (122) appears supplemented by a term of second order in the momentum derivative – a momentum diffusion term – so that the acceleration term takes on the full form of the Fokker-Planck operator. For diffusive acceleration at shock fronts, however, it is quickly shown that the appropriate momentum diffusion coefficient $\left\langle\Delta p^2\right\rangle/t_{\text{c}}$ is of the order of $\Delta u/v$. It must, therefore, be excluded from (122), which already has the form of an expansion in this small parameter.

For a time independent shock ($u_{1,2}$ and $\kappa_{1,2}$ constant), (122) has a simple solution:

$$n(p,t) = \frac{t_{\text{c}}Q}{p_0\mathcal{D}}\left(\frac{p}{p_0}\right)^{2-s}\left[H\left(p-p_0\right) - H\left(p-p_{\text{max}}\right)\right] \ , \tag{126}$$

where

$$p_{\max}(t) = p_0 \exp\left(t\mathcal{D}/t_c\right) \tag{127}$$

and $H(x)$ is the Heaviside function $[H(x) = 1$, for $x > 0$, $H(x) = 0$, for $x < 0]$. The boundary conditions leading to (126) are that the distribution vanishes before the time $t = 0$, when the (subsequently) constant injection is suddenly switched on. The solution is zero for $p < p_0$ and also for $p > p_{\max}(t)$. In between, it is equal to that of the steady state – a power law of index $s - 2$.

The inaccuracy in this solution concerns the sharpness of the cut-off at $p_{\max}$. For the special case of constant $\kappa_{1,2}$ and $\kappa_1/u_1^2 = \kappa_2/u_2^2$ a time dependent analytic solution of the transport equation in the diffusion approximation can be found using the Laplace transform method (Toptyghin 1980, Axford 1981, Drury 1991):

$$n(p,t) \propto \left(\frac{p}{p_0}\right)^{2-s} \left[\left(\frac{p}{p_0}\right)^{4\alpha} E^+ + E^-\right] , \tag{128}$$

where

$$\begin{aligned} \alpha &= \frac{3\left(u_1 + u_2\right)}{4\left(u_1 - u_2\right)} \\ E^{\pm} &= \operatorname{Erfc}\left[\alpha \ln\left(p/p_0\right)\sqrt{t_{\mathrm{acc}}/t} \pm \sqrt{t/t_{\mathrm{acc}}}\right] \\ t_{\mathrm{acc}} &= 4\kappa_1/u_1^2 \qquad (= 4\kappa_2/u_2^2) \ . \end{aligned}$$

Figure 18 shows this solution for a shock front of compression ratio 4 ($\alpha = 1.25$) for various times. For values of the momentum well above that of injection, the solution is a reasonably sharp step-function, in agreement with the behaviour of the approximate treatment (126).

In the case of a supernova, the shock front evolves with time, so that to find the maximum possible particle energy one must integrate the equation

$$\frac{\mathrm{d}p_{\max}}{\mathrm{d}t} = p_{\max}\frac{\mathcal{D}}{t_c} \tag{129}$$

with the initial condition that the momentum at $t = 0$ is that of injection, which is presumably a few MeV/c. However, we can make a simple estimate by putting the acceleration timescale of the most energetic particle equal to the dynamical timescale of the remnant, which, for a self-similar solution, is just the age of the remnant:

$$\frac{p_{\max}}{\mathrm{d}p_{\max}/\mathrm{d}t} = t \ . \tag{130}$$

The minimum plausible value of the diffusion coefficient, is obtained by assuming $D_{\mu\mu}$ in (78) is roughly equal to the gyrofrequency (times the factor $1 - \mu^2$ in the isotropic case). This is known as 'Bohm diffusion':

$$\kappa_{\mathrm{Bohm}} = \frac{2v^2\gamma mc}{15ZeB} \tag{131}$$

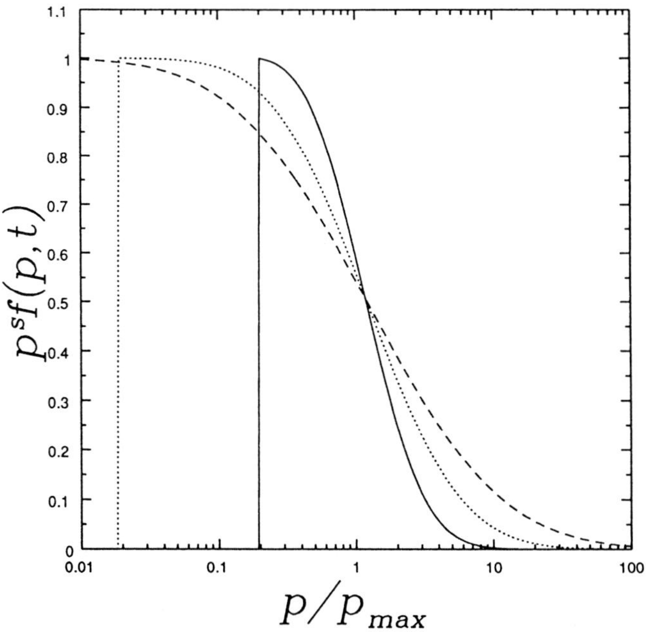

Fig. 18. The distribution function at the shock front in diffusive shock acceleration as a function of momentum at three times: $t = 2t_{\rm acc}$ (solid line), $t = 5t_{\rm acc}$ (dotted line) and $t = 10t_{\rm acc}$ (dashed line). Note that the distribution function is identically zero for momenta lower than that of injection: $p < p_0$. Since the abscissa is $p/p_{\rm max}(t)$, the injection momentum moves to the left as time advances.

(sometimes defined with a slightly different numerical coefficient). Inserting it into (130) gives for the maximum Lorentz factor

$$\gamma \approx \left(u_1^2 t\right) \frac{6ZeB(\rho_{\rm c}-1)}{v^2 mc\rho_{\rm c}(\rho_{\rm c}+1)} \quad . \tag{132}$$

In the Sedov phase, the shock remains strong ($\rho_{\rm c} = 4$), but from (115) we see that the combination $u_1^2 t$ falls off as $t^{-1/5}$. Consequently, the highest energy achieved will not increase significantly during this phase. The sweep-up phase, on the other hand has $u_1 \approx$ constant. Assuming constant magnetic field during this phase implies that the highest energy is reached just at the transition to the Sedov phase. Taking 300 years for the time $t_{\rm trans}$ of transition, $3000\,{\rm km\,s^{-1}}$ as a typical shock speed and $3\,\mu$G for the magnetic field, one finds for a proton:

$$\gamma mc^2|_{\rm max} \approx 2.1\times 10^4 \left(\frac{u_1}{3000\,{\rm km\,s^{-1}}}\right)^2 \left(\frac{t_{\rm trans}}{300\,{\rm yr}}\right)\left(\frac{B}{3\,\mu{\rm G}}\right)\ {\rm GeV} \quad . \tag{133}$$

The result presents a difficulty for the theory (e.g., Lagage & Cesarsky 1983), because the cosmic ray spectrum shows no sign of a departure from the power law index of 4.71 until energies greater than 10^6 GeV/nucleon are reached. Völk & Biermann (1988) have suggested that a systematically stronger magnetic field might be found close to the progenitor of a supernova, which would alleviate the problem. Alternatively, Jokipii (1987) has suggested that an oblique shock should accelerate particles faster than suggested by (129) (see the discussion in Chapter 9).

7 Cosmic Ray Acceleration in Supernova Remnants II

Leaving aside the problem of the maximum energy, the questions one would like answered about the application of diffusive shock acceleration to the production of cosmic rays in supernova remnants are (i) is the remarkable power law form reproduced, at least for energies below 10^{13} eV and (ii) can the mechanism satisfy the stringent efficiency requirements? Important progress in answering both of these questions would follow if we could solve the cosmic ray transport equation (80) and couple it with realistic hydrodynamics. A limited class of analytic solutions is known for self-similar flows (Krymsky & Petukhov 1980, Prischep & Ptuskin 1981, Drury 1983), but this class does not include cases where the acceleration time is comparable to the dynamical evolution time of the flow. Considerable effort has also been invested in numerical approaches to the problem, and there has been some recent progress (Berezhko et al 1993) in this field. Using a more limited approach, one can still hope to gain an understanding of the underlying physics. Two such methods will be described in this chapter. The first concerns itself more with the formation of the spectrum, the second more with the question of the energy budget.

7.1 The Onion-Shell Model

Shortly after the publication of the theory of diffusive acceleration (Axford et al 1977, Krymsky 1977, Bell 1978, Blandford & Ostriker 1978) several groups attempted a more or less realistic calculation of the spectrum of cosmic rays to be expected from a supernova remnant (Blandford & Ostriker 1980, Bogdan & Völk 1983, Moraal & Axford 1983). The physical reasoning was in each case basically the same: cosmic rays should be accelerated at the (unmodified) shock front of a supernova in the Sedov phase of evolution, and, after leaving the vicinity of the shock, should undergo basically just adiabatic expansion. Although the details of the treatments differ the results are comparable.

Let us formulate such a method, starting from the phenomenological equation (122), using the dynamic friction coefficient relevant to diffusive acceleration (96, 99 and 121)

$$\frac{\partial n}{\partial t} + \frac{\partial}{\partial p}\left(\frac{p\mathcal{D}}{t_c}n\right) + \frac{P_{\rm esc}}{t_c}n = Q\delta\left(p - p_0\right) \ . \tag{134}$$

This equation refers to particles undergoing acceleration in the vicinity of the shock front. The precise size of this region need not be specified. On physical grounds one would expect it to extend a few diffusion lengths (κ/u) up and downstream from the shock. The quantity n must then be regarded as the integral of $\mathcal{N}$ integrated over this length (times the area of the shock front). In the microscopic picture, particles escape from the vicinity of the shock into the downstream plasma. Here, acceleration ceases, because we assume all velocity gradients to be unimportant over the particle mean free path. The distribution function is then governed by (54):

$$\frac{\partial f}{\partial t} + \mathbf{u} \cdot \nabla f - \frac{1}{3}(\nabla \cdot \mathbf{u}) p \frac{\partial f}{\partial p} = 0 \ . \tag{135}$$

In order to couple the two solutions, we must express the condition that the particles which escape from the shock enter the adiabatic region. With the help of the expression for the rate at which particles cross an imaginary border in the downstream region (97) we can write:

$$4\pi p^2 u_2 f[p, \mathbf{x} = \mathbf{x}_s(t)] = \frac{P_{esc}}{t_c} n \ . \tag{136}$$

For the function f, we can just as well consider the boundary at $\mathbf{x} = \mathbf{x}_s(t)$ to be the shock front itself, since a diffusion length is by assumption very small compared to the typical distance scale downstream.

The model is now mathematically complete. Equation (134) can be solved once the hydrodynamic solution is prescribed, and the injection is specified. Equation (136) then provides the necessary boundary condition for the subsequent adiabatic expansion, described formally by (135). Physically, we can imagine that the shock front continuously accelerates particles. The newly injected ones are given a power law distribution with an index corresponding to the instantaneous compression ratio (91); the older particles retain their index, if it corresponds to a harder spectrum, or else they too receive the imprint of the instantaneous value of s. As the front progresses it leaves behind it a trail which at each point records the distribution which was present at the shock front when that particular fluid element was overtaken. The neglect of diffusion in the adiabatic region means that there is no blending of the distributions there. In a spherically symmetric situation, the particles in each concentric shell remain locked in it – their spectrum changing only by adiabatic expansion (or compression) and so keeping the same power law index. In this picture, a supernova remnant contains its energetic particles until the expansion has finally finished, and the interior slowly 'unpeels' into the interstellar medium. Figure 19 shows an example of the results presented by Bogdan & Völk (1983) who coined the name 'onion shell' for this model. The most remarkable property is that despite the intricate calculation taking account of the different momentum scale present at each stage of the evolution, the final result is to a good approximation a spectrum which is a single power law. Even the index is in remarkably good agreement with the value required for an accelerator of cosmic rays. The model used in this paper differs slightly from that described above, in that it is assumed that the shock front leaves behind a distribution of particles in each fluid element which is the time asymptotic result of acceleration at a shock of the instantaneous compression ratio, up to a momentum cut-off. There is thus no memory built in, and so no account taken of the fact that particles which reach high energies have accompanied the shock front through a whole range of different conditions. This shortcoming is, however, not serious for particles far below the maximum energy.

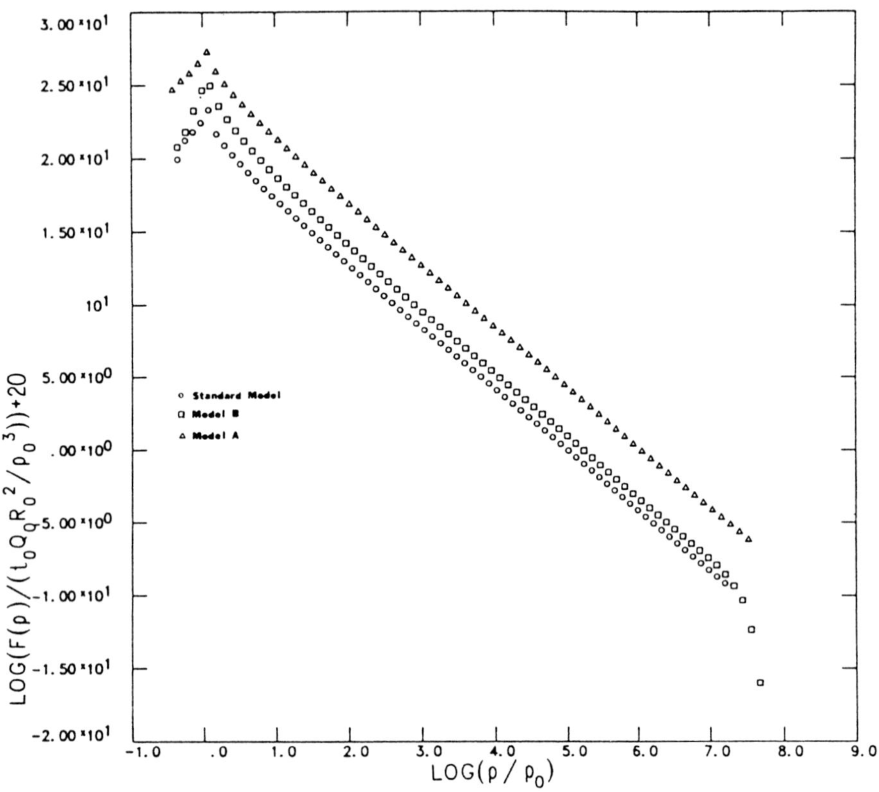

Fig. 19. Results of the onion-shell model from Bogdan & Völk (1983). The ordinate is proportional to the logarithm of the final density of cosmic rays $\mathcal{N}(p)$ produced by a single SNR. The abscissa is the logarithm of the ratio of particle momentum p to a fixed injection momentum p_0. Injection of protons at an energy of a few keV implies $p_0 \sim$ few MeV/c, so that the maximum momentum in these models is roughly 10^{13} eV/c. Three different sets of parameters for the Sedov phase of the explosion have been chosen. The precise values are not important for the purposes of our discussion; each produces a spectrum of power-law type with index close to 2.1, as required for a source of cosmic rays

7.2 Cosmic Rays and Hydrodynamics

Perhaps the main deficiency of the method described in Section 7.1 is that it assumes the shock front is unaffected by the particles it accelerates. In order to produce the power required to sustain the density of cosmic rays in the galaxy, the efficiency of acceleration in a supernova remnant must be of the order of 10%. In this case, one should expect substantial modification of the evolution of a remnant. The effect of the energetic particles on the background plasma is relatively simple to incorporate into the equations of hydrodynamics, because a fluid is described completely by the values of its density, velocity and temperature (assuming we know the equation of state). Thus, because the number density of cosmic rays is negligibly small, we need only calculate the rate at which they exchange momentum and energy with the fluid.

The case treated so far, in which pitch-angle scattering drives the distribution of energetic particles towards isotropy in the rest flame of the fluid, is particularly easy. Measuring, as usual, the pitch angle in this frame we can write the scattering term as

$$\left(\frac{\partial f}{\partial t}\right)_{\rm scatt} = \frac{\partial}{\partial \mu}\left(D_{\mu\mu}\frac{\partial f}{\partial \mu}\right) \quad . \tag{137}$$

The rate at which the background is heated by pitch-angle scattering is then

$$-2\pi\int_0^\infty \mathrm{d}p\, p^2 \int_{-1}^{+1} \mathrm{d}\mu E \left(\frac{\partial f}{\partial t}\right)_{\rm scatt} = 0 \ , \tag{138}$$

where $E = \sqrt{m^2c^4 + p^2c^2}$ is the particle energy. In this approximation, then, no energy is exchanged between the particles and the fluid as seen in the fluid rest frame. The fluid remains adiabatic, and we can continue to employ the usual equation of state. However, if, as is usually assumed, the scattering is due primarily to Alfvén waves, then the small but finite velocity with which these move through the fluid leads to the possibility of an exchange of energy between particles and waves (in fact, there is no single frame of reference in which all Alfvén waves are stationary). This complicates the situation appreciably, since one must in principle introduce a new equation to describe the waves, their damping and growth, as well as their energy and momentum fluxes (McKenzie & Völk 1982). One way out of the problem, but by no means a very satisfactory one, is to assume the wave damping (mainly by non-linear Landau damping) to be so effective that all the energy put into the waves by the particles is immediately and locally transferred to the background plasma as heat. In that case, the level of wave intensity is unimportant, and the energy transfer can be described by an entropy source in the equation of state of the gas, together with an appropriate damping term for the particles.

The deposition of momentum, on the other hand is rather easier to compute, since the equation of motion of the gas just has to support the pressure gradient of the cosmic rays. The fluid equations are thus modified to read, in one dimension,

1. The equation of continuity for the plasma:
$$\frac{\partial \rho}{\partial t} + \frac{\partial}{\partial x}(\rho u) = 0 \ , \tag{139}$$
where ρ is the density, and it is assumed that the number of particles transferred from the thermal population to the cosmic rays (via injection) is negligible.
2. The equation of conservation of momentum
$$\frac{\partial}{\partial t}(\rho u) + \frac{\partial}{\partial x}(\rho u^2 + P_{\rm g}) = -\frac{\partial}{\partial x} P_{\rm CR} \ . \tag{140}$$
Here the left hand side contains the divergence of the momentum flux of the gas ($P_{\rm g}$ being the gas pressure). Neglecting the inertia of the cosmic rays compared to that of the gas, we have on the right hand side only the contribution of the pressure term from the cosmic rays. The pressure is defined as an integral over the distribution function, giving the rate of transport of momentum across unit area perpendicular to the x–axis:
$$P_{\rm CR} = 2\pi \int_0^\infty {\rm d}p\, p^2 \int_{-1}^{+1} {\rm d}\mu\, (\mu p)(\mu v) f(p,\mu,x) \ . \tag{141}$$
In the diffusion approximation, there is a contribution to this term from both the isotropic part of the distribution $f^{(0)}(p,x)$ and the anisotropic part $f^{(1)}(p,\mu,x)$. The latter has not been investigated thoroughly, but would appear to be unimportant when the cosmic ray inertia is negligible (see Webb 1989, Baring & Kirk 1991).
3. The equation of conservation of energy
$$\frac{\partial}{\partial t}\left(\frac{1}{2}\rho u^2 + \frac{P_{\rm g}}{\gamma_{\rm g}-1}\right) + \frac{\partial}{\partial x}\left(\frac{1}{2}\rho u^3 + \frac{\gamma_{\rm g}}{\gamma_{\rm g}-1} P_{\rm g} u\right) = -(u+v_{\rm A})\frac{\partial}{\partial x} P_{\rm CR} \ . \tag{142}$$
Here $\gamma_{\rm g}$ is the ratio of the specific heats of the gas, and $v_{\rm A}$ is the velocity. On the right hand side there appears the work done by the particle pressure gradient against the moving gas and against the moving Alfvén waves.

There have been several numerical solutions of this set of equations, together with the cosmic ray transport equation (80), which reads:
$$\frac{\partial f^{(0)}}{\partial t} + u\frac{\partial f^{(0)}}{\partial x} - \frac{1}{3}\frac{\partial u}{\partial x} p \frac{\partial f^{(0)}}{\partial p} - \frac{\partial}{\partial x}\left(\kappa \frac{\partial f^{(0)}}{\partial x}\right) = 0 \tag{143}$$
(assuming the velocity and distribution functions depend only on distance x from the shock front). The general appearance of a shock transition is changed in the manner shown in Fig. 20. The pressure of the cosmic rays which diffuse ahead of the shock front sets the upstream gas into motion, and compresses it. It is in this 'precursor' that energy is exchanged between cosmic rays and the

kinetic energy of the gas. Subsequently, the gas may pass through a weak sub-shock, downstream of which the cosmic ray distribution and the gas velocity are constant (at least for a planar shock). In principle, it is possible for the gas sub-shock to disappear altogether. Analytic solutions for this kind of structure are well-known for the closely related case of conduction dominated shock fronts (Zeldovich & Raizer 1967), as well as photon, or cosmic ray dominated shocks (Blandford & Payne 1981, Drury et al 1982).

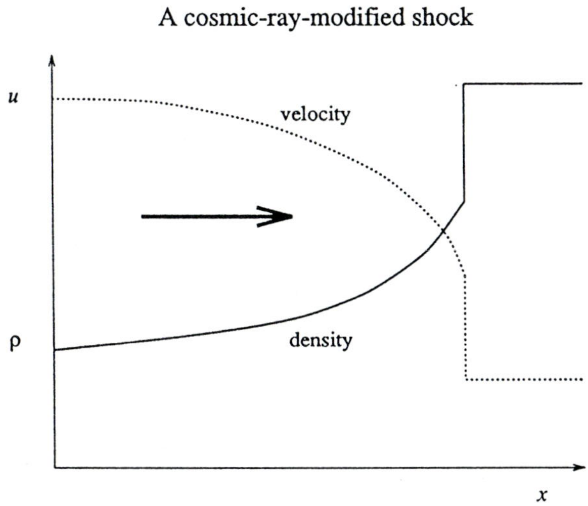

Fig. 20. A sketch of the dependence of the density and velocity of background plasma on position in a shock front which is modified by cosmic rays and develops a precursor upstream of the gas sub-shock.

The most serious difficulty facing a numerical treatment is the very wide range of spatial scales which arises in the precursor, when the diffusion coefficient is an increasing function of momentum. If, for example, we assume Bohm diffusion, then κ is proportional to γv^2, and varies over a range of ten orders of magnitude between a typical injection energy of a few keV and the maximum energy expected in a supernova remnant. Even worse, in the early stages of the sweep-up phase, when the particle energy is still quite low, the velocity is at its highest, so that the length scale typical of the precursor (κ/u_1) is very small. In contrast, towards the end of the Sedov phase, the shock slows down, and particles of the highest energy are present. For this reason, most investigations have used a momentum dependence of κ less severe than that of Bohm diffusion (Falle & Giddings 1987, Bell 1987, Kang & Jones 1991). Furthermore, planar geometry has normally been employed. Only recently has the more realistic problem of a spherical supernova remnant with Bohm diffusion been attacked, although not, as yet with the effects of wave damping or the cooling of the background plasma during the snow-plough phase (Berezhko et al 1993).

7.3 The Two-Fluid Model

At a more primitive level, one can hope to make progress using a cruder representation of the spectrum of the cosmic rays in a supernova remnant. One approach, for example, is to take moments of the angular dependent transport equation (69). Those corresponding to the number density $N_{\rm CR} = 2\pi \int d\mu \int dp p^2 f$ and to the kinetic energy density $E_{\rm CR} = 2\pi \int d\mu \int dp p^2 (E - mc^2) f$ involve an unweighted integration over angles which leads directly to the cosmic ray transport equation (143). We can, therefore, just as well take the moments of this equation, integrating only over p. For the kinetic energy, this leads to

$$\frac{\partial E_{\rm CR}}{\partial t} + \frac{\partial}{\partial x}[u(E_{\rm CR} + P_{\rm CR}^{(0)})] - u\frac{\partial P_{\rm CR}^{(0)}}{\partial x} = \frac{\partial}{\partial x}\bar{\kappa}\frac{\partial E_{\rm CR}}{\partial x} , \tag{144}$$

where

$$\bar{\kappa} = \frac{\int_0^\infty dp\, p^2(E - mc^2)\kappa(\partial f/\partial x)}{\int_0^\infty dp\, p^2(E - mc^2)(\partial f/\partial x)} . \tag{145}$$

Equation (144) has the same general form as the hydrodynamic energy equation (142), with the exception that now the work done by the cosmic rays is subtracted instead of added and that the energy lost to Alfvén waves has been neglected in the transport equation. Another difference is the appearance of the quantity $P_{\rm CR}^{(0)}$, which is just the contribution of the isotropic part of the distribution to the pressure, as discussed above, but the main innovation is the inclusion of a diffusive contribution to the energy flux in the term containing $\bar{\kappa}$.

The hydrodynamic equations, however, do not require knowledge of $E_{\rm CR}$, but rather of the pressure $P_{\rm CR}$. A direct attempt to compute the pressure requires weighting with the factor μ^2, and cannot be performed using (143) as a steppingstone. Returning to the angular dependent transport equation is also futile, because other unknown moments of the distribution function enter the equation for the pressure moment. This situation is a familiar one in many branches of physics – it is usually called a 'closure problem'. Exploiting the analogy with hydrodynamics, we can define as a closure parameter the ratio of specific heats for the cosmic rays $\gamma_{\rm CR}$, which links $E_{\rm CR}$ to $P_{\rm CR}$:

$$P_{\rm CR} = (\gamma_{\rm CR} - 1)E_{\rm CR} . \tag{146}$$

In hydrodynamics, we know the distribution function in a gas is close to that of thermodynamic equilibrium, so that the ratio of specific heats can be evaluated directly by integration. The problem we are faced with here, however, is one of calculating the cosmic ray spectrum, so that $\gamma_{\rm CR}$ is unknown. The other closure parameter, $\bar{\kappa}$, is analogous to a transport coefficient – in hydrodynamics an additional equation would be needed to find the spatial dependence of the distribution function (e.g., the heat conduction equation).

Several attempts have been made to find a reasonable physical basis on which to estimate γ_{CR} and $\bar{\kappa}$ (see Duffy et al 1994). Given that the spectrum in the vicinity of a supernova shock front depends on the entire history of the shock and the rate at which particles have been injected into the acceleration process at it, it is clear that this is a difficult task. The real value of such an enterprise lies in the possibilities it opens up of investigating in detail the evolution of a supernova in the presence of cosmic rays. To illustrate this point, Fig. 21, taken from Dorfi (1993) shows the evolution of a model of a supernova remnant, computed using the 'two-fluid' approach described here. Radiative cooling and heating by Alfvén wave damping have not been included in this example, which, however, illustrates nicely the points made in section 7.1, namely that the maximum energy is determined in the sweep-up phase, whereas the main power input occurs in the Sedov and snow-plough phases. The energy input into cosmic rays in this example is an impressive 60% of the total supernova energy. This quantity, however, is sensitive to the rate at which particles from the thermal plasma are injected into the cosmic ray gas at the shock front. All current models of cosmic ray acceleration are faced with a similar difficulty, usually referred to as the 'injection' problem. At present there is no self-consistent theory which would enable one to compute such a quantity, so that one must resort to a rather arbitrary parameterisation. Because the number density of cosmic rays does not enter the two-fluid formulation, it is necessary to assume a small transfer of energy between the fluids at the shock front. In the case presented in Fig. 21, one part in 10^4 of the kinetic energy flux of the gas entering the sub-shock is put into the cosmic ray gas.

8 Jets and Active Galactic Nuclei

8.1 Introduction

That energetic particles play a fundamental part in the physics of active galaxies has been clear for some time. Recently, this view has been emphatically confirmed by observations of very high energy photons from several such objects. The evidence for the presence of nonthermal particles can be grouped into three categories:

1. In six instances, bright hot spots in the jets of double radio sources display synchrotron radiation extending from the radio to frequencies in the near infra-red or higher (Meisenheimer et al 1989). In addition, the relatively nearby galaxy M87 has a jet which emits synchrotron radiation at optical and possibly higher frequencies. Of course, all radio galaxies show evidence of nonthermal electron populations, but these jets are the most extreme examples in terms of the high energy of the emitting electrons. Standard methods of estimating the magnetic field lead to Lorentz factors of the electrons of up to 10^6.
2. Gamma-rays in the range 100 MeV to 10 GeV have been detected from about 40 AGNs by the EGRET instrument on the Compton Gamma-ray Observatory (Fichtel et al 1993). The flux level implies that gamma-rays play a

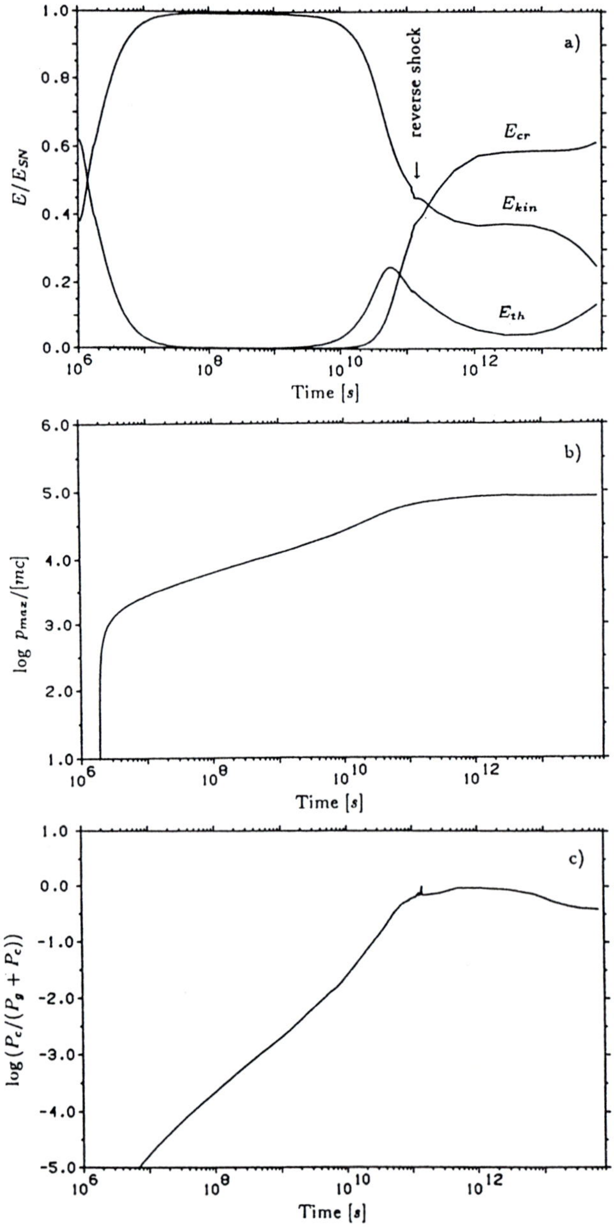

Fig. 21. Results obtained by Dorfi (1993) showing the temporal evolution of a supernova remnant in the two-fluid model. Plotted is the ratio of the energy in cosmic rays to the total energy of the explosion, the maximum momentum of an accelerated particle (in units of mc) and the ratio of cosmic ray pressure to total pressure (i.e., cosmic ray plus gas) at the shock front.

major role in the energy budget of the source. Photons of TeV energy have been detected from one source (Mrk 421) by the Whipple telescope, using the atmospheric Cherenkov technique (Punch et al 1992). Although such gamma-rays can be produced by nuclei of relatively modest Lorentz factor $\sim 10^3$, most current models assume a leptonic origin (inverse Compton scattering), in which case one again requires electrons of Lorentz factor 10^6.

3. The X-ray spectra of Seyfert galaxies have a distinctive power-law form in the energy range 2 – 10 keV (Mushotzky et al 1993). One possible explanation involves the generation of an electromagnetic cascade by relativistic electrons and positrons, which would require Lorentz factors in the hundreds. An alternative picture involves Compton scattering in an electron/positron gas with a temperature of about 50 keV. Although this gas is assumed to have an equilibrium distribution, the explanation requires some form of particle acceleration, because the two-body collision rate is too slow to maintain a Maxwellian spectrum in the face of cooling by Compton scattering.

Unambiguously identifying the presence of accelerated particles does not, however, mean that we can unambiguously identify their origin. The physical conditions in the source are generally too uncertain to permit this. In fact, of the three examples mentioned above, only (1) has seen the application of a reasonably detailed and nevertheless plausible acceleration model, simply because of the availability of high quality, spatially resolved data. In example (2) we have no possibility of locating the source observationally, but must rely on indirect arguments concerning the variability of the flux, or the optical depth to absorption of the central region of the AGN. The X-rays referred to in example (3) almost certainly originate from a very small region (10^{14} – 10^{15} cm) around a super-massive black hole (in the standard scenario), but whether or not the particle acceleration mechanism is associated with shocks is unclear.

The new aspect which arises in a discussion of acceleration in these sources is that of loss mechanisms. Both the magnetic field and the photon field are estimated to be considerably stronger than in the case of supernova remnants, so that losses by synchrotron radiation and inverse Compton scattering (for the leptons) and by pair production and photon-pion production (for the hadrons) are correspondingly more rapid. Two applications will be discussed in this chapter: hot spots in jets, and the nonthermal emission from a 'leaky box' model of the central source.

8.2 Hot Spots in Jets

A relativistic electron emitting synchrotron radiation loses energy at a rate

$$\frac{-\mathrm{d}p/\mathrm{d}t}{p} = \frac{2p}{m^2c^2}\sigma_{\mathrm{T}}(1-\mu^2)\left(\frac{B^2}{8\pi}\right) , \tag{147}$$

where σ_{T} is the Thomson cross section ($= 6.65 \times 10^{-25}\,\mathrm{cm}^2$). This effect is important for electrons which are being accelerated by the first order Fermi process in the vicinity of a shock front if it is comparable with the rate of acceleration given by (129). The necessary modification of the transport equation can be found by returning to the equation expressing the conservation of flux in phase space (2). If we assume pitch-angle scattering is sufficiently rapid to keep the distribution almost isotropic, then we can easily include this term in the phenomenological equation used to describe the differential density of particles in the 'acceleration region' close to the shock front (134):

$$\frac{\partial n}{\partial t}+\frac{\partial}{\partial p}\left[\left(\frac{p\mathcal{D}}{t_{\mathrm{c}}}-\beta p^{2}\right)n\right]+\frac{P_{\mathrm{esc}}}{t_{\mathrm{c}}}n=Q\delta\left(p-p_{0}\right)\ , \tag{148}$$

where

$$\beta=\frac{4\sigma_{T}}{3m^{2}c^{2}}\left(\frac{B^{2}}{8\pi}\right)\ . \tag{149}$$

Clearly there exists an important momentum scale in this problem: $p_{\max} = \mathcal{D}/(\beta t_{\mathrm{c}})$, i.e., the point where acceleration is exactly balanced by losses. It is to be expected that the solution will always vanish for higher momenta. Assuming $p_0 < p_{\max}$, particles are accelerated away from the injection point, so that the distribution vanishes for $p < p_0$. Just as in the problem without losses (126) the distribution also vanishes for momenta to which there has not been enough time to accelerate particles. In this case, however, the upper bound $p_1(t)$ is given by

$$\frac{1}{p_{1}(t)}=\frac{1}{p_{\max}}+\left(\frac{1}{p_{0}}-\frac{1}{p_{\max}}\right)\exp\left(\frac{-t\mathcal{D}}{t_{\mathrm{c}}}\right)\ , \tag{150}$$

so that it tends to $p_{\max}$ as $t \to \infty$, but never exceeds it. Once again, the distribution between p_0 and p_1 is independent of time, and has the shape

$$n(p,\infty)\propto p^{-2}\left(\frac{1}{p}-\frac{1}{p_{\max}}\right)^{(P_{\mathrm{esc}}-\mathcal{D})/\mathcal{D}} \tag{151}$$

(Kardashev 1962). We see from this that the slope is unaffected by losses for $p \ll p_{\max}$, but as the critical value is approached, the distribution either steepens (for $s = 3 + P_{\mathrm{esc}}/\mathcal{D} > 4$) or flattens, tending to infinity at $p_{\max}$ (for $s < 4$).

If we want to explain a hot spot in a jet by this acceleration mechanism, we must calculate the synchrotron emission from the distribution of particles. Not only would we expect a contribution from particles in the vicinity of the shock, but also from each of the 'onion shells' left behind. In the case of a jet of constant cross section, the adiabatic losses included in (135) are absent, but we must instead take account of synchrotron losses. The distribution function downstream of the shock front is then determined by the equation

$$\frac{\partial f}{\partial t}+\mathbf{u}\cdot\nabla f-\frac{1}{p^{2}}\frac{\partial}{\partial p}\left(\beta p^{4}f\right)=0\ , \tag{152}$$

together with the boundary condition at the shock

$$\pi v p^{2}f[p,\mathbf{x}=\mathbf{x}_{\mathrm{s}}(t)]=\frac{P_{\mathrm{esc}}}{t_{\mathrm{c}}}n\ . \tag{153}$$

The solution of this equation is again simple. The main properties can be found by qualitative considerations and are illustrated in Fig. 22. In interpreting an observation, we must take account of the fact that the instrument integrates the total synchrotron emission originating from a region of finite size. In the case of a hot spot, we can assume this region includes the shock front and a finite length of the downstream plasma. It is this length, or more precisely, the time taken by a fluid element to traverse it, which defines a characteristic momentum p_{br} to which a particle starting off at the shock front with $p = p_{max}$ will have had time to cool. (Provided the region is not too small, this 'break momentum' is almost independent of p_{max}.) Well below p_{br}, the particle spectrum remains $f \propto p^{-s}$ throughout the whole emission region, so that the synchrotron spectrum is given by the usual formula, for frequencies below ν_{br}, which is that typically radiated by a particle of momentum p_{br}.

$$I_\nu \propto \nu^{-(s-3)/2} \qquad \text{for } \nu < \nu_{br} \ . \tag{154}$$

However, as shown in Fig. 22, the distribution above p_{br} has a cut-off, in the neighbourhood of which the spectrum steepens or peaks (depending on s). The frequency of this structure moves gradually down in p-space as the fluid element progresses further and further from the shock, arriving at p_{br} at the edge of the emission region. The number of radiating particles at any given p is therefore proportional to the magnitude of the distribution at the shock, times the length over which such particles can avoid cooling. From (150) one sees that this length is proportional to $1/p$, so that the overall effect is to steepen the 'injected' distribution by unity. Thus, for a frequency range above ν_{br}, the emitted spectrum is steepened by 0.5:

$$I_\nu \propto \nu^{-(s-2)/2} \qquad \text{for } \nu_{max} > \nu > \nu_{br} \ , \tag{155}$$

where ν_{max} is the characteristic frequency emitted by a particle of momentum p_{max}. This argument breaks down if the peak which forms close to the upper cut-off contains an important fraction of the radiating particles, which occurs for $s < 3$. Independently of s, the spectrum then tends towards that emitted by monochromatic cooling particles, which have $I_\nu \propto \nu^{-1/2}$ in the range $\nu_{max} > \nu > \nu_{br}$. The result (154) remains valid unless the low frequency emission emitted by energetic particles overwhelms that emitted by lower energy ones i.e., unless $s < 7/3$.

This type of spectrum has been used to model the emission from hot spots in the six cases mentioned above (Meisenheimer et al 1989). In general, the model spectra fit well, and have enabled constraints to be placed on the jet parameters. In particular, the frequency of the observed cut-off together with an estimate of the magnetic field fixes the value of the diffusion coefficient

$$\kappa \sim 10^4 \kappa_{Bohm} \ . \tag{156}$$

Clearly, this model does not run into the difficulty which besets cosmic ray acceleration in SNR, namely the lower limit $\kappa > \kappa_{Bohm}$. However, we must remember that it is electrons rather than ions which are being accelerated here. Figure 23

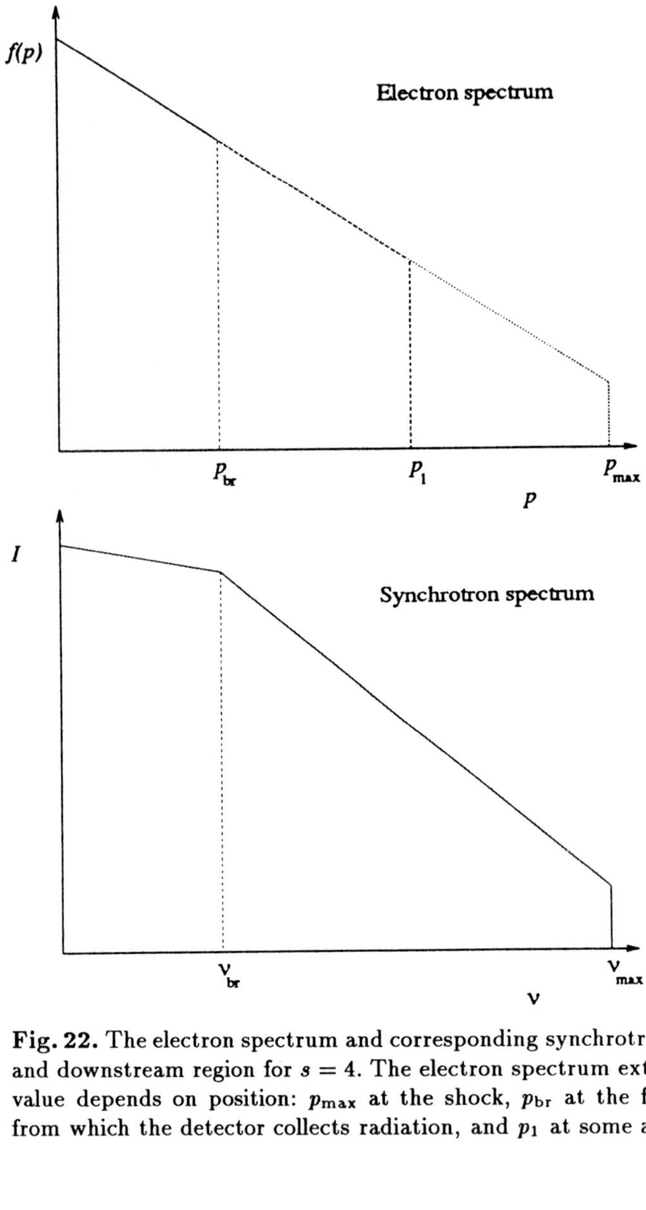

Fig. 22. The electron spectrum and corresponding synchrotron emission from the shock and downstream region for $s = 4$. The electron spectrum extends up to a cut-off whose value depends on position: $p_{\rm max}$ at the shock, $p_{\rm br}$ at the furthest point downstream from which the detector collects radiation, and p_1 at some arbitrary point in between.

(from Meisenheimer et al 1989) displays the data on hot spot 'A' in the jet of the radio galaxy 3C111. Two theoretical models have been plotted, one in which the electron distribution is assumed to be a power law $f \propto p^{-3.9}$ up to a maximum, above which there is a sharp cut-off ($f = 0$ for $p > p_{\rm max}$) (dashed line) and one in which the distribution is an approximation based on an exact solution of the cosmic ray transport equation, including synchrotron losses (Heavens & Meisenheimer 1987). Clearly, the most sensitive part of the fit is close to the upper cut-off. Unfortunately, it is precisely here that the phenomenological model used above fails. The reason is that the model treats the acceleration region as homogeneous. Near to the cut-off, however, particles lose a significant amount of energy whilst still inside the acceleration region. In fact, the higher the energy of a particle, the thinner is the sheet around the shock front in which they can be found. Thus, in order to calculate the emission reliably, one must return to the transport equation (79) and solve it at a shock front including the synchrotron loss term (Webb et al 1984, Bregman 1985, Heavens & Meisenheimer 1987). These analytic approaches give a cut-off which is much too gradual to fit the observations shown in Fig. 23, but there remain two possibilities of saving the model: the use of a momentum dependent diffusion coefficient (not taken account of in the analytic treatments) and the generalisation to relativistic flows. The first of these seems to offer good prospects of obtaining very sharp cut-offs (Fritz 1989, 1989), the ability of the second to do so is more controversial (Kirk & Schneider 1987b, Krülls 1992)

8.3 The Central Source

A fascinating property of the X-ray emission of Seyfert galaxies and quasars is its rapid variability. Occasionally, these objects have been known to change their luminosity by a factor of two within about one hour. To quantify this behaviour, observers define the 'compactness', which is proportional to the ratio of the X-ray luminosity to the fastest timescale of variability. Denoting this timescale by Δt, the dimensionless compactness is

$$\ell_X = \frac{L_X \sigma_T}{4\pi \Delta t\, mc^4} , \tag{157}$$

where L_X is the luminosity in the X-ray band, calculated on the assumption that the source radiates isotropically, and lies at a distance indicated by its redshift. This quantity is typically in the range 10 – 100 for variable sources. The presence in (157) of σ_T indicates that ℓ_X is connected with the reaction rate or optical depth of an electromagnetic process. The minimum value of the energy density of radiation in the source occurs if photons escape freely and is given by $L_X/(4\pi R^2 c)$, which translates into a photon number density when divided by the energy $\langle x \rangle mc^2$ typical of an X-ray photon. If we consider now a gamma-ray of energy sufficiently high that it can create an electron positron pair on collision with an X-ray photon, we can use the fact that the cross section for this process is roughly $\sigma_T/4$ to write for the optical depth of the source

$$\tau_{\gamma\gamma} > \frac{\ell_X}{4\langle x \rangle} . \tag{158}$$

For X-rays in the band 2 – 10 keV, the average photon energy lies in the range $1/50 > \langle x \rangle > 1/250$, so that for variable sources $\tau_{\gamma\gamma} \gg 1$. This observation has led to the generally accepted view that electron positron pairs are present and are important factors in forming the spectra of sources of high compactness. This

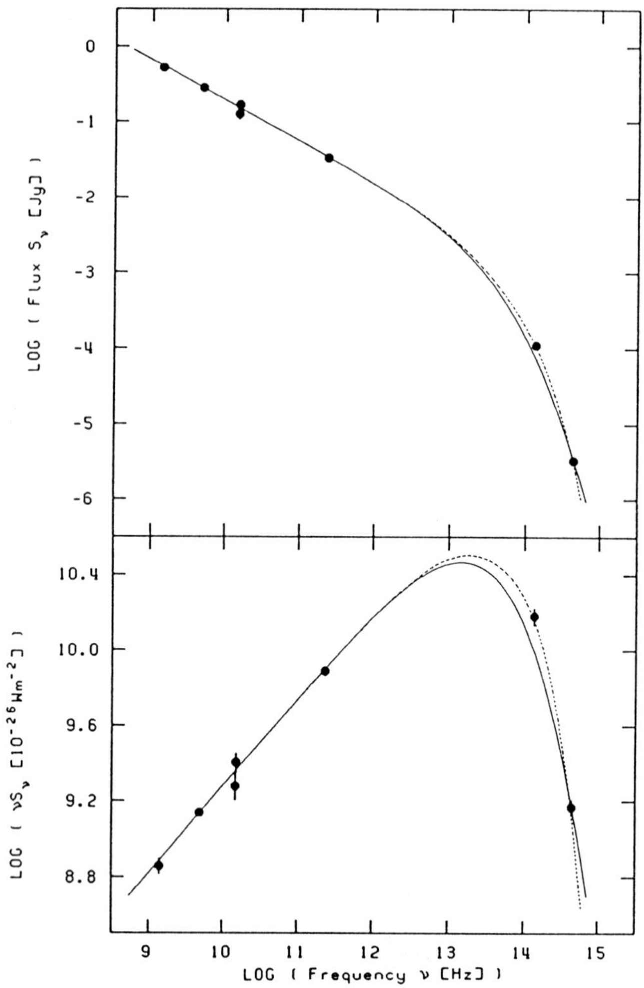

Fig. 23. The synchrotron spectrum of the hot spot 3C111 East, from Meisenheimer et al (1989). Both the flux $S(\nu)$ and $\nu S(\nu)$ (which is proportional to the luminosity per frequency decade) are shown. Two models are drawn on each plot: the solid line results from an approximate solution of the electron transport equation, the dashed line from the idealised distribution shown in Fig. 22. This source does not display a spectral break at low frequency.

is probably true, but it is as well to remember that even in sources of very high compactness, no pairs are created unless there are pair producing gamma-rays. Caution is not out of place, because recent observations by the SIGMA telescope and by the OSSE instrument on the Compton Gamma-Ray Observatory have rather unexpectedly failed to confirm the presence of electron positron pairs in these sources (Jourdain et al 1992, Maisack et al 1993). Another point concerning the compactness which is often overlooked is that it gives only a lower limit to the photon column density. There may be many sources of low compactness which have very high photon column densities, but do not vary rapidly. In fact, the X-ray luminosity of Seyferts and quasars seems to be positively correlated with the timescale of variability, which could even mean that low compactness sources are those with the highest photon column densities (see Mushotzky et al 1993).

Although basically an observed quantity, the term 'compactness' has been widely adopted as a name for related parameters in theoretical models. Thus, the rate at which electrons or photons are injected into a source model is usually given as a dimensionless compactness. In this case, the definition uses the actual source dimension R instead of the quantity $\Delta t\, c$ in (157). Given that most theoretical models compute stationary spectra, the observed compactness is predicted to be zero! Another convenient quantity related to the observed compactness (and usually called by the same name) is defined in terms of the photon energy density $U_{\rm rad}$ in the source:

$$\ell = \frac{U_{\rm rad} R \sigma_{\rm T}}{mc^2} \ . \tag{159}$$

To link this with the observed compactness one must not only relate Δt to R, but also $L_{\rm X}$ to $U_{\rm rad}$. In the same spirit, one can define a magnetic compactness:

$$\ell_B = \frac{B^2 R \sigma_{\rm T}}{8\pi mc^2} \ . \tag{160}$$

From the synchrotron loss-rate of an electron, Eq (147) one quickly sees that the energy loss timescale measured in units of the light crossing time R/c is just $3/(4\ell_B\gamma)$. If, as is frequently assumed (see, for instance Protheroe & Kazanas 1983) , the energy density in the magnetic field in the inner regions of an AGN is of the same order of magnitude as the energy density in photons, then we can conclude that electrons or positrons will cool long before they are able to leave the source. The same estimate applies also to the cooling time of a relativistic electron by inverse Compton scattering off soft photons. In the case of a Seyfert or quasar, the most numerous soft photons are presumably from the spectral maximum in the UV region. Interpreting $U_{\rm rad}$ as referring to these, one arrives at $3/(4\ell\gamma)$ as the inverse Compton cooling time in units of R/c. The inescapable conclusion is that the variability timescale is much longer than (and therefore unlikely to be related to) the loss timescale of the electrons/positrons. Since in any model the acceleration rate must exceed the loss rate at least up to the maximum energy, variability cannot be connected with the acceleration mechanism either.

In view of the extremely rapid cooling rate for leptons, it has been suggested that the basic process of production of nonthermal particles is not one of electron acceleration, but primarily one of proton acceleration (Protheroe & Kazanas 1983, Sikora et al 1987). Electrons then arise as the products of pair and pion producing interactions of the energetic protons with the background plasma, and, especially, with the photons. Of course, the physical conditions in the central region of an AGN are not known with any degree of certainty, so that it is perhaps wisest to adopt a crude approach to modelling particle acceleration. In the context of first order Fermi acceleration, the obvious first step is to regard the acceleration region as a box, within which particles are accelerated at a constant rate, and out of which they can escape with a certain probability. Formally, this is just the same as the phenomenological model of Chapter 6 (122). One has, however, the freedom of choice, at least in principle, between allowing the escaping particles to cool and further contribute to the emitted radiation, or losing them once and for all into the black hole (e.g., Mészáros & Ostriker 1983).

One can fairly generally estimate the maximum possible energy of a proton by comparing the acceleration rate in a particular model with the loss rate. To order of magnitude, the rate of energy gain in first order Fermi acceleration is (123)

$$t_{\mathrm{acc}}{}^{-1} \approx u^2/\kappa \; . \tag{161}$$

With the assumption of Bohm diffusion, one finds

$$t_{\mathrm{acc}}{}^{-1} \approx \frac{u^2}{c^2}\Omega \; , \tag{162}$$

where Ω is the gyrofrequency. This estimate is based on the diffusion approximation, and thus depends on the shock being nonrelativistic. Numerical investigations of the relativistic case have given results ranging from a factor of 3 to 13 faster (Quenby & Lieu 1989, Ellison et al 1990). It is interesting to note that the acceleration rate for shock-drift acceleration, which makes use of the electric field induced in the plasma by motion across the magnetic field at speed u, is approximately $t_{\mathrm{acc}}{}^{-1} \approx \Omega u/c$; almost the same (to order of magnitude) as that given in (162).

The important loss processes, on the other hand, are threefold

1. Nuclear collisions with the background plasma

$$p + p \rightarrow p + p + \pi' \mathrm{s} \tag{163}$$

$$p + p \rightarrow p + n + \pi' \mathrm{s} \; , \tag{164}$$

which result in a total energy loss rate:

$$t_{\mathrm{pp}}^{-1} = 6.7 \times 10^{-16} n_{\mathrm{p}} \; [\mathrm{secs}^{-1}] \; , \tag{165}$$

where n_{p} is the proton number density of the background.

2. Pair production

$$p + \gamma \rightarrow p + e^+ + e^- \tag{166}$$

3. Photo-pion production

$$p + \gamma \rightarrow p + \pi^0 \tag{167}$$

$$p + \gamma \rightarrow n + \pi^+ \tag{168}$$

The relative importance of processes 2 and 3 depends on the photon spectrum: 2 has a larger cross section but a very low inelasticity. However, it enjoys the advantage of a lower threshold, giving it many more target photons. For a power law spectrum with intensity inversely proportional to frequency, 2 and 3 turn out to be almost equally effective, resulting in an energy loss rate:

$$t_{\mathrm{pair}}^{-1} \approx t_{\mathrm{pion}}^{-1} \approx 3 \times 10^{-15} \gamma \bar{U}_{\mathrm{rad}} \ [\mathrm{secs}^{-1}] \ , \tag{169}$$

where $\bar{U}_{\mathrm{rad}}$ is the radiation energy density per frequency decade (measured in $\mathrm{erg\,cm}^{-3}$). If the background plasma density is low, we can confine ourselves to a consideration of the photon losses, in which case equating rates (162) and (169) leads to an estimate of the maximum proton Lorentz factor:

$$\gamma_{\max} \approx 10^{9} \left(\frac{B}{10^{3}\mathrm{G}}\right)^{1/2} \left(\frac{\Delta t}{10^{4}\mathrm{s}}\right) \left(\frac{10^{42}\mathrm{erg\,s}^{-1}}{L}\right)^{1/2} \left(\frac{u}{c/10}\right) . \tag{170}$$

This is, of course, an upper limit, because of the identification we have made of the radiation energy density with luminosity and of the source size with the variability timescale. Nevertheless, it hints at the possibility of producing very energetic protons in AGNs.

A closer consideration of the reactions (165) and (168) leads to the conclusion that the energy they remove from the protons is largely put into other potentially interesting forms such as neutrons and neutrinos (Kazanas & Ellison 1986, Biermann & Strittmatter 1987, Kirk & Mastichiadis 1989, Begelman et al 1990, Atoyan 1992, 1992). The fact that the interaction cross section of the neutrino rises as a function of energy, means that the highest energy neutrinos produced in the decay of the muons which, in their turn, come from the pions produced in reactions (165) and (168) could be observable using proposed neutrino detectors (Stecker et al 1991).

Another possible effect of these reactions is a contribution to the observed cosmic ray flux between 10^{15} and 10^{19} eV. The formulation of the acceleration as a stochastic process with quite specific values of escape probability and acceleration rate lends itself readily to a Monte-Carlo simulation. Processes which result in a very small change in particle energy per event (such as pair production losses or acceleration on crossing a shock front) can be treated as quasi-continuous and incorporated into the particle orbit, whereas discrete jumps in energy (as in pion production) are simulated stochastically. Protheroe & Szabo (1992, 1994) have recently performed such simulations including a careful treatment of the loss processes, and have calculated the neutron spectrum emitted by the source. These particles escape ballistically from the central region and decay into protons at a distance which depends on the Lorentz factor: $d \approx 3 \times 10^{13}\gamma[\mathrm{cm}]$, thus contributing to the cosmic ray density in the host galaxy.

The two largest uncertainties in such a model are the speed of the acceleration process (equivalent to the strength of the magnetic field, since it is assumed that $\kappa \sim \kappa_{\mathrm{Bohm}}$) and the strength of the radiation field in the source. The latter can be parameterised by the value of the maximum proton energy, provided a realistic estimate can be made of spectral distribution of the radiation. Finally, in order to compute the contribution of all AGN to the cosmic ray spectrum measured at Earth, one must estimate the escape probability of energetic cosmic rays from the host galaxy of an AGN and the average number density of AGN in the universe as a function of redshift. Then, provided the characteristics of the intergalactic medium (i.e., the diffusion coefficient) allow cosmic rays to diffuse over a distance larger than the average intersource distance, one can estimate the AGN contribution to the local density. Using models intended to span a plausible range for these quantities, Protheroe & Szabo arrive at the prediction shown in Fig. 24. The closeness of the predicted flux with that observed is impressive, especially if one remembers that the normalisation of the model flux is not a free parameter. This model also makes a clear prediction of the composition, because only neutrons can leak out of the central parts of an AGN. Thus, if AGNs are important for cosmic rays of energy around 10^7 GeV, one should expect mostly protons in this range. Recent results from the Fly's Eye (Gaisser et al 1993), however, seem to indicate a different picture.

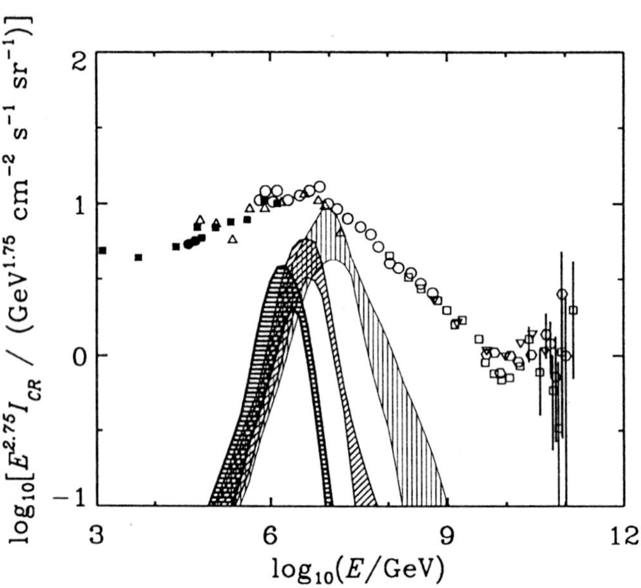

Fig. 24. The contribution to the local cosmic ray flux by AGNs predicted by Protheroe & Szabo (1994). The shaded bands give the range of results found using different AGN luminosity functions and shapes for the radiation spectrum in the source. Each pair of curves corresponds to a different assumption about the strength of the acceleration process: $\kappa = \kappa_{\mathrm{Bohm}}$ (strongest flux), $10\kappa_{\mathrm{Bohm}}$, and $100\kappa_{\mathrm{Bohm}}$ (weakest flux). The points are from compilations of the cosmic ray flux at Earth

The Monte-Carlo calculation of Protheroe & Szabo is, of course, linear, in the sense that the photon field responsible for the energy loss of the protons is taken as fixed. In fact, it is to be expected that the relativistic particles themselves supply the energy which subsequently appears in the X-ray band.

In this case, different parts of the photon spectrum originate in different spatial regions, and a simple estimate of the radiation density in the source is likely to be unreliable. The only way to improve on the calculation is then to attempt to include the feedback between energetic particles and photons in a self-consistent manner. To date, only preliminary investigations of this difficult problem have been attempted (e.g., Kirk & Mastichiadis 1992).

9 Radio Supernovae

9.1 The Radio Emission of Supernovae

Supernova *remnants* are prominent and well-studied sources of nonthermal radio emission. But, until fairly recently, none of the 'modern' supernovae (i.e., those discovered after 1885, when the first extragalactic supernova was found in Andromeda) had been detected at radio wavelengths. There are now over a dozen examples of supernovae which are also radio sources, including SN1987A in the Large Magellanic Cloud (for a review see Weiler & Sramek 1988). The emission, which has a nonthermal spectrum, does not always start at the same time as the optically detected explosion, but has, in some cases, been detected for the first time several months afterwards, and in one case was seen about ten days before 'maximum light' (i.e., the optical maximum). As in the applications we have considered in previous chapters, the acceleration of nonthermal particles (in this case electrons) accompanies a phenomenon which produces a shock front. Therefore, it seems natural to attempt once again to apply the best developed model of particle acceleration – diffusive acceleration – to these sources.

There are several reasons why theorists have been cautious in doing so. Synchrotron emission, as we have seen in Sect. 3.1 is a very sensitive indicator of the hydrodynamics: a minor fluctuation in the magnetic field strength – which is quite likely to be unimportant dynamically – suffices to make a particular region stand out on radio maps. Furthermore, in the case of young supernova remnants such as Cas A it is known that the magnetic field in the radio emitting region is too strong to be just the interstellar field compressed by the shock front (Anderson et al 1991). Turbulent motion is thought to be responsible for enhancing the field, and where there is turbulence, there will be particle acceleration, but not necessarily of the diffusive kind. Even if most of the energy going into nonthermal particles does so by means of diffusive acceleration at the shock front, the synchrotron emission may still be dominated by those electrons located in (and possibly accelerated in) regions of amplified magnetic field. Another difficulty with the application of diffusive acceleration is that the spectrum of electrons responsible for synchrotron emission in both radio supernovae and supernova remnants is not the simple power-law of index $s = 4$ which the diffusive process predicts for test particles at a strong shock front. Whereas this problem is not so severe for the case of electron acceleration in the hot spots of jets (Sect. 8.2), supernovae and supernova remnants sometimes display spectra which are very much too steep (e.g., $s = 5$) to be accommodated in a simple picture.

Because of these difficulties, work on the radio emission of SNe and SNRs has concentrated on the hydrodynamical aspects, rather than the particle acceleration model. A widely accepted and well-developed model of radio supernovae has been proposed Chevalier (1982). In it, it is assumed that the supernova shock front propagates in a medium whose density falls of inversely as the square of the radius, such as in a stellar wind of approximately constant velocity. In fact, it seems from observations that only massive stars explode into supernovae which emit in the radio, and these stars are indeed thought to drive a dense stellar wind in their pre-explosion phase. In the model, the emission switches on when the absorbing screen of wind material becomes sufficiently thin to let out the radio waves. The dominant absorption mechanism is free-free absorption, which is stronger at low frequency, so that high frequency emission is predicted to switch on first, in agreement with observation. The underlying, unabsorbed emission always decreases with time, and its exact time dependence can be calculated using a self-similar solution for the free-expansion phase, in which the expanding shell of shocked material is treated as being geometrically thin (the 'mini-shell model'). The time dependence of the shock radius in such a model is given by

$$r_{\rm s} \propto t^m \ , \tag{171}$$

where m is related to the density profile of the outer edge of the ejecta, and is very close to, but slightly less than unity. Thus, the shocked ejecta suffer slight deceleration by the less dense surroundings – a situation analogous to that of a heavy fluid supported against gravity by a light one. This configuration is subject to the Rayleigh-Taylor instability. The growth time turns out to be of the same order as the age of the supernova, making it plausible that turbulence will develop, amplifying the magnetic field and accelerating particles. The total thermal energy $E_{\rm th}$ scales as the product of volume and pressure of the shocked material, i.e., as $r_{\rm s}^3 \times (\dot{r}_{\rm s}/r_{\rm s})^2$, so that

$$E_{\rm th} \propto t^{3m-2} \ . \tag{172}$$

In the absence of a detailed theory describing how acceleration occurs, the model assumes that a fixed fraction (typically 1%) of $E_{\rm th}$ is converted into magnetic field energy, and an equal amount also goes into relativistic electrons, which leads to a magnetic field scaling

$$\begin{aligned} B &\propto E_{\rm th}^{1/2} r_{\rm s}^{-3/2} \\ &\propto t^{-1} \ . \end{aligned} \tag{173}$$

Given a time independent power law index for the electron spectrum as defined in (41), one finds $C \propto t^{3m-5}$, which leads via (42) to a synchrotron flux which, in the absence of absorption, has the time dependence:

$$\mathrm{I}_\nu(\mathrm{t}) \propto t^{3m-3-\alpha} \ , \tag{174}$$

where $\alpha = (s-3)/2$ is the spectral index of the underlying radio emission. Equation (174) describes the decay of the radio flux which sets in once the source becomes optically thin. When combined with the variable absorption controlling the switch-on phase, a description of the emission results with four (or, including internal free-free absorption, five) free parameters. Extensive fits of this model to the data can be found in Weiler et al (1986). Although remarkably good for some RSNe, there are several cases in which the model has difficulty.

9.2 Supernova 1987A

Studies of this remarkable object have given new impetus to many branches of research into supernovae. This may well turn out to be the case for the theory of particle acceleration in radio supernovae too.

There have been two phases of radio emission from SN1987A: an initial 'prompt' radio burst which started about two days after explosion, and lasted for a couple of weeks, and the current, steadily rising emission, which was first detected just over three years after explosion, in July 1990. The prompt burst has been the subject of a considerable amount of theoretical work (Storey & Manchester 1987, Chevalier & Fransson 1987, Benz & Spicer 1990, Bisnovaty-Kogan 1990, Kirk & Wassmann 1992) involving various acceleration models, but this is not the topic of the present discussion. Perhaps the most interesting and at the same time least controversial conclusion one can draw from the prompt emission is that the stellar wind material allowed radio waves to escape from close to the star as early as two days after explosion. Consequently, the switch-on of the second phase of emission after three years cannot be due to the emergence of the shock front from an absorbing screen, and a straightforward application of the mini-shell model fails.

Not just the existence of the prompt emission, but also the details of the switch-on of the second-phase emission itself show the inadequacy of the absorbing screen explanation. The emission was first detected at low frequency (843 MHz) and only later at high frequency (4.8 GHz). This is particularly interesting from a theoretical point of view. Applying an 'onion-shell' type model to the particles in the acceleration zone around the supernova shock front (122) one arrives at a solution for the distribution of these particles given by (126):

$$n(p,t) = \frac{t_c Q}{p_0 \mathcal{D}} \left(\frac{p}{p_0}\right)^{2-s} \left[H\left(p-p_0\right) - H\left(p-p_{\rm max}\right)\right] \tag{175}$$

where

$$p_{\rm max}(t) = p_0 \exp\left(t\mathcal{D}/t_c\right) \ . \tag{176}$$

This solution applies for time independent flow speeds, such as is expected in the early (sweep-up) phase of the explosion. If we assume that the magnetic field did not change much in the emission region between detection at low frequency ($\nu=\nu_1$, $t=t_1$) and at high frequency ($\nu=\nu_2$, $t=t_2$), the maximum particle energy must have changed during this time by a factor

$$\begin{aligned}\frac{p_{\rm max}(t_2)}{p_{\rm max}(t_1)} &= \sqrt{\frac{\nu_2}{\nu_1}} \\ &= \exp\left[\mathcal{D}(t_2-t_1)/t_c\right] \ . \end{aligned} \tag{177}$$

Recalling the definitions of t_c (121) and $\mathcal{D}$ (125), one finds that given the speed of the shock front, the time delay $t_2 - t_1$, which is observed to be about 30 to 60 days, yields the spatial diffusion coefficient directly (Ball & Kirk 1992). The speed of the material immediately behind the shock front $(u_1 - u_2)$ can be estimated to be about $25,000\,\mathrm{km\,s^{-1}}$ from the observations of fast moving line emitting plasma seen immediately after the explosion, since we do not expect the ejecta to have decelerated appreciably. The compression ratio of the shock follows from the observed synchrotron spectrum and is $u_1/u_2 \approx 2.7$. Taken together, these lead to

$$\kappa \approx 2 \times 10^{24}\,\mathrm{cm^2\,s^{-1}} \ . \tag{178}$$

Other parameters can be derived if we use an estimate of the magnetic field. Thus, taking a value of 10^{-3}G, which can be found by extrapolating the (unfortunately model dependent) estimates of the magnetic field strength during the prompt burst, one can deduce that the Lorentz factor of the radio emitting electrons (4.8 GHz) is roughly

$$\gamma_{\mathrm{max}} = 2000 \ . \tag{179}$$

Diffusive acceleration takes about 300 days to accelerate an electron from an energy of 500 keV up to a Lorentz factor of 2000, given the values of u_1, u_2 and κ derived here. We can conclude that the acceleration process must have begun operating about two years after explosion.

In quasilinear theory, the coefficient of pitch-angle scattering is proportional to the square of the magnitude of the magnetic field fluctuations $(\delta B/B)^2$ (see, for example, Blandford & Eichler 1987), which enables us to approximate the spatial diffusion coefficient of a relativistic electron as:

$$\kappa \approx \frac{1}{3}\lambda_{\mathrm{g}} c \left(\frac{B}{\delta B}\right)^2 , \tag{180}$$

where $\lambda_{\mathrm{g}} = \gamma mc^2/eB$ is the gyroradius. This expression can be turned around to estimate the amplitude of the magnetic fluctuations present around the shock front of SN1987A. Inserting the value of κ from (178) one finds:

$$\left(\frac{\delta B}{B}\right) \approx 0.003 , \tag{181}$$

which is similar to the level of fluctuations required in the local interstellar medium for cosmic rays of this energy, but significantly below that estimated at the shocks of young supernova remnants (Blandford 1992, Achterberg et al 1994), indicating that the level of turbulence in SN1987A is as yet rather modest.

These arguments, concerning the switch-on time at different frequencies, depend solely on the upper limit to the energy of particles in the acceleration zone around the shock front. The onion-shell approach, however, is capable of giving the entire history of the synchrotron emission including that emitted by electrons left behind in the expanding shocked material. All we must do is specify the rate at which particles are injected into the acceleration process, and give the radial dependence of the magnetic field and density. The shock, we can assume, is essentially piston driven, and maintains a constant velocity. If it moves out into the undisturbed stellar wind of the progenitor, then it is reasonable to assume the density falls off as $1/r^2$ and the magnetic field, which is likely to be wound up into a toroidal direction by the rotation of the progenitor, varies inversely with r. The number of particles overtaken by the shock front per second is then constant, so that a constant injection rate is suggested. Under these assumptions, the model is so simple that an analytic expression for the emitted synchrotron flux as a function of time can be found (Ball & Kirk 1992). The time dependence is shown in Fig. 25. Initially, the emission rises on the acceleration timescale. This slows as adiabatic losses of the electrons downstream of the shock make themselves felt. Finally, even though injection continues at a constant rate, the light curve turns over. This indicates that the freshly accelerated electrons are unable to compensate for the adiabatic losses of their predecessors. It arises because the magnetic field decreases monotonically outwards, so that the freshly accelerated electrons emit progressively less synchrotron radiation.

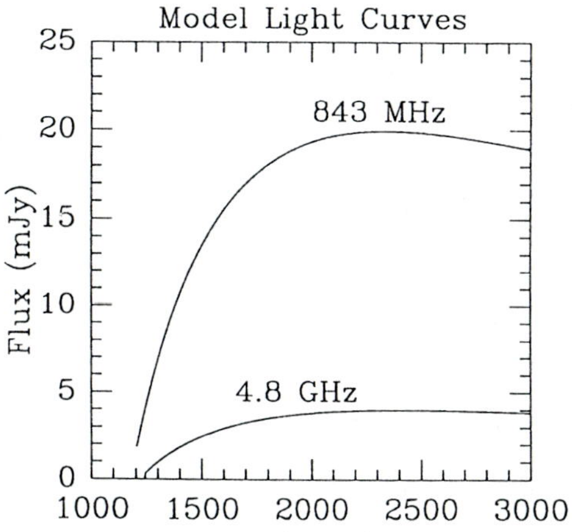

Fig. 25. The light curve of a single clump at low and high frequency. (From Ball & Kirk 1992)

Observations of the supernova paint a different picture. Firstly, fluctuations on a timescale of several days were observed shortly after the emission was detected (Staveley-Smith et al 1992), indicating that the emitting region is smaller than a spherically symmetric shell of radius equal to that of the shock front. Secondly, the flux did not turn over, but increased sharply after nearly one year. One is thus driven to the conclusion that one or more small clumps of material are contributing an important part of the emission. The model described above can accommodate such clumps, provided they take part in the radial flow pattern, so that one can hope to model the emission using several components similar in appearance to that in Fig. 25. The result obtained with two clumps is shown in Fig. 26, superimposed on the data up to early 1992.

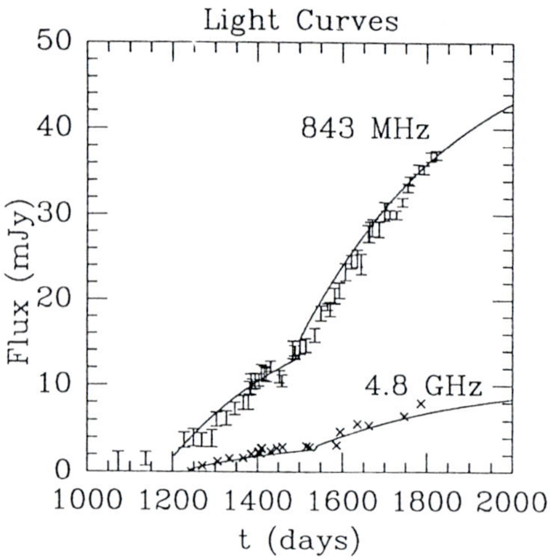

Fig. 26. The predicted light curves at 843 MHz and 4.8 GHz from two clumps of emitting electrons superposed on the observations up to day 1800. (From Ball & Kirk 1992)

9.3 Future Prospects

In this section, I would like to depart from the approach adopted hitherto and present some opinionated speculation about problems associated with much of the material I have presented, and about the trends of future research which I hope will address them. I have chosen the context – the application to SN1987A – deliberately: a wealth of data should be forthcoming on the radio emission of this object and on other RSNe in the next few years, giving us a chance to gain a much better grasp of the physical processes involved.

The model calculations shown in Fig. 26 agree well with the data. The first order Fermi process at a shock front is, it would seem, not too unreasonable a model for the radio emission of SN1987A. However, it is important to note that the predictive power of the simple model described here is limited. In fact there are three aspects of the model which are not dealt with self-consistently, and these are closely connected with fundamental problems in the theory of diffusive acceleration.

The first, and perhaps least serious, concerns the compression ratio of the shock front. In the simple model, this is taken from the observed power-law index of the synchrotron emission according to the well-known formulae connecting the synchrotron index α with the electron index s (appropriate to the phase-space density): $s = 2\alpha + 3$, and the electron index with the compression ratio: $\rho_c = s/(s-3)$ (see 91). The data give $\rho_c \approx 2.7$, whereas a strong shock front in an ideal gas of adiabatic index 5/3 has a compression ratio of 4. However, if the shock front accelerates not only electrons, but also protons, and if the pressure in the energetic protons becomes significant, then a structure of the type depicted in Fig. 20 establishes itself. Electrons of GeV energy, which have a much lower mean free path than the energetic protons, move adiabatically in the precursor region of this structure. Their diffusive motion 'feels' only the sharp compression of the sub-shock. Because the plasma in front of this sub-shock is compressed and heated by the protons, and also because it is accelerated in the direction of the flow, the Mach number of the sub-shock is reduced. This effect, suggested as an explanation of the range of spectral indices observed in supernova remnants by Bell (1987) has also been investigated by Ellison & Reynolds (1991) in the same context.

In principle, it is possible to compute the modification of the shock structure for a radio supernova using the techniques described in Chapter 7. However, close to the site of the explosion the progenitor can be expected to have modified its environment substantially. An undisturbed stellar wind such as assumed in the simple model is one possibility, but SN1987A is known to be more complex, possessing not only a dense ring of material at a radius of about 0.2 pc, but also most probably a region interior to this in which the wind stagnates (see McCray 1993 for a recent review). A computation of the self-consistent evolution of the compression ratio of the sub-shock in this environment would clearly supply valuable information for the interpretation of the radio data. However, numerical difficulties due to the strong magnetic field and the correspondingly short acceleration times close to the position of the progenitor must first be overcome. Preliminary calculations are reported in Kirk et al (1994).

The second problem concerns injection. This is an issue which I have avoided in most of these lectures. The main reason is that injection involves the transport properties of particles of thermal energy, for which the structure of the collisionless shock itself, complete with the effects of electrostatic potentials, magnetic overshoots and non-coplanar field components, is crucial. In view of this, the most promising attack on the injection problem seems to me to be full-scale numerical simulation (cf. the section on simulations in Zank & Gaisser 1992). However, this is as yet of little use in direct astrophysical application. An alternative approach is to ignore nonstochastic effects and insist that the transport is governed by pitch-angle diffusion. Unfortunately, this also does not get us very far. A self-consistent analytic treatment of the resulting anisotropic distribution is lacking. Monte-Carlo simulations are successful (see Jones & Ellison 1991), but are limited to a very specific pitch-angle diffusion coefficient (isotropic) and to stationary solutions. Thus, the only practical course at present is to use the cosmic ray transport equation (80) for energetic particles and to adopt a more or less plausible prescription for injection. In the application to SN1987A, the injection problem arises for both electrons and protons, but fortunately the observations of both light curves and spectra are sufficiently detailed that one can still hope to constrain the models.

The third problem is that of the transport process of energetic particles. The overall structure of the magnetic field in the simple model of SN1987A is toroidal. The spatial diffusion coefficient found in (178) thus refers to transport across the direction of the average field. In simple scattering theory, the spatial diffusion coefficient parallel to the magnetic field is given in terms of the mean-free path for scattering λ by

$$\kappa_{\|} = \frac{1}{3}\lambda v \tag{182}$$

and is related to the perpendicular diffusion coefficient by

$$\frac{\kappa_{\perp}}{\kappa_{\|}} = \frac{1}{1+(\lambda/\lambda_{\mathrm{g}})^2} , \tag{183}$$

so that $\kappa_{\perp} < \lambda_{\mathrm{g}} v/6$. The value of κ given in (178) violates this limit by about five orders of magnitude. Clearly, something is wrong either with the model, or with the picture of scattering outlined above.

Anomalous transport across magnetic field lines is a problem which has been studied for some time in laboratory plasmas. Basically, although the direction of the average magnetic field is well-defined, each individual field line can wander around and diverge systematically from its neighbours. Even if a particle undergoes no scattering in the usual sense, and remains tied to a single field line, it will nevertheless undergo stochastic transport across the direction of the mean field (Rechester & Rosenbluth 1978). Clearly, such an effect could be important in giving the electrons at the shock front of SN1987A such a large perpendicular diffusion coefficient (Achterberg & Ball 1994).

9.4 Concluding Remarks

In assembling the material for these lectures I have followed a fairly straight and narrow course along the path of diffusive shock acceleration, with occasional excursions to consider shock-drift acceleration and relativistic flows. The reason for this is partly fashion, since the theory has undergone rapid development over the last twenty years. But, more importantly, this theory provides us with a fairly simple means of making useful statements about very exotic objects. The plasma physics which lies behind it is, ignoring cross-field diffusion for a moment, fairly well understood. It is also robust – we can rely on streaming cosmic rays to provide their own scattering centres (see Melrose, Lecture 4, this volume), and do not have to assume special properties of the turbulence, as in most other stochastic acceleration models. The theory has its problems, as I have tried to point out, but these do not seem to be insuperable. In any case, they are an indispensable part of any theory worth further research.

Acknowledgments

I would like to thank Arnold Benz and Thierry Courvoisier for their excellent organisation of this Advanced Course. Several figures in the text have been taken from the work of others; for permission to do this, my thanks are due to Rob Decker, Ernst Dorfi, Klaus Meisenheimer, Ray Protheroe and Heinz Völk. Finally, a thorough and constructively critical reading of the text was undertaken by Lewis Ball, to whom I am especially grateful.

References

Abramowitz, M., Stegun, I.A. 1972 *Handbook of Mathematical Functions* (Washington DC: National Bureau of Standards)

Achatz, U., Steinacker, J., Schlickeiser, R. 1991 *Astron. Astrophys.* **250**, 266

Achterberg, A., Ball, L.T. 1994 *Astron. Astrophys.* **284**, 687

Achterberg, A., Blandford, R.D., Reynolds, S.P. 1994 *Astron. Astrophys.* , in press

Anderson, M., Rudnick, L., Leppik, P., Perley, R., Braun, R. 1991 *Ap. J.* **373**, 146

Atoyan, A.M. 1992a *Astron. Astrophys.* **257**, 465

Atoyan, A.M. 1992b *Astron. Astrophys.* **257**, 476

Axford, W.I. 1981 *Proc. 17th. Int. Cosmic Ray Conf.* (Paris) **12**, 155

Axford, W.I., Leer, E., Skadron, G. 1977 *Proc. 15th. Int. Cosmic Ray Conf.* (Plodiv) **11**, 132

Baade, W., Zwicky, F. 1934 *Phys. Rev.* **45**, 138

Balescu, R. 1988 *Transport Processes in Plasmas*, North-Holland (Amsterdam)

Ball, L.T., Kesteven, M.J., Campbell-Wilson, D., Turtle, A.J., Hjellming, R.M. 1994, *M. N. R. A. S.* , in press

Ball, L.T., Kirk, J.G. 1992 *Ap. J. Letters* **396**, L39

Baring, M.G., Kirk, J.G. 1991 *Astron. Astrophys.* **241**, 329

Begelman, M.C., Kirk, J.G. 1990 *Ap. J.* **353**, 66

Begelman, M.C., Rudak, B., Sikora, M. 1990 *Ap. J.* **362**, 38

Bell, A.R. 1978 *M. N. R. A. S.* **182**, 147

Bell, A.R. 1987 *M. N. R. A. S.* **225**, 615

Benz, A.O., Spicer, D.S. 1990 *Astron. Astrophys.* **228**, L13

Benz, A.O., Thejappa, G. 1988 *Astron. Astrophys.* **202**, 267

Berezhko, E.G., Yelshin, V.K., Ksenofontov, L.T. 1993 *Proc. 23rd. Int. Cosmic Ray Conf.* (Calgary) **2**, 354

van den Berg, S., Tammann, G.A. 1991 *Ann. Rev. Astron. Astrophys.* **29**, 363

Biermann, P.L., Strittmatter, P.A. 1987 *Ap. J.* **322**, 643

Bisnovaty-Kogan, G.S. 1990 *Proc. Joint Varenna – Abastumani – ESA – Nagoya – Potsdam Workshop on Plasma Astrophysics* Eds: T.D. Guyenne, J.J. Hunt, Telavi, Georgia.

Blandford, R.D. 1992 in *Particle Acceleration in Cosmic Plasmas*, AIP conference procedings #264, eds: G.P. Zank, T.K. Gaisser, page 430

Blandford, R.D., Eichler, D. 1987 *Physics Reports* **154**, 1

Blandford, R.D., Ostriker, J.P. 1978 *Ap. J. Letters* **221**, L29

Blandford, R.D., Ostriker, J.P. 1980 *Ap. J.* **237**, 793

Blandford, R.D., Payne, D.G. 1981 *M. N. R. A. S.* **194**, 1041

Bogdan, T.J., Völk, H.J. 1983 *Astron. Astrophys.* **122**, 129

Bregman, J.N. 1985 *Ap. J.* **288**, 32
Chevalier, R.A. 1982 *Ap. J.* **259**, 302
Chevalier, R.A., Fransson, C. 1987 *Nature* **329**, 611
Cox, D.P., Reynolds, R.J. 1987 *Ann. Rev. Astron. Astrophys.* **25**, 303
Decker, R.B. 1988 *Space Sc. Rev.* **48**, 195
Decker, R.B. 1990 in *Particle Acceleration in Cosmic Plasmas*, AIP conference proceedings #264, eds: G.P. Zank, T.K. Gaisser, page 183
Dolginov, A.Z., Toptyghin, I.N. 1966 *Sov. Phys. JETP* **51**, 1771
Dorfi, E.A. 1993 in 'Galactic High-Energy Astrophysics High-Accuracy Timing and Positional Astronomy' (Lecture Notes in Physics #418) Eds.: J. van Paradijs, H.M. Maitzen, Springer-Verlag, Berlin
Drury, L.O'C. 1983 *Rep. Prog. Phys.* **46**, 973
Drury, L.O'C. 1991 *M. N. R. A. S.* **251**, 340
Drury, L.O'C., Axford, W.I., Summers, D. 1982 *M. N. R. A. S.* **212**, 413
Duffy, P., Drury, L,O'C., Völk, H.J. 1994 *Astron. Astrophys.* , in press
Duin, R.M., van der Laan, H. 1975 *Astron. Astrophys.* **40**, 111
Eichler, D. 1985 *Ap. J.* **294**, 40
Ellison, D.C., Jones, F.C., Reynolds, S.P. 1990 *Ap. J.* **360**, 702
Ellison, D.C., Reynolds, S.P. 1991 *Ap. J.* **382**, 242
Falle, S.A.E.G., Giddings, J.R. 1987 *M. N. R. A. S.* **225**, 399
Fermi, E. 1949 *Phys. Rev.* **75**, 1169, also *Collected Papers vol. II* page 656 (1965 Chicago: University of Chicago Press)
Fermi, E. 1954 *Ap. J.* **119**, 1, also *Collected Papers vol. II* page 970 (1965 Chicago: University of Chicago Press)
Fichtel, C.E., Bertsch, D.L., Hartman, R.C., et al. 1993 *Astron. Astrophys. Suppl.* **97**, 13
Fritz, K.D. 1989a *Astron. Astrophys.* **214**, 14
Fritz, K.D. 1989b *Ap. J.* **347**, 692
Gaisser, T.K. 1990 *Cosmic rays and particle physics* Cambridge University Press, Cambridge.
Gaisser, T.K. et al 1993 *Comments on Astrophys.* **17**, 103
Ginzburg, V.L., Syrovatskii, S.I. 1965 *Ann. Rev. Astron. Astrophys.* **3**, 297
Gleeson, L.J., Axford, W.I. 1967 *Ap. J. Letters* **149**, L115
Green, D.A. 1984 *M. N. R. A. S.* **211**, 433
Green, D.A. 1992 *Proc. 22nd. Int. Cosmic Ray Conf.* (Dublin) **2**, 412
Hall, D.E., Sturrock, P.A. 1967 *Phys. Fluids* **10**, 2620
Han, X., Hjellming, R.M. 1992 *Ap. J.* **400**, 304
Heavens, A.F., Drury, L.O'C. 1988 *M. N. R. A. S.* **235**, 997
Heavens, A.F., Meisenheimer, K. 1987 *M. N. R. A. S.* **225**, 335
Hjellming, R.M., Johnston, K.J. 1988 *Ap. J.* **328**, 600
de Hoffmann, F., Teller, E. 1950 *Phys. Rev.* **80**, 692
Holman, G.D., Pesses, M.E. 1983 *Ap. J.* **267**, 837
IAU Colloquium #142 1994 *Ap. J. Suppl.* ,
Jokipii, J.R. 1987 *Ap. J.* **313**, 842
Jones, F.C., Ellison, D.C. 1991 *Space Sc. Rev.* **58**, 259
Jourdain, E., Bassani, L., Bouchet, L. et al., 1992, *Astron. Astrophys.* **256**, L38
Kang, H., Jones, T.W. 1991 *M. N. R. A. S.* **249**, 439
Kardashev, N.S. 1962 *Sov. Astron. J.* **6**, 317
Kazanas, D., Ellison, D.C. 1986 *Ap. J.* **304**, 178
Kennel, C.F., Coroniti, F.V. 1984 *Ap. J.* **283**, 694

Kirk, J.G., Duffy, P., Ball, L.T. 1994 *Ap. J. Suppl.* **90**, 807
Kirk, J.G., Heavens, A.F. 1989 *M. N. R. A. S.* **239**, 995
Kirk, J.G., Mastichiadis, A. 1989 *Astron. Astrophys.* **213**, 75
Kirk, J.G., Mastichiadis, A. 1992 *Nature* **360**, 135
Kirk, J.G., Schlickeiser, R., Schneider, P. 1988 *Ap. J.* **328**, 269
Kirk, J.G., Schneider, P. 1987a *Ap. J.* **315**, 425
Kirk, J.G., Schneider, P. 1987b *Ap. J.* **322**, 256
Kirk, J.G., Schneider, P. 1988 *Astron. Astrophys.* **201**, 177
Kirk, J.G., Wassmann, M. 1992 *Astron. Astrophys.* **254**, 167
Krülls, W.M. 1992 *Astron. Astrophys.* **260**, 49
Kruskal, M. 1962 *J. Math. Phys.* **3**, 806
Krymsky, G.F. 1977 *Sov. Phys. Dokl.* **22**, 327
Krymsky, G.F., Petukhov, S.I. 1980 *Sov. Astron. Lett.* **6**, 124
van der Laan, H. 1962a *M. N. R. A. S.* **124**, 125
van der Laan, H. 1962b *M. N. R. A. S.* **124**, 179
van der Laan, H. 1966 *Nature* **211**, 1131
Lagage, P.O., Cesarsky, C.J. 1983a *Astron. Astrophys.* **118**, 223
Lagage, P.O., Cesarsky, C.J. 1983b *Astron. Astrophys.* **125**, 249
Laing, R.A. 1981 *M. N. R. A. S.* **195**, 261
Leroy, M.M., Mangeney, A. 1984 *Ann. Rev. Geophys.* **2**, 449
Luhmann, J.G. 1976 *J. Geophys. Res.* **81**, 208
McCray, R. 1993 *Ann. Rev. Astron. Astrophys.* **31**, 175
McKenzie, J.F., Völk, H.J. 1982 *Astron. Astrophys.* **116**, 191
Maisack, M., Johnson, W.N., Kinzer, R.L. et al. 1993, *Ap. J. Letters* **407**, L61
Mastichiadis, A., Kirk, J.G. 1992 in 'High-Energy Neutrino Astrophysics', eds: V.J. Stenger, J.G. Learned, S. Pakvasa, X. Tata, (World Scientific, Singapore), page 63
Meisenheimer, K., Röser, H.-J., Hiltner, P.R., Yates, M.G., Longair, M.S., Chini, R., Perley, R.A. 1989 *Astron. Astrophys.* **219**, 63
Melrose, D.B. 1969 *Astrophys.& Space Sci.* **4**, 143
Melrose, D.B. 1980 *Plasma Astrophysics Vol. II* (Gordon & Breach, New York)
Melrose, D.B., Dulk, G.A. 1987 *Physica Scripta* **T18**,29
Mészáros, P., Ostriker, J.P. 1983 *Ap. J. Letters* **273**, L59
Moraal, H., Axford, W.I. 1983 *Astron. Astrophys.* **125**, 204
Miller, J.A., Buessoum, N., Ramaty, R. 1990 *Ap. J.* **361**, 701
Mushotzky, R.F., Done, C., Pounds, K.A. 1993 *Ann. Rev. Astron. Astrophys.* **31**, 717
Parker, E.N. 1958 *Phys. Rev.* **109**, 1328
Parker, E.N. 1965 *Planet. Space Sci.* **13**, 9
Prischep, V.L., Ptuskin, V.S. 1981 *Sov. Astron. J.* **25**, 446
Protheroe, R.J., Kazanas, D. 1983 *Ap. J.* **265**, 620
Protheroe, R.J., Szabo, A.P. 1992 *Phys. Rev. Letts.* **69**, 2885
Protheroe, R.J., Szabo, A.P. 1994 submitted to *Astroparticle Physics*
Punch, M. et al 1992 *Nature* **358**, 477
Quenby, J.J., Lieu, R. 1989 *Nature* **342**, 654
Rechester, A.B., Rosenbluth, M.N. 1978 *Phys. Rev. Letts.* **40**, 38
Schatzman, E. 1963 *Annales d'Astrophys.* **26**, 234
Schlickeiser, R. 1984 *Astron. Astrophys.* **136**, 227
Schlüter, A. 1957 *Zeitschr. Naturforsch.* **12a**, 822
Sedov, L.I. 1959 'Similarity and Dimensional Methods in Mechanics' Academic Press, New York
Shklovskii, I.S. 1960 *Sov. Astron. J.* **4**, 243

Sikora, M., Kirk, J.G., Begelman, M.C., Schneider, P. 1987 *Ap. J. Letters* **320**, L81
Skilling, J. 1975 *M. N. R. A. S.* **172**, 557
Staveley-Smith, L., et al 1992 *Nature* **335**, 147
Stecker, F.W., Done, C., Salamon, M.H., Sommers, P. 1991 *Phys. Rev. Letts.* **66**, 2697
Storey, M.C., Manchester, R.N. 1987 *Nature* **329**, 421
Swann, W.F.G. 1933 *Phys. Rev.* **43**, 217
Taylor, G.I. 1950 *Proc. Roy. Soc.* **A201**, 159
Toptyghin, I.N. 1980 *Space Sc. Rev.* **26**, 157
Völk, H.J., Biermann, P.L. 1988 *Ap. J. Letters* **333**, L65
Webb, G.M. 1989 *Ap. J.* **340**, 1112
Webb, G.M., Axford, W.I., Terasawa, T. 1983 *Ap. J.* **270**, 537
Webb, G.M., Drury, L.O'C. 1984 *Astron. Astrophys.* **137**, 185
Weiler, K.W., Sramek, R.A., Panagia, N., van der Hulst, J.M., Salvati, M. 1986 *Ap. J.* **301**, 790
Weiler, K.W., Sramek, R.A. 1988 *Ann. Rev. Astron. Astrophys.* **26**, 295
Whipple, E.C., Northrop, T.G., Birmingham, T.J. 1986 *J. Geophys. Res.* **91**, 4149
Wu, C.S. 1984 *J. Geophys. Res.* **89**, 8857
Zank, G.P., Gaisser, T.K. 1992 (editors) *Particle Acceleration in Cosmic Plasmas*, AIP conference procedings #264.
Zel'dovich, Ya.B., Raizer, Yu.P. 1967 *Physics of Shock Waves and High-temperature Hydrodynamic Phenomena* Academic Press, (New York)

Appendices

A Physical, Astronomical and Plasma Constants

physical quantity	symbol	value
speed of light	c	$3.0 \times 10^{8}\,\mathrm{m\,s^{-1}}$
		$3.0 \times 10^{10}\,\mathrm{cm\,s^{-1}}$
fundamental charge	e	$1.6 \times 10^{-19}\,\mathrm{C}$
		$4.8 \times 10^{-10}\,\mathrm{esu}$
electron mass	m_e	$9.1 \times 10^{-31}\,\mathrm{kg}$
		$9.1 \times 10^{-28}\,\mathrm{g}$
proton mass	m_p	$1.67 \times 10^{-27}\,\mathrm{kg}$
		$1.67 \times 10^{-24}\,\mathrm{g}$
mass ratio	$m_\mathrm{p}/m_\mathrm{e}$	1837
Boltzmann constant	κ_B	$1.38 \times 10^{-23}\,\mathrm{J\,K^{-1}}$
		$1.38 \times 10^{-16}\,\mathrm{erg\,K^{-1}}$
electron volt	eV	$1.6 \times 10^{-19}\,\mathrm{J}$
		$1.6 \times 10^{-12}\,\mathrm{erg}$
gravitational constant	G	$6.67 \times 10^{-11}\,\mathrm{N\,m^2\,kg^{-2}}$
		$6.67 \times 10^{-8}\,\mathrm{dyne\,cm^2\,g^{-2}}$
(Planck's constant)/2π	$\hbar$	$1.05 \times 10^{-34}\,\mathrm{J\,s}$
		$1.05 \times 10^{-27}\,\mathrm{erg\,s}$
classical e^- radius	r_0	$2.8 \times 10^{-15}\,\mathrm{m}$
		$2.8 \times 10^{-13}\,\mathrm{cm}$
Thomson cross section	σ_T	$6.65 \times 10^{-29}\,\mathrm{m^2}$
		$6.65 \times 10^{-25}\,\mathrm{cm^2}$
critical B field	B_c	$1.44 \times 10^{9}\,\mathrm{T}$
		$1.44 \times 10^{13}\,\mathrm{G}$
permittivity of free space	ε_0	$8.85 \times 10^{-12}\,\mathrm{F\,m^{-1}}$
permeability of free space	μ_0	$1.23 \times 10^{-6}\,\mathrm{H\,m^{-1}}$

astronomical quantity	symbol	SI units	gaussian units
astronomical unit	AU	$1.5 \times 10^{11}\,\mathrm{m}$	$1.5 \times 10^{13}\,\mathrm{cm}$
parsec	pc	$3.09 \times 10^{16}\,\mathrm{m}$	$3.09 \times 10^{18}\,\mathrm{cm}$
light year	ly	$0.95 \times 10^{16}\,\mathrm{m}$	$0.95 \times 10^{18}\,\mathrm{cm}$
angstrom unit	Å	$10^{-10}\,\mathrm{m}$	$10^{-8}\,\mathrm{cm}$
solar radius	$R_\odot$	$7.0 \times 10^{8}\,\mathrm{m}$	$7.0 \times 10^{10}\,\mathrm{cm}$
Earth's radius	R_E	$6.4 \times 10^{6}\,\mathrm{m}$	$6.4 \times 10^{8}\,\mathrm{cm}$
Jupiter's radius	R_J	$7.1 \times 10^{7}\,\mathrm{m}$	$7.1 \times 10^{9}\,\mathrm{cm}$
solar mass	$M_\odot$	$2.0 \times 10^{30}\,\mathrm{kg}$	$2.0 \times 10^{33}\,\mathrm{g}$
solar luminosity	$L_\odot$	$3.9 \times 10^{26}\,\mathrm{W}$	$3.9 \times 10^{33}\,\mathrm{erg\,s^{-1}}$

SI and Gaussian Units

The following plasma formulas give expression in SI units and gaussian units when the physical quantity is given in the relevant units. Thus the formula for v_A in SI units (upper value) involves B in tesla and n_e per cubic meter, formula in gaussian units (lower value) involves B in gauss and n_e per cubic centimeter. Formulas involving the mass density contain a factor μ defined by $\eta = \mu m_p n_e$. Temperature is in kelvin.

astronomical quantity	symbol	value
plasma frequency	ω_p	$56.4 n_e^{1/2}\,\mathrm{s}^{-1}$
		$5.64 \times 10^4 n_e^{1/2}\,\mathrm{s}^{-1}$
	$\omega_p/2\pi$	$9.0 n_e^{1/2}\,\mathrm{Hz}$
		$9.0 \times 10^3 n_e^{1/2}\,\mathrm{Hz}$
electron gyrofrequency	Ω_e	$1.76 \times 10^{11} B\,\mathrm{s}^{-1}$
		$1.76 \times 10^7 B\,\mathrm{s}^{-1}$
	$\Omega_e/2\pi$	$2.8 \times 10^{10} B\,\mathrm{Hz}$
		$2.8 \times 10^6 B\,\mathrm{Hz}$
proton gyrofrequency	Ω_p	$0.98 \times 10^7 B\,\mathrm{s}^{-1}$
		$0.98 \times 10^4 B\,\mathrm{s}^{-1}$
Alfvén speed	v_A	$2.2 \times 10^{16} B(\mu n_e)^{-1/2}\,\mathrm{m\,s}^{-1}$
		$2.2 \times 10^{11} B(\mu n_e)^{-1/2}\,\mathrm{cm\,s}^{-1}$
sound speed	c_s	$1.5 \times 10^2 T^{1/2}\,\mathrm{m\,s}^{-1}$
		$1.5 \times 10^4 T^{1/2}\,\mathrm{cm\,s}^{-1}$
ion sound speed	v_s	$91 T_e^{1/2}\,\mathrm{m\,s}^{-1}$
		$9.1 \times 10^3 T_e^{1/2}\,\mathrm{cm\,s}^{-1}$
Debye length	λ_{De}	$69 T_e^{1/2} n_e^{-1/2}\,\mathrm{m}$
		$6.9 T_e^{1/2} n_e^{-1/2}\,\mathrm{cm}$
thermal electron speed	V_e	$3.9 \times 10^3 T_e^{1/2}\,\mathrm{m\,s}^{-1}$
		$3.9 \times 10^5 T_e^{1/2}\,\mathrm{cm\,s}^{-1}$
collision frequency	ν_0	$1.37 \times 10^{-5} (\ln \Lambda) n_e T_e^{-3/2}\,\mathrm{s}^{-1}$
		$13.7 (\ln \Lambda) n_e T_e^{-3/2}\,\mathrm{s}^{-1}$

B Conversions Between Units

Conversion factors between different systems of units are made most conveniently by introducing conversion factors and regarding units as algebraic symbols. For example, consider conversion from meters to centimeters, or vice versa. Given that $10^2\,\mathrm{cm} = 1\,\mathrm{m}$, one may introduce conversion factors by writing $1 = 10^2\,\mathrm{cm\,m}^{-1}$ or $1 = 10^{-2}\,\mathrm{m\,cm}^{-1}$, both of which follow directly from the basic relation. Then if one is given a formula in meters, as $L = x\,\mathrm{m}$, where x is a number, then the quantity in centimeters, that is, $L = y\,\mathrm{cm}$ where y is another number, is given by $L = x\,\mathrm{m}(10^2\,\mathrm{cm\,m}^{-1}) = \mathrm{y\ cm}$, so that one identifies $y = 10^2 x$.

quantity	SI/gaussian	gaussian/SI
length	10^{-2} m/cm	10^{2} cm/m
mass	10^{-3} kg/g	10^{3} g/kg
energy	10^{-7} J/erg	10^{7} erg/J
power	10^{-7} W/erg s^{-1}	10^{7} erg s^{-1}/W
force	10^{-5} N/dyne	10^{5} dyne/N
charge	$\frac{1}{3} \times 10^{-9}$ esu/C	3×10^{9} C/esu
electric field	3×10^{4} V m^{-1}/esu	$\frac{1}{3} \times 10^{-4}$ esu/V m^{-1}
current	$\frac{1}{3} \times 10^{-9}$ A/emu	3×10^{9} emu/A
current density	$\frac{1}{3} \times 10^{-5}$ A m^{-2}/emu	3×10^{5} emu/A m^{-2}
magnetic induction	10^{-4} T/G	10^{4} G/T

Use of Conversion Factors

The detailed formulas in these notes are given in SI units. To convert any formula to gaussian units involves no change if no electromagnetic quantities are involved. If electromagnetic quantities appear, no change is made to the following quantities

charge q, electric field $\mathbf{E}$, current density $\mathbf{J}$.
charge density ρ, electric potential Φ,

The following changes are made

magnetic field $\mathbf{B} \rightarrow \mathbf{B}/c$, permittivity of free space $\varepsilon_0 \rightarrow 1/4\pi$,
vector potential $\mathbf{A} \rightarrow \mathbf{A}/c$, permeability of free space $\mu_0 \rightarrow 4\pi/c^2$.

Other conversion factors include the following.

quantity	factor	inverse
temperature	$8..6 \times 10^{-5}$ eV/K	1.16×10^{4} K/eV
X-ray energy	4.1×10^{-15} eV/Hz	2.4×10^{14} Hz/eV
angle	2.06×10^{5} arcsec/rad	4.85×10^{-6} rad/arcsec
time	3.16×10^{7} s/yr	3.17×10^{-8} yr/s

Subject Index